Ambassadors from Earth

Outward Odyssey
A People's History of Spaceflight

Series editor: Colin Burgess

AMBASSADORS
FROM EARTH

Pioneering Explorations with Unmanned Spacecraft

JAY GALLENTINE

UNIVERSITY OF NEBRASKA PRESS • LINCOLN AND LONDON

Library of Congress Cataloging-
in-Publication Data
Gallentine, Jay.
Ambassadors from Earth:
pioneering explorations with
unmanned spacecraft /
Jay Gallentine.
p. cm.—(Outward odyssey:
a people's history of spaceflight)
Includes bibliographical
references and index.
ISBN 978-0-8032-2220-5
(cloth: alk. paper)
1. Space probes—Popular works.
2. Outer space—Exploration—
Popular works. I. Title.
TL795.3.G35 2009
629.43′5—dc22
2009017239

Set in Adobe Garamond
and Futura.
Designed by R. W. Boeche
and Shirley Thornton.

Dedicated to those citizens of Planet Earth
who have labored to bring the solar system
just a wee bit closer to all of us.

We shall not cease from exploration
And the end of all our exploring
Will be to arrive where we started
And know the place for the first time.

T. S. Eliot

"Now, Voyager, sail thou forth
to seek and find."

Walt Whitman

Contents

Illustrations

Acknowledgments

How wrong I was to ever consider the writing of this book to be a solo effort. So many wonderful people have so unselfishly contributed to the project.

To my loving wife, Anne, provider of endless understanding and support during the slow years of compiling information, who put up with constant probe talk and let me go enough to get this thing done. I love you to the ends of the solar system. To my sons Ben and Matt, you guys are my heroes, and I am distinctly privileged to spend a lifetime being your dad.

To the publisher, the University of Nebraska Press: Gary Dunham, Rob Taylor, Sara Springsteen, Elisabeth Chretien, Ray Bouche—all are magnificent folks who have been of immense help. Thanks for having me do this book!

To Colin Burgess, author extraordinaire. Through endless e-mail exchanges, he coached me through a million little day-to-day issues. Colin, thanks so much!

Jeremy Hall skillfully polished the text with his professional copyedit.

To all of my interview subjects, who gave selflessly of their time and memories and helpful feedback for the sake of history. Without you all, there *are no* ambassadors; there is no story. Nearly every one reviewed the manuscript sections that involved themselves and were kind enough to make corrections, offer feedback, and add valuable personal anecdotes. In particular, Charley Kohlhase went *way* beyond the call of duty. He was kind enough to critique the entire work from start to finish and help me chase down the seemingly endless factual concerns. James Van Allen initially refused an interview, only to later set aside three priceless days to receive me in his office for one of the most memorable chats I might ever have. Robert Cesarone assisted me with the intricate machinations of spacecraft trajectories and supplied personal materials relating to his work and the directions that *Voyagers 1* and *2* are headed. Michael Minovitch, at his own expense, sent me box after box of primary source documents related to his work at UCLA and JPL, along with many photographs. Thank you! Candice Hansen reviewed the Voyager section, pointed out some inconsistencies, and added warm stories. Jim Burke, George Ludwig, Charley Kohlhase, Bud Schurmeier, Candy Hansen, Anita Sohus, Michelle Thomsen, and Jon Lomberg graciously dug through their personal photographic collections to provide readers with seldom-seen imagery.

By sheer dumb luck, I encountered Kay Ferrari of the Jet Propulsion Laboratory during a 2004 Lab tour; over the last four years, she's been a constant source of assistance, suggestions, and amusing insight. Kay, you're amazing!

Thanks to Philip Corneille, Fellow of the British Interplanetary Society, who was an endlessly encouraging, helpful reviewer of early drafts. He also kept needling me to find better and better pictures and even opened his entire photographic collection up to my cherry-picking.

Thanks to Paolo Ulivi, who diverted time from his own writing on unmanned spacecraft to review pretty much the entire manuscript and comment in such helpful detail. Much gratitude also goes to Bruce Moomaw, who supplied a treasure trove of material on the early air force probes. This is an element of space history that would be just about gone if not for people like Bruce and his dedicated research.

Down in Iowa City, filmmaking friend David Gould more or less implored me to come stay with him for my interview with Van Allen. Over the course of the week there, David served not only as host but also as consultant, sitting down with me each evening to review the substance of that day's discussions, locate holes, and suggest follow-up questions.

Kathy Kurth of the University of Iowa Department of Physics and Astronomy was most gracious in assisting me with the sourcing of pictures. Many thanks also to the JPL Archives for their generosity and expertise: Julie Cooper, Charlene Nichols, Charles Avis.

Cargill Hall endured many questions from me concerning Ranger details, including more than five of the who-did-what-and-when variety. Asif Siddiqi kindly responded to several e-mails seeking to clarify obscure details of the early Soviet space efforts. Former JPL worker Henry Richter added much weight to the argument that "the Deal" originally referred to poker.

For one highly interesting year, I enjoyed the company of a Russian neighbor, Alex Fomenko. Think six-foot colossus with a moustache befitting his stature. From him I learned much about Russian culture and history, and about the Soviet military. Before moving away, I received a warm arm around my shoulder: "Russian or American, we are all the same," he said. "We are people."

And yes of *course*, double-special thanks to Ms. Ilene Kelly and Ms. Kathie Williams, Millard North High School English mavens. Both were eternally frustrated with my inability to diagram even basic sentences, yet encouraging of the "little stories" I continually dribbled up on the side. Hope you like this one.

Introduction

If the neighborhood has already gone to bed, then it's easier for stargazing. Or for checking up on things, which is what learned men do. So on balmy summer nights in Tullahoma, Tennessee, Gary Flandro sometimes wanders outside after a good day and sights up his thumb at the darkened sky.

A select few constellations always get the man's attention, and he knows right where they are . . . not a lot of searching involved. Gary's thumb extends in the general direction of the Serpent Handler, which makes for something of an arresting choice of nicknames. Astronomers describe this faint stellar grouping as Ophiuchus, a fairly ancient member of the skies. It's snuggled between Hercules and Sagittarius and is typically not the first constellation on the tip of anyone's tongue. Except maybe Gary's.

Around 150 AD the Greek astronomer Ptolemy was affected by folkloric inspiration as he attempted to make sense of the heavenly pinpricks of light rotating above him. Ptolemy catalogued some 1,080 stars and approximately forty-eight groupings of them based on figures and patterns from the Greek stories of times gone past. One of them dealt with the Serpent Handler. It appears in the earliest of Greek mythology and stands as a symbol of medicine and longevity.

However, this is *not* what Gary Flandro thinks about when he looks up at Ophiuchus.

Maybe it's something more tangible than Greek mythology—something like the unusually vast collection of astronomical curiosities that just happen to fall within the constellation's border. It contains no less than seven globular clusters—obscenely huge, dense groupings of very old stars, up to a million in some cases. And in Ophiuchus, they are all reasonably bright, making this constellation a popular target for the binocular set.

Although this is pretty fun stuff, it isn't why Gary looks up to the sky like he does. Hmm, what about the nova? Ophiuchus contains a recurring nova, or stellar explosion, which occasionally flares up to naked-eye visibility. Is that it? Sorry. What about the *other* oddities inside our Serpent Handler? Barnard's Star then? Or Kepler's Supernova of 1604?

Nope.

Yet *something* out there needles Gary into thinking about the time he got his picture taken. It wasn't on his birthday . . . or that one family camping trip. In fact, he's not sure exactly when it was taken, which may perhaps be considered a bit odd. Sometime around Valentine's Day 1990 . . . The picture (probably of his good side) is a matter of fact; he's seen it. But a buddy wasn't snapping from nine feet in front of him. The camera wasn't on the ground, which may help explain Gary's lack of details.

No, the picture isn't of some backyard picnic, shot from his second-floor deck. It is not his class group picture from long ago, either. Gary is tough to make out in this image, so keep going up. The roof of the Sears Tower? Farther still . . . climb.

Ancient Greeks wrote with magnificent artistic license. But even they might not have been prepared for the true source of Flandro's picture. The night air in Tennessee gets nippy sometimes. Gary twists his thumb a little to the left to block the minor boundary separating Ophiuchus from Hercules. Swiveling his wrist ever so slightly, he makes the star Rasalauge wink on, then off . . . then on again from behind his thumb. It gets him smiling. And he brings down his arm, but not before squinting past it one last time: *There. She's right about . . . there.*

Plenty of folks have seen the aforementioned photos. None of them are the same one.

A woman named Candice Hansen had lobbied in favor of the shot for years. So had Carl Sagan, the Cornell University astronomer who brought science to the masses on his television show *Cosmos* and from whom the tall idea originally gushed. Every time the subject came up, they'd get knocked down. Both Hansen and Sagan took a lot of lumps: The picture wasn't scientific; it wasn't necessary. It would be too expensive. It probably wouldn't even be *in focus.*

Sagan and Hansen conceded that all those arguments were true.

But why not try?

Because what it was, was *human.*

People who do this work spend a lot of time getting knocked down in general, and it doesn't take long at all to figure out what's involved in picking yourself up again. So the pair kept banging on doors, preaching their message, until someone finally waved a green flag. The shot was on. Gary Flandro would get to see himself after all.

When everything dribbled through, Candy Hansen plunked down before her computer screen, eyeballing one lackluster image after another—mostly depicting familiar patterns of dust against the rich blackness of outer space. She had become well acquainted with dust specks on the camera lenses and began playing "spot the difference" for one that struck her as unfamiliar.

A little while later, she found it.

Shortly thereafter, Carolyn Porco made a print of that image and looked it over. The shot was mostly black with a scattered few dinky colored blobs. Fresh out of the processing chemicals, photographs tend to be a bit tacky. So Carolyn held up her print sideways in the light to peruse all the bits of dust stuck to the top. She brushed the back of her hand across the face trying to clean it all off, then again and again. One tiny speck wouldn't budge, and that's when Carolyn realized what it was.

Not long after at a press conference, Carl Sagan hoisted his shabby wooden pointer and gestured at the middle of a television screen. It showed the same thing his associate had been trying to brush off. "This may look like just a dot," he commented, "but it's actually *less than a pixel*. You will notice in the color image, a faint blue."

Nobody spoke, because they couldn't grasp it. Some people still can't. Sagan kept going, "And *this* . . . is *where we live*."

It was Earth.

The picture Carl and Candy and Carolyn were looking at, the one Gary thinks fondly of, is of nothing less than Mother Earth taken from 4 billion miles out. On that day, an old camera shot a family portrait of our solar system. Most every planet appears together with the Sun in one panoramic vista. While the image itself is blurred and unremarkable, the very nature of it is revolutionary. The idea that any human-made camera could be so far off as to *take a picture of our solar system* is an electrifying, numbing concept. It eclipses common sense. The fact that it was actually taken and sent back to Earth is positively stupefying.

Out at the very edges of our solar system, this camera was not floating by

1. You are here: Earth as seen from 4 billion miles. Everyone you know is on this pale blue dot. (Courtesy of NASA/JPL-Caltech)

itself. It was attached to one of the most meticulously designed and carefully built machines ever created. *This* camera was on an unmanned spacecraft, gunning for the Handler, its direction indicated by Gary's thumb. A machine unglamorous yet dripping with exotic, jeweled beauty in the eyes of any mortal fortunate enough to have collaborated in her being. Such participants can put their hands up and be recognized: *I, Bud Schurmeier; I, Charley Kohlhase; I helped make her go. I cultivated an endangered species.* It was an apparatus put on more weight-reducing diets than some desperate prom queen with a tight dress. Considered fodder and filler in the waning pages of any given Sunday paper by the apathetic members of Planet Earth who were more concerned with dishwashing liquid sales than the nature of their own existence. *Oh, this machine they had wrought.* Covertly designed to do what it wasn't allowed to and built on an ungodly pathetic shoestring, it was finally sent off with universal greetings of peace and welcome. For all time. So for a little afternoon shutterbugging, by golly, she was only too happy to oblige.

Somebody once made the joke that after a nuclear war the only things left would be bugs and Keith Richards. They forgot about a magnificent creation that left Earth in 1977 to learn our solar system's dirty little secrets, which no one has ever been able to go and discover in person. With an odometer flipping ten miles every second, the trip has gone billions of miles . . . it counts on.

Her name is *Voyager 1*, and she will outlast us all.

Voyager 1 didn't go up just for a picture. She went to see the outer planets, specifically those beyond Mars. Laden with instrumentation, the ship left Earth with the idea of scoping out Jupiter, Saturn, and a particularly curious moon of Saturn named Titan. Anything beyond that counted as extra credit.

Flandro's picture-taker continues receding, but curiously enough, it is not alone. Down in Tullahoma, there is less to listen to; most of the traffic is gone. Right now he can't make out any southern constellations, but Gary sure wishes he could settle his thumb over one called Pavo, also known as the Peacock. He steps a few feet toward the horizon, bringing himself infinitesimally closer. This doesn't really accomplish anything, but Gary can't help himself. That's where *Voyager 1*'s sister is going.

Armed with an identical complement of toys, *Voyager 2* was named second but left Earth first; some might argue that she trumps even her sister as far as expeditions are concerned. *Voyager 2*'s baggage tags would humble the most seasoned international business traveler. Sure, *Voyager 1* makes the Tennessean smile and think fondly, but everyone has a favorite. And Gary Flandro's favorite is *Voyager 2*, because *she* made it past Jupiter and Saturn, out to Uranus, then Neptune . . . and she wouldn't have done so—indeed, *nothing* in this lifetime would have done so—if it hadn't been for Gary.

Planet Earth has a lot of gadgets lying around. We have plastic-fish wall hangings that play recorded music, Sports Illustrated Sneaker Phones, and machines that reset bowling pins. And how about that Easy-Bake Oven? These ingenious marvels of engineering are no doubt vital for civilized life as we know it. But some folks like to learn more about what's far away in the heavens, and luckily there are machines for that, too. The Newtonian telescope is a great example. Looking through a twelve-inch Newtonian telescope is like having an eyeball one foot across. And with a relatively inexpensive, modest device such as this, it's possible to watch the clouds move on Saturn. But the telescope itself doesn't do any looking. Same deal with unmanned spacecraft like Voyager: if nobody's back at home base, then nothing gets discovered.

So to call them "unmanned spacecraft" is a gross injustice. The term implies that they are dumb boxes of wires launched in vague, ballistic guesstimates to their intended locations, operating solely on autopilot while running some tight computer program the whole way. Not so at all. These fantastic machines are piloted and controlled as much as any craft with beating hearts aboard. The people aren't on the ship, but they are still along for the ride. This sort of arrangement means that, in many ways, the entire planet can go along on the discovery. In the case of Voyager, these ships have discovered the true end of our solar system. They are there right now, today, chatting with home on a radio signal needing twenty-eight hours for the round trip.

The structures of Voyager were forged in the fires of high-altitude exploration already more than twenty years old at the time. Their molds are cultivated in planning the ships of today and those of our great-grandchildren. They exist on a continuum; they will always evolve. Everything has some kind of heritage, so where are these things from? How did humanity get from a stone wheel to interstellar spacecraft? What sorts of people made these vessels, how did they do their work, what troubles did they face? How did they learn to make Voyager?

The answer doesn't begin with a picture of our solar system, or Gary Flandro, or even the ancient Greeks. The story actually begins late in 1957, on a bobbing ship in the Pacific Ocean.

1. Aboard the *Glacier*

Obviously the Russians had the capability,
and I knew that already.

James Van Allen

Larry Cahill sardined himself into the communication shack with James Van Allen, who had headphones popped over his ears. It was quite hot and late at night. Both were thankful that they could be near the top of the ship, because that was farthest from the blistering engine room.

"Well, I wonder if that could really be it," Van Allen said to nobody in particular. There was this inexplicably goofy beep-beeping sound in the headset. Something didn't add up here. He glanced around. On the assembled faces were looks of confusion, like something didn't fit. Like a game of Clue—everybody knew it couldn't be Colonel Mustard with the rope, but all the evidence said otherwise.

What bugged him was the frequency, this 20-megahertz business. It was a common enough spot on the dial for amateurs. The National Institute of Standards and Technology's radio station wwv used it too. And it *never just beeped*. If this new signal was indeed genuine and its creators were playing by the rules, it would be coming through at 108 megahertz, like everybody agreed upon.

Van Allen asked, "Do you think we're getting some kind of weird transmission effect from a terrestrial station?" He wasn't at all sure what to think, so there was a dissonance ricocheting through his brain. He had heard the news. The news was out there, but they were out *here*, in pretty much the middle of nowhere. And learning more wasn't as easy as calling the neighbors. They had to take what came through, and right now there wasn't much.

"Do you think this is really true, Larry?"

Three hours earlier a bearded, grimy, forty-two-year-old James Van Allen had been stuffing his face in the ship's mess. They were a few days west of the Panama Canal, which coincidentally enough was also forty-two years old. Van Allen and Cahill were passengers on the USS *Glacier*—a naval ice-breaking ship ultimately bound for the Antarctic Ocean and perhaps the last place to find two physicists from the Midwest. On paper the vessel was working *Operation Deep Freeze III*, aiming to pulverize ice for supply ships traveling between New Zealand and Antarctica. But that wasn't everything that the ship did; it also carried some thirty-odd secondhand rockets stashed down in the hold, which figured into Van Allen and Cahill's plans along the way.

Two guys with rockets? Did the captain even know? How on earth did they get aboard? For others, it might not have been so easily done. Over the preceding eleven years, James Van Allen had come to be something of an authority in the slightly obscure field of high-altitude research and had gotten to this point by playing a brilliant series of cards.

It had been a regular day, that October 4. Van Allen was familiar enough with creaky government ships to think of himself as an old hand while living on them for months at a time. He'd been in the navy in World War II through two tours—plenty long enough to demonstrate how the radio proximity fuze he coinvented could help bring down lots more enemy aircraft. That was his first card, and a professor of Van Allen's had guided him into the position of being able to lay it down.

Antiaircraft artillery is most effective when it's able to hit something—like an airplane you'd rather not have flying around shooting at you. And the Pacific theater's strategy at the time was to load up the skies with shrapnel and hope to God that enemy craft ran into some of it. Unfortunately, this didn't prove to be any kind of effective, long-term means of conducting antiaircraft operations, and it sure tended to gobble up the ammo. Timing was everything. The guns had a muzzle velocity in excess of half a mile a second. Real fast. So if the gunner's trigger finger was off by even a second . . . well then, there went the bullet, whizzing harmlessly by that Nakajima Gekko half a mile away.

Van Allen and his brilliant cocreators tended to think outside the box, so the group took things one step further and came up with a way to have themselves a smart bullet. They built a dinky radio emitter that could withstand twenty thousand g-forces, and they capped it on top of a five-inch artillery shell. Van Allen himself was responsible for one of those back-of-a-napkin inspirational moments, cooking up an elegant way to protect the fine filaments inside the transmitter tubes from snapping after such an intense force. The final product emitted a continuous radio signal, and if the sensor sensed that it was getting close to something—like a troublesome Kawasaki Ki-61—then boom went the projectile. It was a hit, literally and figuratively, as the U.S. military scooped up the idea and sent it into production. Tens of *millions* were assembled and used with great success. Their effectiveness was estimated to be about five times better than plain old naked rounds. Van Allen spent months at sea training U.S. Naval crews; so by the time he walked aboard the *Glacier*, ships, sailing, ship ways, and even ship food were no big deal. "I was quite at home on ships, yes sir," he said. He was even at home on this one, which featured no keel and commonly rolled twenty degrees in the lightest of winds.

Back in the *Glacier's* sauna of a radio room, Van Allen took off the headphones and passed them to Larry Cahill. They'd walked up there together after a rather abysmal movie was shown over on the helicopter deck following dinner. Cahill was one of Van Allen's graduate students in the University of Iowa physics department. More prestigious-sounding institutions probably exist in which to chase a physics education. Van Allen was once directly asked why anyone should go to Iowa for this path of study, and there was no hesitation before his response, like he'd been asked a million times: "Physics work the same in Iowa City as they do anywhere else!"

David Armbrust shifted on his feet—the *Glacier's* radioman was generally unaccustomed to such commotions and goings-on in his little world of a room way up here. But he knew there'd be a wave of visitors after he first heard the news. John Gniewek—a passenger on this trip who was ultimately headed to the nontouristy Antarctic to operate a magnetometer station there for the next year or so—then came in to find out what the latest was. Gniewek had found out about the occurrence around the same time as Cahill but wondered if Armbrust had been able to tune in the signal itself. He listened to the sweaty headphones passed to him, and what he heard was *beep-beep-beep-beep-beep*—short, staccato tones, a little like Morse code.

"Hey!" Van Allen blurted out. "Why don't we try and tape-record it?"

"Yeah," Cahill agreed. The signal faded some, increasing the tension in the room. Everyone looked at Armbrust, who delicately massaged his gear and tuned it back in.

"We've got that Ampex recorder in the hold," Van Allen began, then immediately stopped himself. True, they had a tape recorder way down in their lab with the rockets, but it was so god-awful heavy that they might not make it back up in time. It also would've been a shame to rip out all the wiring Cahill had already done for their telemetry system.

The signal faded again; Armbrust had to play with the dials some more to bring it back. An urgency came over the men. The beeping would surely be gone for good soon; nobody knew if it would return. They had to make a recording, to get a record of the signal, to find something that could do that. Gniewek spoke up—he had a portable recorder in his bunk and scampered off to get it. About the same time, Van Allen darted out of the radio room, hoofing it down to the lab after all. For a second, his eyes roamed the myriad stacks of gear, then he plucked out an oscilloscope and hauled it back up.

T plus nine minutes and counting. It had been nine minutes since Cahill and Van Allen had walked through the radio room door to peek at the news teletypes and Armbrust had turned to them saying, "I think that I have it!"

Van Allen strung out the cord to his machine, plugged in a few wires, and flicked the thing on. Tube electronics always take a minute to warm up. He crouched down and hovered over the gear, trying not to let any sweat drip on the mechanics. *Gosh it's hot.* The small greenish circular screen finally pittered and then began displaying a graphical representation of the beeping. Within a minute Gniewek's tape recorder was operational too, and it began laying down loop after loop of the curious tones. And then the beeps vanished. Gniewek squished both lips together and hit Stop, but his fingers hovered over the controls. Everyone in the room stayed glued to their respective equipment, not knowing for sure what would happen. They started stiffening up.

About an hour later, the beeps came back. Van Allen's wide eyes went back to his scope, and Gniewek punched the recorder on again. They repeated that cycle once more before standing down their watch, moving the gear to one side, and finally settling down in the room to discuss what was happening. By that time, the ship's captain, B. J. Lauff, had wandered up to join the impromptu gab session. His inquiry made Van Allen kick himself: *Jim, what about the Doppler shift? Was there any?*

Of course, that's why the signal kept fading!

Van Allen and the others jumped up and stood in front of the radio gear, back on full alert. "In order to keep the strongest signal, we kept continuously tuning it," he explained, "and I measured how much we had to tune it, from when going in to coming out." He jumbled some numbers on paper, scratching figures, based on a crude calibration of the radio dial and how much it had had to be adjusted each time in order to coherently reacquire the signals. *Add those, carry the two.*

Van Allen looked at his results and finally knew what they meant. "It checked very well with the velocity required to be in orbit." So he dashed out a telex, which Captain Lauff wired to the U.S. National Committee of the International Geophysical Year (IGY) in Washington DC:

USS GLACIER

IGY WASHINGTON DC

RECEIVED SIGNAL . . . BELIEVED RUSSIAN SATELLITE TRANSMITTER . . .

DISCOVERED BY DAVID ARMBRUST RM3 AND CONFIRMED AND RECORDED

BY IOWA SCIENTISTS . . .

For better or worse, it was *Sputnik.*

2. Problem Child

Van Allen is and has always been,
in my experience, a calm, unexcitable person.

Larry Cahill, describing his boss's temperament

For more than the last half of the twentieth century James Van Allen researched and taught physics at the University of Iowa—a terribly oversimplified description. To appreciate the man's impact, it is necessary to understand that for the last forty-plus years before his death his office was in the *Van Allen Building*, which is just off Jefferson Avenue in Iowa City and two short walking minutes from the bagel shop.

Van Allen was a man scourged by problems. They raced around the corners of his mind, little tigers pacing in cages, honeybees in the hive. Percolating. Tying up brain cells. These problems were not bad things— they were evidence of his unyielding desire to resolve as many unknowns as possible about the natural world. His problems inquired, "What exactly is happening there?" or, "Why did that change?" And no matter how many of them Van Allen reconciled, he always seemed to find more and want to answer those, too. They kept him up after the rest of his family went to bed—kept him hunched over the battered little desk in the corner of his living room, working by the sixty-watt bulb of a small lamp. The condition existed even in his elementary school days. "If the instructor assigned five problems at the end of the chapter, out of twenty, I'd work the whole twenty," he said, darn excited just to talk about it. "I mean, I *loved* to see how they came out!"

Born in September 1914 as the second of four boys, James Alfred Van Allen grew up on what he termed a "junior-grade farm" in Mount Pleasant, Iowa, in the southeastern part of the state. The town's population was only about

2. James Van Allen working at his desk, late 1950s. Note the sophisticated filing system. (Courtesy of the University of Iowa)

three thousand. Its colorful environment was perfect for a young, eager boy like him. "We had a big orchard and a huge garden—a one-acre garden— flocks of chickens, and a big barn," he explained of his world at that age. "So it was kind of a hybrid between an in-town [home] and a farm."

Van Allen's father, Alfred, was a lawyer—and by his son's admission a real domineering sort of fellow. He had been an only child and presumably grew quite accustomed to getting his own way. You didn't cross Van Allen's father. Even so, Dad parked his kids every night after supper for at least an hour of undivided attention. This was his time to read to them. *National Geographic* . . . the *Atlantic Monthly* . . . whatever they had around. Sometimes *An Illustrated History of the Civil War*. After that, the children were off to their homework.

Before she married Alfred, Van Allen's mother, Alma, taught at the country schoolhouse in Eddyville, Iowa, down near Ottumwa, where she managed to conduct the first grade through the sixth all in one single room. "My parents were both very dedicated to education in general," Van Allen said of his folks, which might help explain their son's path in life.

Some of Van Allen's early antics scared poor Alma half out of her wits. In an issue of *Popular Mechanics*, ten-year-old Jimmy read about how to build a tesla coil from spare parts. He laid his hands on a Model T spark coil and fastened it to a cardboard tube wrapped in lengths of wire, which he in turn connected to a huge battery. "My older brother was a coconspirator in this," Van Allen shrugged, deflecting responsibility. "We stood on a dry board and pushed our hand over to it, and we'd get big sparks." Big sparks indeed: tesla coils generate what amounts to lightning. "Our hair would sort of stand up on end. We gave a demonstration for my mother, and she was pretty appalled by the whole operation!" He cracked a grin at the memory. "It was really quite safe!" At least that spark coil didn't come from his family's vehicle. The Van Allens owned a car but seldom drove it, even in the summer. When the Iowa winters came through, Alfred stuck it up on blocks "to save the tires."

Despite his electrical hijinks, the natural world didn't much appeal to Jimmy right from the start. "I don't remember having any grand aspirations at that time," Van Allen offered of his childhood. "I was mostly preoccupied with trying to do well in school and learning everything they were teaching." Yes indeed, solving those problems. At the time, he definitely wouldn't have counted Earth's atmosphere among his prevalent mysteries. "The atmosphere was just something we breathed, as far as I was concerned!"

Then in high school Van Allen discovered physics and chemistry. "Those were the things that really fascinated me," he indicated with much enthusiasm of the serendipitous find. Van Allen graduated as class valedictorian and delivered the commencement address. *Not* attending college was entirely out of the question for one of Al and Alma's boys. They paid forty-five bucks a semester so Jim could attend Iowa Wesleyan College right there in Mount Pleasant. He lived at home, graduating summa cum laude with a BS in physics. He took one astronomy course. "I still have the book as a matter of fact," Van Allen said. "Still a good book!" The guy never did throw much of anything away.

When the time came in mid-1934 to scout graduate colleges, Van Allen and his father lowered the car off its blocks for an extended road trip. They checked out the University of New Mexico, went on through to the California Institute of Technology (Caltech), and then looked at Stanford. But Van Allen decided to stay relatively close instead, choosing the State

University of Iowa—as it was then called—only fifty miles to the north. "Close to home and very inexpensive." Undertaking a difficult course load, he earned his master's degree in physics in 1936, followed by a PhD in nuclear physics.

During this time, James Van Allen first happened on the topic of cosmic radiation. "Well, I'd hardly ever heard of a thing!" he exclaimed. The matter was brought to his attention in the form of a raging controversy between two physicists named Arthur Compton and Robert Milliken over what exactly caused it. See, atmospheric scientists had a problem of their own: more background radiation existed on Earth than could be accounted for by natural processes. In 1912 the plot thickened when an Austrian researcher went up to seventeen thousand feet in an open balloon and lugged along a radiation detector to keep himself company. It measured *lots* of very high-energy particles cascading down. The higher he went, the stronger the counting rates became, leaving Compton and Milliken in a rather vociferous disagreement as to where in the bejesus all of it came from.

"So that sort of got me interested in cosmic rays," Van Allen explained.

Marginally understood and definitely feared right through the 1940s, these particles were capable of anything from inducing headaches to outright killing a man—depending on whom you asked. Even as late as 1955 and after Van Allen and his colleagues had begun to unravel exactly what the rays could affect and what they couldn't, magazines like *Popular Science* issued grave warnings: "What these blockbusters may do to the body is conjecture—no one has ever been exposed to them. Space-medicine experts foresee that they may perforate the body with such wounds as sterilized needles would leave—and cause flashes of light as they strike the eye's retina, perhaps with permanent damage to vision."

In the spring of 1939 Van Allen's PhD work so impressed the folks at the Department of Terrestrial Magnetism that they offered him a gig up in Washington DC. Van Allen moved out there to focus his attention on the recently discovered principle of nuclear fission, a barely understood concept. So much of what the department did, though, involved natural magnetic fields and solar radiation, which were more his style. Within a year, Van Allen admitted to himself that he'd rather study *cosmic* physics than *nuclear* physics, but he didn't know how to make the change. Soon thereafter the bombing at Pearl Harbor precipitated a slew of meetings between the armed forces and Van Allen's boss. The guy's name was Merle Tuve, an absolutely brilliant experimental physicist. He'd wanted to help. After peppering the establishment with a billion questions about how American physicists might best contribute to the war effort, Tuve sorted the answers and reached clarity. What everybody seemed to be telling him was that, of all things, *antiaircraft weaponry* fell inexcusably short of being able to clinch a victory for the red, white, and blue. And in Tuve's head solidified the idea of his department somehow uniting to kiss that issue goodbye. With a flourish, Tuve redirected almost his entire staff toward the effort. Initially Van Allen remained out of the loop, but the number of problems increased. Tuve then spoke with Van Allen about it.

Problems? *Yeah . . .* he was in.

If such a thing can be said, war had been good to James Van Allen. After finishing his naval duties, he returned to Washington DC and picked back up where he had left off. Building the proximity fuze had so taxed the Department of Terrestrial Magnetism that halfway through the war Tuve essentially shoveled most of the work over to Johns Hopkins University. The college set up a special wing in a rented Chevrolet garage and called it the Applied Physics Laboratory (APL). At the time, the one and only application of physics carried out there was developing a certain hush-hush antiaircraft gizmo.

Such was the environment Van Allen now returned to—something of a different place in which to hang his hat. Merle Tuve ran the whole show. He looked quite kindly upon Van Allen's wartime contributions and achievements leading to V-J Day. And so a most wonderful question thus came forth: *what would you like to do now, young man?*

"He sort of gave me a blank check," was how Van Allen put it. The question dredged up all those old longings, and the kid from Mount Pleasant now suddenly found himself able to follow his heart. Cosmic radiation. Definitely. No matter what, those questions kept darting about inside his head.

"It's not practical reasons that I was interested," Van Allen later offered. "It was purely scientific. Where do these rascals come from? Where do they get their energy? How do they interact with matter?"

At long last the man was on his way to finding out answers to his questions. And to do that, he knew the *real* answers could only be found by ascending to visit the rays in their most native state—before dilution by Earth's atmosphere.

Van Allen smiled at his boss. "I'll organize a small group," he pledged.

With Tuve's blessing—and money—Van Allen immediately switched gears to plan for high-altitude research, using what then promised to be a hearty supply of captured German v-2 rockets. "So that's sort of how I got into it." He thought he *could* do it, too. It was the radio-bomb effort that saddled him with confidence: blending ruggedness, miniaturization, and low power drain, not to mention classified project work in the unpleasant face of bureaucracy. Van Allen put together a band of like-minded souls and went to work. Another card expertly played.

What went into the nose of a v-2 was about to change. It used to be high explosives, typically earmarked for a populous area of London. But Van Allen wasn't interested in rockets so much as he was in their capability for delivering science experiments to heretofore unattainable altitudes. The v-2 provided an opportunity to further understand these mysterious cosmic ray particles that as of late had begun consuming most of Van Allen's waking hours. *What were they made of? Where the heck were they coming from?* A few were hitting his body each minute he just stood there thinking about them.

Whump. The driver in front had backed into him.

Van Allen sat in his car at a stoplight on the way to work. When the light changed, the car ahead of his shifted into reverse for whatever reason and crunched his bumper a little. Whoever was behind the wheel didn't move; she just sat there. So Van Allen pulled around and shot her a frustrated glance.

Van Allen had left this trouble behind after parking outside the lab, but seconds later he ran into the same woman. She really got in his face. "Who do you think you are? Giving me dirty looks!" came the accusation. And then the figure strode toward a nondescript entryway. Uh-oh, she *also* worked at the Applied Physics Lab.

Van Allen tried to put *that* trouble behind him, too, but he kept coming back to the woman. She was a real striking damsel and kind of a firebrand; so he did a bit of investigating and found out her name was Abigail Halsey. A mathematician, no less. He got up the gumption to ask her bicycling.

"When he came to see me," she remembered, "he dreaded having to talk to my roommates while he waited for me. He'd walk in, look wildly around for a magazine, and bury his face in it just to avoid making small talk." Later on, while rolling around town, they smiled and got to know each other and had a gay old time. "Well, he claims he wasn't brilliant, and I agree with him," she said. "But he was very, very smart."

Their relationship matured to the point where the two exchanged nuptials late in 1945. "When we finally decided to get married, the girls thought I was crazy," Abigail continued. "They asked, 'How can you marry a guy like that?'"

The happy couple knew they would probably settle down someplace else, but the wedding itself was in Southampton, New York, where her parents had a summer home. Van Allen was on cloud nine. "The marriage business is a great business!" he proclaimed with a big grin. "I'm all for marriage!"

Afterward, they returned to the APL in Silver Spring, Maryland, and to gobs of research. "Boy," Abigail Van Allen opined years later, "I think when he married me he thought he was going to get somebody who was gonna reduce his data. And I did it for a while, but then I had children!" She laughed. "I liked that better."

By the summer of 1946, Van Allen had fallen in with a small gathering of scientists at the Naval Research Laboratory. They had no endorsed or official agenda; they didn't even have a budget or some weird acronym as a name. But everyone there had a similar passion involving high-altitude research on natural phenomena. John Hopfield was dying to know more about the distribution of ozone. Others were after a better understanding of topics like the Sun's ultraviolet spectrum. Clyde Holliday wanted to figure out a way for a rocket to take pictures while it was so far up there.

U.S. Army Ordinance had figured out how to get their hands on all these v-2 rockets and move them seemingly overnight from war-torn Europe to the deserts of New Mexico. But they hadn't spent much time deciding what to do with them after that, except maybe shooting one off every now and again. Otherwise, the stuff was all just lying around inside corrugated tin shelters. So Van Allen and his cohorts enamored themselves to those watching over it all and became entrusted with developing a program of scientific research. They picked a grand name: the v-2 Upper Atmosphere Panel, which sounded formal enough to please anyone at Army Ordinance.

With the meetings and general organizational tasks over with, it was time to produce. In collaboration with others, Van Allen envisioned a simple cosmic ray experiment. And applying the lessons of the fuze, he painstakingly fashioned it to be undersized, rugged, and simple. It had to fit in the nose of the rocket. It had to share space with other devices. It had to cooperate with the telemetry system provided by the New Mexico College of Agriculture and Mechanic Arts.

The guys finally got one of their widgets operational, so Van Allen packed it up and bought a train ticket. He carried the thing on his lap, rattling all the way down to White Sands Range—the Land of Enchantment's mostly secret venue where the U.S. government had stashed their load of v-2s along with most of the Germans who had built them.

When his train reached New Mexico, Van Allen hopped off in Alamogordo and went the rest of the way by truck. It was a pretty sparse place, with a hangar for assembling the v-2s, some mechanical shops, and bare-bones launching facilities. General Electric was hired to manage the actual rocket preparations, while German engineers peeked over everybody's shoulders to help them figure out what was what. For anyone getting bored, there was also lots of wind and heat.

Van Allen walked into the assembly hangar and had his first in-person assessment of Germany's former v-2, which stands for Vengeance Weapon Number Two. The rocket was almost fifty feet high, and Van Allen had trouble grasping the scale. He gazed up and down its length and finally said, "Uh, gee. That's really something. I mean that's a huge rocket."

Another man was down there already. He was a guy by the name of William Pickering, who was attached to some outfit called the Jet Propulsion Laboratory (JPL) up in Pasadena. Apparently, JPL was doing a lot of secret work for the U.S. Army on developing military rockets. They had been dying to get their hands on a v-2. As an offshoot, Pickering fought the foibles of readying payload enclosures for things like Van Allen's instruments. They didn't just go inside a tin box, of course. The enclosure supplied power, thermal control, and data relays back to the ground—enough to keep a Caltech-educated New Zealander like Pickering on his toes.

This being his first time in town, Van Allen located the man in charge and handed off his part of the puzzle, which had been vigilantly cradled with love and attention the entire trip. General Electric installed it inside the next available v-2. The rocket was trucked into position and launched. Even the air was rumbling. From the blockhouse, Van Allen watched it climb skyward and then blow up. Nobody said anything to him. He looked around, unsure of what he should do. Pieces began raining downrange; some of those pieces represented whatever was left of Van Allen's machine. Back in the hangar, there were lots of v-2s, but Van Allen hadn't brought along a spare cosmic ray experiment in his briefcase. There was no overnight delivery in those days—and even if there had been, there *still* wasn't another finished apparatus back on the shelf at the Applied Physics Laboratory. "We felt lucky to get one working," he said. All that effort just turned into fireworks.

"It was a reality check."

Ultimately the v-2 opportunity never panned out like it was supposed to. The rockets were worse than overbred dogs: scarce, fidgety, and ungodly expensive. The promise of easily accessible v-2s wasn't coming true by anybody's estimation—especially Van Allen's. After a number of similar do overs, he began shopping around for other ways to get his work done. Before leaving New Mexico, Van Allen always made a point to surf the local Indian craft fairs to pick up a little something for Abbie.

The problem with the launch at White Sands Range wasn't Van Allen's experiment, which had been carefully engineered and prepared to survive just about anything—save an exploding booster loaded with fuel. The problem was the booster itself. So Van Allen called up a guy he knew named Rolf Sabersky at the defense contractor Aerojet, and the two of them designed a modest rocket (really a toy compared to the v-2's towering immensity) that was twenty-five and a half feet tall and one and a quarter feet in diameter.

Of the 1,600 pounds making up this so-called Aerobee, 68 of them were reserved for payload. This rocket was more user-friendly than a v-2, incorporating a solid booster and liquid upper stage. It cost a whole lot less money. And the thing tended to not blow up. Early tests at White Sands went amazingly well, and Van Allen was hooked.

Most planets are surrounded by some variety of magnetic field generated from deep within the planet itself. They're not visible to human eyes and do not take the form of perfect circles. Fields bob and weave like the rim of a mildly traumatized bike wheel. Just exactly how they're shaped depends partly on latitude. Patterns near the equator differ from those up north and down south. So after two years and about a dozen Aerobee flights from White Sands, Van Allen figured he had enough data from New Mexico latitudes to begin scouting around for other options.

One of them seemed to lie within the Office of Naval Research. They had been in and out of his field of view since the v-2 days, and Van Allen felt he was in good company with them. Sharing similar goals, they arranged for him and a couple others from Applied Physics to tag along with the USS *Norton Sound*, an aging tender for seaplanes that had tons of supplies to deliver across the ocean. From Maryland the physicists took a train all the way across the Lower 48 to Port Hueneme in California and hopped the railing with Aerobees in tow.

Van Allen was on the water yet again. No better match existed for his endeavor: ships could steam in different directions to effectively cancel out the wind, and they were hell and gone from cities or towns or anyone who might be set off by an unscheduled shower of debris landing in their privet hedges.

He managed two successful flights off the coast of Peru, near Earth's true geomagnetic equator. On another trip a year later, two more rocketed from the *Norton Sound*'s helicopter deck up in the Gulf of Alaska. Each voyage took weeks, and Van Allen developed the tradition of letting his beard grow the entire time at sea. Typically not a whimsical man, he nevertheless engaged in this minor bit of whimsy.

All this data from high altitudes brought him slightly closer to solving his problem. The habitat of true primary cosmic radiation lay well above sixty miles in altitude—the radiation's natural state before slamming into Earth's atmosphere. Plenty of good data from this height was coming back, although it did nothing but whet Van Allen's appetite for more: *more* readings, *higher* altitudes, *more* data. But the Aerobee could only make eighty miles, tops. He wanted to go all the way.

In July 1948 Van Allen drafted up a paper for a colleague to deliver at a major geophysical symposium in Oslo, Norway. The colleague in question was Ralph Gibson, who had recently taken over from Merle Tuve as director of APL in its entirety. One day, Gibson drifted by Van Allen's area and casually mentioned he wanted to discuss the paper's content.

When Van Allen showed up in the boss's office later on, Gibson explained his position. A part of the paper really bugged him. It was the part that mentioned how "serious consideration is being given to the development of a satellite missile which will continuously orbit around the Earth at a distance of, say, 1,000 km." The concept hadn't even been Van Allen's idea, but he was about to get in trouble for it.

"That's too speculative," Gibson indicated, sniffling.

Drafts had already gone around to many colleagues at the APL. None of them had a problem with it, Van Allen protested.

Gibson said, "Yeah, well, I really don't want to put that in the paper. It's too speculative."

Van Allen groaned.

"Why don't we strike that out," Gibson said in a very leading way. A discrete puff of tension rose up. Van Allen could tell his boss's mind was really made up. Silence. It was his turn to speak.

He finally conceded, "*Hmm*. Okay, I'll take it out." More kvetching wouldn't have done any good. Gibson wasn't budging and might have just pulled rank if he had to. So Van Allen took it out, and Gibson delivered the paper in a slightly truncated form.

James and Abigail Van Allen held a dinner party at their home in Silver Spring early in 1950. While visiting the APL, prominent British geophysicist Sydney Chapman had mentioned to Van Allen that he'd be interested in getting together with other Washington DC–area folks in the same line of research. An inspired Van Allen worked the phones and rounded up nearly a dozen commitments to break bread. American scientist Lloyd Berkner gladly came that night. Over the course of the evening many appetizers vanished in discussion of the thrilling Aerobee shots. Stories of progress were always a welcome delight to the other researchers.

During a break in the discussion, Berkner tossed out, "Sydney, isn't it about time we had another Polar Year?"

Chapman lit up in a smile. "Well, I've been thinking along the same line!"

A grand total of two International Polar Years had previously occurred: the first in 1882 and another in 1932. The whole idea had been a globe-circling festival of extended scientific research focused on the earth's polar regions. Born there in the Van Allen living room, the International *Geophysical* Year was formally proposed by Berkner in 1952 and was then embraced worldwide as some sixty-seven nations agreed to collaborate on an extended series of observations and experiments to deepen their understanding of the planet. If it happened on Earth, they would study it: auroras, geomagnetism, gravity, meteorology, oceanography, seismology—oh yeah, and cosmic rays, for which Van Allen had a few plans in the works. The party wouldn't start until mid-1957 though, in order to coincide with a period of high solar activity. A full participant, the Soviet Union also harbored some plans of *its* own.

As if all that wasn't enough, more change was to come. Late in December 1950 Van Allen accepted an invitation with a familiar ring to it: Iowa City. A former professor of his still worked there at the State University; when its physics department head left for greener pastures, one particular man's name quickly materialized. Van Allen packed up his house, his Abigail, their two rapidly growing daughters, plus a ton of raw research data and books, and prepared to leave Silver Spring. He was off to the old alma mater.

The Chinese military celebrated the 1951 new year by collaborating with North Koreans to overrun United Nations boundaries in South Korea. The Van Allen family celebrated by reaching snow-packed Iowa City in their ancient station wagon, pulling a trailer very much on its last legs. Abbie had been west of the Mississippi only once before.

For the time being, home would unfortunately not be a house. To assist married students and visiting professors—or professors just new to town—Iowa City made intelligent use of the temporary barracks still left over from World War II. Shelling out thirty-five bucks a month, the Van Allens in return got a half-cylinder of corrugated tin on the west end of town, warmed only by a cast-iron stove dominating the living room. The stove burned fuel oil dripping in from a fifty-five-gallon drum perched outside on stilts. The family unpacked and hung a few pictures; then Van Allen began teaching physics classes and running the department a couple miles east toward downtown. He also wasted no time in pitching a familiar theme to his old pals at the Office of Naval Research: the shipboard launching of rockets at unique latitudes.

To be sure, Van Allen enjoyed shooting off his Aerobees like that, but the weeks at sea typically netted only a couple of launches. Even Aerobees were just too darn complicated to rig up and make go. He needed something simpler yet, costing less money. The grants he could lay his hands on didn't pay for much.

The quest led him to recall a discussion he'd had at sea over a year earlier. He'd been in the *Norton Sound*'s wardroom in March 1949, gabbing it up with fellow traveler Lee Lewis, who said a few things that really made sense. The lower atmosphere is the rough part of the trip, Lewis pointed out. It's where the air is densest and the rocket has more work to do.

"Wouldn't it be easier," Lewis thusly concluded, "to lift a rocket on a balloon above most of the atmosphere, and then fire it?"

Lewis's idea made gobs of sense. Tons of surplus Loki antiaircraft rockets were to be found since the war had ended. Van Allen could probably get a few from that Pickering fellow out in Pasadena. A few weather balloons wouldn't break the bank. It all sounded so cheap and easy, which is the kind of thing university professors truly appreciate. It also sounded like balloons could escort the rocket to higher altitudes than Aerobees, returning cleaner and better data.

Intrigued, Van Allen poked into the idea and found himself succumbing to its charms. He christened these apparatuses "rockoons." Dutifully, the physicist prepared grant applications and logged correspondence with the Office of Naval Research and continued teaching all at the same time.

Finally, Van Allen pitched the idea to some of his more promising graduate students, who ate it up. They set sail on the first rockoon expedition in August 1952, completing nearly twenty-four flights over two excursions within the span of thirteen months. Like with the Aerobees, raw data streamed down by way of a simple radio transmitter to be recorded on audio tape. After returning home, all of it would then be hand reduced onto a graph in order to lay out patterns of activity.

More trips went down in '54 and '55. Data flowed like cold beer at a baseball game. They could practically bathe in it. Each individual launch cost $750. Hurrah for cost-effective research—and recycling. To boot, Van Allen's students relished the opportunity to earn PhDs based on truly unprecedented work.

The more readings came down, the more Van Allen's curiosity perked up. It was all so puzzling—the readings kept changing, growing, the higher they went. Van Allen needed to go higher still. No matter what that wag Ralph Gibson thought, he would get there.

On the very first expedition, however, almost nothing seemed to be working. The balloon filled with no problems, and it then was simply released from the aftermost deck of the ship with the rocket dangling below on a line. That much went okay. Short-roped underneath the rocket's tail hung its firing box. This brick-sized enclosure contained Iowa's trigger setup—an onboard barometric switch inducing the fireworks at a predetermined altitude. It shared space with batteries and a low-tech backup— nothing more than a wind-up kitchen timer—in case the fancy altimeter didn't work.

Over the course of several years and 109 total attempts, the Iowans learned how a good sixty minutes typically elapsed for the rig to ascend past firing altitude, tens of thousands of feet up. In all those launches, they only got to see one actually ignite. Every other time, it drifted into heavy clouds; they'd be stuck waiting like slugs as the crew, with great effort, maneuvered the boat back on course. But this first time, nobody knew the drill. The rockoon floated off, and an hour later . . . nothing. Very quiet skies. It should've been plenty of time for even the backup kitchen dial to go, but no luck. There was a flat line on the radio receiver.

They stood there—a grubby Van Allen and two even-grubbier grad students—on the deck of a Coast Guard cutter in the middle of the frigid Arctic trying to troubleshoot a dud rocket one hundred thousand feet above them.

"Why *was* that?" someone asked of the failure.

Van Allen scratched the lengthening whiskers on his face and pondered his clouds of breath and had an idea. "I wonder if it's too cold?" That was an awful factor to consider, however, seeing as how the idea was to get readings from a variety of latitudes—even the frosty ones. Van Allen needed to see if that was it—and if not, he needed to cross it off the list and try something else.

The next rocket was portaged up from below. Somebody found a plastic bag with which to cover the whole thing up. Somebody else unearthed a few blocks of Styrofoam and packed them all around the rocket's firing box. Now the working parts were insulated but still not warm. Where to get a portable heater in the mid-Arctic? Van Allen hustled down to the ship's galley and raided it of a large orange-juice can, filling it to the brim with scalding water. Then he dashed back up and stuffed that steaming hot potato into the firing box. Van Allen let things stew for a few minutes, then stripped off everything to let 'er go. An hour later, *whoosh*.

"By George, the next one did work, so we did that from then on," Van Allen remembered of his pregame warm-up.

With activities like these, the Iowa physics department began to command little pockets of media attention. The locals definitely paid notice. Just west of Iowa City, Van Allen drove to the flyspeck town of Tiffin one morning to be on a small radio program literally broadcast from a farmhouse breakfast table. The show *Fresh from the Country* was hosted by a onetime Evangelical United Brethren minister and schoolteacher. Partway through the interview the host stopped asking about rockoon flights to introduce a family member: "My son is just back from the air force," the man proudly explained. "Might you have something for him to do?"

"Well, I might," Van Allen offered. He turned to the boy, who admitted an enthusiasm for electronics. In fact, the kid had built the radio show's amplifier. Van Allen told him, "Come into my office on Monday, and we'll talk it over." Next week, the boy went and Van Allen hired him on the spot to clean up test equipment, just like that. It's the way he was.

Another successful rockoon expedition happened in the summer of '57, lasting from late July through mid-August. Van Allen brought along two of his students, Larry Cahill and undergrad Gary Strine, on an Arctic cruise aboard the USS *Plymouth Rock*. "I thought he was a reasonably friendly, low-key, intelligent person," Cahill said of Van Allen's disposition. And Cahill considered himself very fortunate indeed to even be on these kinds of trips. As he later attested, "From a faculty point of view, one never knows how good or useful a student will be until he has worked through a variety of challenges for a year or two."

Van Allen, Cahill, and Strine returned from their voyage grubby yet safe and extremely well fed. Shortly after arriving back in Iowa City, they repacked and headed out again to Boston Harbor on another excursion. It would be a different ship this time, the USS *Glacier*, taking them through the Panama Canal and down into the Antarctic, before dropping off the Iowa contingent in New Zealand.

During a stopover in Balboa, Van Allen was racing up a set of the *Glacier*'s steps when he caught his leg on the sharp tread of the metal stairs. "It was a very minor little thing," he said, "and just made a little cut and bled a little bit. I didn't think anything of it." He continued his business of inspecting helium tanks for the balloons, and soon they were out in open water.

Ocean-bound ships are not exactly havens of cleanliness. Two days later Van Allen was flat on his back in the ship's medical bay, with his core body temperature through the roof and an IV bag dripping into his arm. The guy couldn't talk; he couldn't even open his eyes.

James Van Allen's entire bloodstream had gone septic. That means it was thick with rampant infection, which is deadly serious in any location at any latitude. And it sure isn't good on a naval ship plowing through the South Pacific. If Van Allen had died on the *Glacier*—which was entirely possible—virtually every facet of unmanned space exploration *to the present day* would have been affected.

"Injuries are common on board ships," is how Larry Cahill positioned the uniquely undulating environment that was the keelless *Glacier*. "At one dinner, a large bowl of hot stew slid down the table and into the lap of an unfortunate person."

Van Allen stoically clung on while his graduate student temporarily assumed control of the rockoon launches. "I stopped by to see him and report progress to him frequently during the day," Cahill mentioned. Over the next five days this routine persisted as Cahill incessantly pressed for updates from the ship's doctor while successfully keeping the launches on schedule. And soon Van Allen resumed work.

"The *Glacier* was a great ship," Cahill later indicated. "We were able to launch the rockoons from the helicopter deck, at the stern." Like Van Allen, Cahill was no stranger to life at sea. His father and grandfather both had been worldwide sailors on navy and commercial ships.

"The food was okay. The movies were okay, also, as far as I was concerned. I didn't go to very many—I wasn't there for movies or great food!"

He wasn't there to die either. The first launch off the *Glacier* didn't quite go as planned. "It was essential that the ship be cruising with the wind," Cahill explained, "so that there was nearly zero wind on the launch deck. Otherwise, the large, helium-filled balloon was subject to being pushed horizontally by the wind, while we were attempting to let it lift the rockoon." So right before launch time, Cahill or Van Allen had to go make requests to the bridge crew for which way to turn and go. Quite often, this took the ship in a direction completely opposite that of its final destination.

Cahill said, "I got the impression that the *Glacier* officers were not particularly happy to have us putting requirements on them." And the number one thing he really needed was zero wind on the deck. "Could you please make sure there's zero wind on the deck?" Cahill would ask. "Zero wind on the deck? Okay?"

With that accomplished, he marched down to the helicopter deck and prepared for their first attempt of the cruise. The balloon was inflated and leashed to the rocket. Perfect. The balloon began to rise, and Cahill initially cradled the rocket, guiding it up. Before he let go, the balloon needed to ascend higher; otherwise, the rocket would clatter down on the deck and break everything. But during that first time, crosswinds played havoc with the rig: "We couldn't seem to get the bridge to turn the ship downwind and adjust the speed to *zero wind*," Cahill recalled. The balloon took off sideways, rocket tethered behind, Cahill gamely hanging on tight with both arms and his legs splayed out, knees locked, feet skidding across the deck, hoping for the love of Mary that the crew would figure it out. Headed right for the edge.

"ZERO WIND ON THE DECK!!" Cahill shrieked. "ZERO WIND ON THE DECK!!" Spittle flying, he barreled toward the edge of a two-story drop before letting go. *Ploink* went their rockoon into the ocean.

"Van Allen thought I should have held the rocket longer and waited for the bridge crew to get organized," Cahill remarked. "He wasn't holding it."

It was on this very trip that Larry Cahill and James Van Allen heard those eerie beeps in the middle of the hot night on October 4. Down in the yawning bowels of the ship, Cahill had been laboring over the preparation of more experiments. "I probably heard about *Sputnik* on the radio," he remembered of the saga unfolding that night. "The navy had a news radio program, and I listened to the radio while I was working on my rocket payloads."

And when the reports began coming through, Cahill stopped working for a minute. "I wasn't unhappy that the Russians had an experiment up before we did—just interested."

Naturally Van Allen was interested too. So interested, in fact, that he calmly finished dinner and then went up to see a movie on the flight deck. He *hated* movies. Not until *three hours later* did Van Allen finally collect Larry Cahill and go check out the radio signal.

What caused that absolutely interminable delay? Could it have been sheer jealousy? When the question came up, Van Allen spent an awkwardly long time staring off into space. "Well, I don't think so . . . uh, no, I don't think so," he began. Then more staring off, looking down, comparing his thumbs. Swallow.

"But you know, I suppose you could say, maybe some kind of disappointment, 'cause we were still . . ." He trailed off, looked at the ceiling again, and took a breath.

Then, it finally came out.

"The Soviets had beat us to it."

3. The Convict

> He could be categorical, demanding, abrupt,
> even coarse, and often was simple, friendly,
> tactful, ready and willing to render real aid.

Physicist Pavel Kostyukevich, describing Sergei Korolev

At that very same time Van Allen and Cahill stood in the *Glacier*'s radio room, on the other side of the globe, a twice-married Russian father named Sergei Korolev (pronounced "Ko-rol-yov") was having a heck of a good day. He was directly responsible for the patterns of beeps those American sailors were listening to, and little did *any* of them grasp how an innocuous thing like a beeping machine was already in the process of reshaping world events.

For Korolev, *Sputnik*'s launch marked the culmination of years of frustration with the Soviet government's nearsightedness, including his close brush with death at their hands. He'd studied hard through school and worked to improve the results of his amateur rocketry group's endeavors, only to get thrown in prison and very nearly killed. The man had pulled himself through endless personal trials, then risen to the top of his game via sheer tenacity and grueling labor. It didn't net him much in the way of material wealth, but one thing Korolev did own was a set of undeniably long legs. In the months before *Sputnik*, he would easily have qualified as a star in his country's expanding space exploits. Now he was the *biggest star*, but in a cruel twist, he would never see it formally celebrated.

Sergei Pavlovich Korolev considered himself a lucky man, and truer words were never spoken. In the fifties, James Van Allen sat at space exploration's gambling table laying down straight flushes every few years. If he wasn't speaking at some formal international symposium about his latest high-altitude revelations, he was out busting his tuckus on an old ship discovering even more new information.

But the deck was noticeably stacked against Korolev right from the outset. The man who would end up responsible for the first intercontinental ballistic missile (ICBM), the first Earth satellite, the first dog in space, the first man in space, the first woman in space, along with the first probes to Venus, Mars, and the Moon was the product of an arranged marriage. Custom dictates that Russians address each other by their first name and patronymic, which is a form of the father's name. Sergei Pavlovich is then understood to be "Sergei, son of Pavel." And "Pavel" translates to the English "Paul." Last names are rarely used in Russian culture except under very official circumstances. His parents divorced when he was three; little Sergei Pavlovich found himself continually shuffled around until his mother remarried.

Sergei Pavlovich finally made it through a school door for the first time at age fourteen. The kid loved aeronautics, earning his way into the Moscow Higher Technical School by 1926. Among others there, he came into contact with Andrei Tupolev, a professor who would emerge as one of the Soviet Union's most prized aircraft developers—and years later take personal responsibility for yanking Korolev's butt out of a nasty situation.

After graduation Korolev went to work for the government as a mechanical engineer on an aircraft bomber project. He loved the work but kept thinking higher, faster. Aeronautics make for an easy leap into the fields of rocketry and space flight; so by 1931 Korolev ended up joining a new, informal group with their minds in the heavens. An obscure wing of the government subsidized amateur-level endeavors in such areas as hot-air ballooning and car racing, and here is where the group took root. The whole intent of *this* particular outfit was to discuss and explore the fascinating arena of rocketry. Its name was long-winded in typical Russian bureaucratic style: Group for the Investigation of Reactive Engines and Reactive Flight, which was shortened from the Russian name to GIRD. Friedrich Tsander—a wizard who was the driving force behind this assemblage—felt so patently over-enamored with worlds beyond Earth that he named his two daughters Mercury and Astra. The group was a good thing for young Sergei, who used the club and Tsander's lack of organization to begin developing managerial skills that benefited him greatly later on down the road.

One year later Tsander fell into bad health and gave Korolev an easy in.

Korolev took over the group and methodically began connecting dots that would lead to the Soviet Union's first liquid-fueled rocket launch, just over a year later. He had long been considered one of the group's natural leaders anyway by other personalities there who would eventually make their own significant contributors to the space program.

Korolev also got married. Lyala Vincentini had been the object of the rocketeer's affections, as is evident from several years of love letters. He was in Moscow and she, in Kharkov; but the distance between them didn't seem to matter. She packed a bag and made the occasional trip to Moscow. He went to see her in Kharkov when he could. Once he arrived during a rather nasty outbreak of dysentery and typhoid fever, which required her undivided attention as a deputy inspector of the local health department.

Once the courting phase wrapped, they often talked of settling down. *He* wanted to move the wedding up, but *she* wasn't in any big hurry, which complicated matters. Everything popped on August 6, 1931, at what could only be termed a micro-wedding in Moscow. The affair was straight out of Las Vegas. Two guests came and the foursome shared champagne. The new bride then departed in a cab, returning home solo to make preparations for moving to Moscow.

Careerwise, things progressed for Sergei Korolev at a respectable rate throughout the 1930s as the Soviet government looked favorably upon his endeavors as being potentially assistive to the defense arsenal. They were interested in concepts like the prospect of producing a rocket that burned solid fuel, which would make for stable, storable artillery. Most any kind of work in this engaging new field was appreciated.

During this period, another young engineer appeared, named Valentin Glushko—whose voracious appetite for space exploration was nurtured largely by an incessant reading of Jules Verne. At one point in his adolescence, Glushko penned an article titled "Conquest of the Moon by the Earth." He was only eighteen at the time. Over the years, his relationship with Korolev would blossom, grow dysfunctional, then clash and burn to cinders. The pair seemed to get off on the wrong foot and stay there. But until things devolved to the point of intimidation and lies, they had a lot of history to write. Glushko wasted no time, plunging himself into development work for the military that involved liquid-fueled rocket engines. Despite funding Korolev's efforts, the government also underwrote Glushko

3. Sergei Korolev with daughter Natasha (at left), plus a niece, 1938. (Author's collection)

in parallel. It would not be the first time the Communist state found itself bankrolling multiple approaches to the same goal.

Then something bad happened. It was 1938, in the month of June. After Lyala and Sergei had gone to bed, others in town were just getting ready for work. In the middle of the night on June 27, two men crashed through the couple's apartment and hauled Korolev off. They were with the Communist secret police, which at that time had been in the process of acquiring an unfavorable reputation by detaining suspected enemies of Josef Stalin. Two "witnesses" tagged along that night, neither of whom came down on the young rocket engineer's side. Korolev didn't have time to say goodbye to his wife or three-year-old daughter or even to grab a fresh pair of underpants. He was thirty-one years old and had been anonymously accused of a number of crimes. At the top of the list was the noticeably vague "subversion in a new field of technology."

This incident was another product of the demons in Josef Stalin's brain. Many of the country's most celebrated engineers, scientists, academicians, and writers found themselves on the wrong side of pretty much everybody.

History laments what the Soviet populace could have accomplished, in general, if not for Stalin's complete paranoia about the country he was running. To be sure, many in the nation did not agree with Stalin's policies, and some undoubtedly organized in opposition to him. This garden-variety breed of disillusionment exists at some level in most any nation. But in Stalin's eyes, the issue boiled down to a simple yet flawed bit of logic: if any Soviet citizen wasn't expressly working in support of the government's goals, then he or she must have been against them. Stalin thought everybody had his number.

Stalin loyalists encircling their ruler *believed* this rhetoric and guided into motion one of history's most tragic events: the Stalin purges. Practically anyone was subject to arrest, trial without representation, torture, imprisonment, murder. Name it. Later, a major contributor to Soviet space exploration grew up knowing his own father had been sentenced to ten years in prison for failing to inform on some colleague who had made a joke about Stalin.

The carnage was widespread, as Ukrainian farmers were starved into submission and even top leaders in the Red Army were executed for imaginary collusion with Germany. Many who were not slaughtered outright wound up in gulags, forced labor camps that tended not to be in resort locations. One of the worst camps was a gold mine in Kolyma, up in the Arctic. Nike sweatshops had nothing on Kolyma. People were transported there in railway cars, on tracks built by other prisoners. Then it was into the hold of a ship for the final leg. Once there, death hung low. Best estimates offered any given prisoner only a 10 percent chance of surviving one full year. Bitter cold, malnutrition, dismal living conditions, overwork—there could be any number of reasons why a man wouldn't rise the next morning.

Plunged into this abstract system of destruction was none other than Sergei Pavlovich Korolev, rocket engineer, project manager, father. "S.P." to his close friends, Korolev was a man whose achievements were already on par with American Robert Goddard's—even though Korolev was some twenty-five years younger. Kolyma—or any gulag for that matter—was the last place he expected to be after all his efforts up to that point. But Korolev didn't go down without a fight. It took two harsh interrogations—punctuated with beatings and torture—before he succumbed to the accusations. A filthy and diseased Korolev finally crumpled into a chair, signing his name to whatever charges lay on the papers before him. At one point before his transfer,

Korolev mustered enough reserve to get a letter off to Stalin himself explaining how there must have been some kind of mistake. Many years later they would meet in person to discuss the developing missile program. For some strange reason, Stalin never mentioned letters, mistakes, or gulags.

Filling his days with manual labor, digging and cutting and lifting, Sergei Korolev existed in this death camp for five months. He could have been hunched over a drafting table, polishing his rocket designs. Or maybe playing with his daughter. Scurvy came to visit instead, taking out half of Korolev's teeth. His jaw was broken by a guard at some point, a condition to later hasten his end.

The entire time, Korolev remained aware of his accuser's identity, perhaps visualizing him, pondering the words to be exchanged the next time they met. Later years would see Korolev in faithful collaboration with Valentin Glushko, knowing since day one of his incarceration how he'd been betrayed. Glushko had come under accusation himself and personally implicated Korolev, trying to save his own hide—which, in turn, had sent the secret police to Korolev's door that fateful night.

Shoveling away in the mines, Korolev returned to his filthy barracks one day to find that some papers had come through the mail. With conciliatory tones, they notified officials in the loop that a reinvestigation of Korolev's case was underway. He'd be leaving, but that didn't mean freedom. Just another prison.

Nobody schlepped a departing prisoner into town; he just walked out the gate and was expected to show up. Or else. Korolev made his way carrying tatters of clothing some other prisoners had collected for him. At one point, he halted a truck and traded one sweater for a ride into the town of Magadan. There was bad news when he arrived: the last ship of the season had already set out across the Okhotsk Sea. The rest of the winter would have to be frittered away somehow in this barren, icy town. Despite everything that had already happened, Sergei Korolev undoubtedly had an alert guardian angel that day: a bad storm came across the Okhotsk three days later and took down that ship with everyone on it.

Two days later his angel took a while in reaching him. It was still fifty below, and Korolev hadn't eaten since getting off the truck. There had to be someplace to catch a few winks around here. At one point, he woke up in an abandoned army barracks shelter and couldn't get up. His clothes were frozen to the floor. Even so, he got kicked out of there, then got kicked out of another set of barracks later that same day. Walking along through the frigid wind, alone with no place to go, Korolev came across a golden and impossibly *warm* loaf of bread. It looked like someone had just laid it on the snow and run away. He would spend the rest of his life wondering how it got there.

In Magadan the rocket engineer finally stumbled onto some work. Various menial yet paying jobs filled his time until spring. Sergei Korolev—general laborer, shoe repairman—was finally able to board an early season ship to begin his long journey to the next way station.

In September of 1940 Korolev reached Central Design Bureau No. 29 on Radio Street in Moscow. The gates swung open, and he moved into what was known as a *sharaga*. It was still prison, but there was a juicy twist: Korolev would be housed with like-minded Soviet researchers and engineers, all of them working as a team on projects. The idea lay in forming a critical mass of talent able to produce substantial, all-encompassing results in short order. It would be like going to university, only with armed guards, meal rations, head counts, and a really tall fence going all the way around campus. Only uncomfortable dormitory furniture held consistent.

Now Korolev began work in the sharaga environment. He fell under the direction of aviation-great Andrei Tupolev, himself a prisoner. Rumor has it that, upon arriving at the sharaga, Tupolev was asked to make a list of two hundred people he would enjoy having around to help him. Whomever he requested was collected from the other prisons, and one of the first names he scribbled down was that of Sergei Pavlovich.

The environment commingled a peculiar mix of incarceration and social networking. A fellow inmate in Korolev's sharaga later recalled, "There were so many well-known, friendly faces . . . the elite, the cream of Russian aircraft technology. . . . It was impossible to conceive that they had all been arrested, and they were all prisoners—this meant a catastrophe for Soviet aviation!" Korolev soon burrowed himself a place in the midst of these great personalities. Among other projects, he worked to realize concepts such as propellants for aerial torpedoes and rocket boosters for a dive-bomber aircraft.

Interviewed survivors have largely told the same tales—involving good food, good company, and productive work. It was all serious research and nobody lounged around. How ironic in light of the ongoing confinement and restricted family visitations. These men still loved the country they lived in and remained driven by their interests and the flickering flame of achievement through socialized labor. These ideals of united workmanship and solidarity oozed from every engineering drawing and page of calculations. Most could also work knowing they hadn't *really* done anything wrong. Innovation might lead to a timely release. The renowned Soviet physicist Andrei Sakharov told one anecdote about a prisoner who professed his innocence to the head of the secret police. The response was said to have been, "My dear man, I know you're not guilty of anything. Get the plane in the air and you'll go free."

No question, the work is what saved Korolev. One day he was consulting with another prisoner when a violin concerto came on the radio. They ceased talking, caught up in the swollen strings, a trigger of happy memories. To them it signified home. Tears welled up in both their eyes. The second prisoner broke down, but not Korolev. He wiped himself off and returned to the tasks before him.

Not every waking minute was spent at research. Prisoners were given rest breaks and time to walk outside in a small yard. Korolev also kept up his letter-writing campaign, machine-gunning envelopes virtually one after another toward the prosecutor responsible for his case. He repeatedly professed his innocence, categorizing the accusations as wrong and unfair. Korolev was also quick to point out his involvement in "the creation of rocket planes which significantly improved the flight and technical performances of the best modern propeller-driven types."

Wasn't that helping our country? None of his letters got any attention.

Finally, in late 1942, Sergei Korolev moved into a different sharaga— established to address such "emerging" concepts as liquid-fueled rocketry. Emerging? It was identical to the kind of research the GIRD club had been doing more than ten years earlier. Had everyone not spent all this useless time having confessions extracted and being incarcerated, the nation may well have been a decade further down rocketry's yellow brick road than they already were.

Korolev unpacked and sorted his things. It wasn't long before he ran into a familiar face. Lately this new bunkmate had been producing concept drawings for jet and rocket engines. He would scratch up the basics and dole them out for other inmates to refine or otherwise improve. The guy also had a whopper of a pet project up his sleeve: the world's most powerful and efficient rocket engine yet designed. This man was none other than Valentin Glushko.

Over the course of twelve hectic months, the group's efforts culminated in live firings of Glushko's RD-1 liquid-fuel engine: rich, angry heat blasting through four combustion chambers to jointly deliver 2,600 pounds of thrust. Throughout experimentation it was shown to dramatically speed up an aircraft in flight. And it didn't burn for mere microseconds, like most rocket engines of the day tended to do before destroying themselves. This one fired for two solid minutes. Someone high in the ranks took a liking to this unique group that snared results, as all were released from the sharaga in June of 1944.

Korolev, however, was not strolling home just yet. The cage door was now open and he could have waltzed out. But another year later, having *voluntarily* stayed on to work at Design Bureau No. 29, Sergei Korolev was promoted to colonel in the Red Army. Then he flew to Germany to involve himself firsthand in the matter of a certain something called the V-2 rocket.

4. Light Fuse, GET AWAY

I lived in the office and slept on the table.
I would resume my work in the morning.

Engineer Georgi Grechko

In the beginning . . . there was the v-2. God didn't create it, but that hasn't stopped the v-2 from being a sort of holy father. This fabled machine of Germany's Third Reich, conceived in hate for the explicit purpose of murder, is nothing less than the combined adam-and-eve of viable, large-scale rocketry. A Russian by the name of Konstantin Tsiolkovsky wrote of his propulsion theories and designs. An American named Robert Goddard surely produced operational rockets that matured through his years. Yet any lineage of sturdily dependable, practical, rocket-based launch vehicles leads directly back to the v-2.

Frequently, the v-2 is associated with a charismatic man named Wernher von Braun. To say it was von Braun's rocket is technically inaccurate and at the very least a stinky injustice to the hundreds of other Germans who worked to develop it and to the thousands of foreign slave laborers who involuntarily gave themselves to build it. But the v-2 was von Braun's as much as it was anyone else's in the close-knit group of German engineers who decidedly transformed rocketry.

To clarify, von Braun was an engineer—not a scientist. He knew how to build launch vehicles and to amazingly manage and promote his methods for doing so, but he cared much less about what they actually carried. As James Van Allen put it, "He was interested in the *achievement* of the rocket." And that's about as far as it went.

Germany wasn't ready to start firing their doohickey until late in World War II, which is nothing shocking considering the slow-motion developmental path of complicated things. Nobody else was having much luck getting *their* rockets to fly correctly either. Working from the German town of Peenemünde, von Braun's whip-smart design group recorded a different outcome. Blooming from a foundation of sound, organized engineering and management, they nattily conquered most of the seemingly insurmountable problems faced by American and Soviet engineers alike. The v-2 worked. It was real. British citizenry were not particularly fond of them, as over three thousand dive-bombed into their nation, fired mostly from Germany and Belgium.

Even so, the whole Nazi rocket program—through no fault of von Braun's—ended up as sort of a last-ditch Götterdammerung by the ailing Reich, a wartime Hail Mary lacking some coordinated program of tactical use. Truth be told, the combined explosive payload of every flown v-2 was less than the average Messerschmitt bombing raid. Still, unlike aircraft, the v-2s came in silent and fast and with absolutely no warning—no hum of prop engines in the sky, no gigantic formations to be spotted by lookouts while dozens of miles out. When and where a v-2 would plunge in and explode were entirely unknown. By the end of the war, some twelve thousand people would lose their lives due to these missiles. A hundred slaves a day died while building them. Germany still conceded.

Wernher von Braun knew the end was coming, so before the Reich fell he loaded nearly his entire operation—including major v-2 assemblies, blueprints, and over a hundred engineers—onto a ramshackle convoy of trucks and rail cars. He painted "vzbv" up the sides—which actually meant nothing—and literally bet his hide that the paranoid ss agents wouldn't risk halting the exodus because of it. Then they disappeared to the United States.

Shortly thereafter, Soviet engineers alighted in the world's most sophisticated rocket facility and shook the travel from their clothes. Peenemunde. U.S. forces never actually came there.

Not a whole heck of a lot remained behind. What did remain had been intentionally ruined for the most part. Debris everywhere, the dust was practically still settling. It had been five days since Hitler put a gun to his head. The Reds assessed the ghost town and then set up camp with a hundred of their men. Everybody was dog tired. A few officers made themselves at home 377 miles away in a mansion called the Villa Franka. It had been von Braun's residence in Bleicherode, outfitted with four bathrooms and a grandiose master bedroom—including mirrors on the ceiling. One of the

4. Where's MacGyver when you need him? Only one example of v-2 debris left behind by the German rocketeers. (Author's collection)

men threw aside the perfectly starched white comforter and plopped down on the bed in his filthy uniform. "You know," he said over a cigarette, "it's really not all that bad in this fascist beast's den of iniquity."

What lay scattered about thrilled those who comprehended what they were looking at. During one early reconnaissance, a young and enthusiastic Soviet engineer earnestly jammed his entire upper half into the nozzle of a v-2's rocket engine. Emerging, he bemusedly exclaimed to the others, "This is what cannot be!" Knowledgeable propulsion engineers were already agog over Valentin Glushko's designs: liquid-fuel engines attaining a couple thousand pounds of thrust. But all that was an order of magnitude removed from the German accomplishments. Here in front of them lay an abandoned yet magnificently engineered power plant *in production* capable of *twenty-five tons* of thrust. They just had to figure out how it all went back together.

After the place had been explored, it was time to focus on the arduous task ahead. The Soviets had come with such immediacy, numbers, and expertise, because their goal was nothing less than reassembling a v-2 and actually making it fly. Not a single intact example remained in Germany. Complicating things, the Soviet group lacked top dogs like von Braun and his inner circle. Straightaway, they began efforts to round up any midlevel German specialists who had dispersed themselves into the local villages.

For over two years, engineers and technicians stayed in Germany, vainly struggling to mend some representative form of the original v-2 to get it working. To do so, they had crumpled metal, scraps of parts, perhaps a complete assembly here and there, chopped-up electrical cables, and what remained of any documentation.

Still, the men kept at it. A contingent of volunteers meticulously combed through the trash. Scouts discovered over a dozen abandoned railway cars the retreating Germans forgot to destroy; those cars were loaded with test equipment. They found other parts hidden in underground mines. More locals joined the ranks: nearly three hundred Germans were now on-site laboring for the Soviets. When a critical piece of equipment *was* acquired, the game plan involved documenting what could first be gleaned from observation and jiggling the moving parts. Next, they disassembled the item down to its last bolt. German subordinates completed the reverse engineering by drafting new blueprints. Soviet engineers, like Boris Chertok, then tried to fabricate a new one. "We solved puzzles every day," he said.

Grand-scale Soviet occupation of Peenemünde began in May of 1945. Sergei Korolev reached Bleicherode that September and ended up staying for two years. Only a couple weeks had elapsed between his release from confinement and departure for Germany. In between, he raced back to Moscow to rejoin the family he hadn't seen in four years, not knowing when he might see them again. Luckily, his wife and daughter were able to join him in May 1946 for the remainder of the summer.

Korolev arrived at Bleicherode in style. Roaring into town behind the wheel of a commandeered Opel, he approached the lavish house of a German electrical contractor, which was to serve as his home away from home. Automotive-handling skills aside, Korolev liked to drive very fast. Inside the grandiose study, Major Chertok and his new assistant had been straightening up in anticipation of Korolev's arrival. They couldn't miss the thumping engine and squeal of brakes. Presently Sergei Pavlovich entered and immediately inquired as to why Chertok got to have such a beautiful German secretary. Chertok said she could type. It was the first exchange between two men who would partner for the next twenty years.

"Korolev looked very healthy," Chertok recalled, "even though he had been in prison. He had a prominent forehead, lively black eyes. He looked directly through you as though X-raying you."

Korolev clacked his heels on the tile. "I would like to have a brief overview of the structure and operation of your institute," he requested of Chertok. Working with GIRD had taught him how to manage people and their tasks over the course of an intricate and unprecedented technological program. Now, after long hibernation, those skills could be unpacked again. And from this fertile brew of innovation and manpower, Sergei Pavlovich Korolev emerged as a true leader.

On September 6, 1945, a patched-up, calico v-2 finally launched, in no small part due to the contributions of Korolev . . . and, of course, his nemesis Valentin Glushko. The argument can be made that without Glushko's intimately focused knowledge of engines, and temporary relocation to Germany, the flight never would have happened. Unfortunately, Glushko's inhumane, micromanaging, and perfectionist attitude didn't exactly endear the guy to his subordinates.

For the other man, gratitude awaited. Nearly a year later, in August of 1946, Korolev was recognized as chief designer for long-range ballistic missiles by the Central Committee of the Communist Party. Nearly as old as the Soviet Union itself, the title of chief designer was, in general, less about prestige or accomplishment and more about identifying the person in charge of a given area of work. If a facility distinguished itself in the production of tractor engines, for example, its head might well become chief designer of heavy power plants. Still, Russian culture thinks highly of titles, and that of chief designer generally evoked a certain level of respect.

To land the honor, Korolev had to come in swinging. One of his bosses, a so-called Minister of Armaments, wanted the rocket program to follow the same development track as aircraft—that is to say, the development and delivery of a product with a fixed use. Get the thing working and hand it over; then *poof*, like that it's done. But that kind of thinking didn't apply here.

"You and your deputies are trying to make me a designer of just a missile!" Korolev shrieked at the minister. "Not even a missile, but a very large automated cannon shell, to be more precise! Listen, if I work as the aircraft designers do, our whole business would collapse very soon. You need to understand: I have to be chief designer of the whole system!"

For the next many years, seven capable and influential Soviet engineers

5. The original Council of Chief Designers, 1959. Korolev sits at center with Glushko to his immediate left. Feel the love? Aleksei Bogomolov is at far left. (Author's collection)

functioned as the Council of Chief Designers, tasked by the Communist Party to open new avenues into this perplexingly modern arena of missiles and space. Each man professed a specialization that contributed to the benefit of all involved. Already, despite a lackluster résumé of trade school and college work, Valentin Glushko had become chief designer of rocket engines. There was a chief designer of launch facilities, and so on.

As the Soviets finally began packing in the summer of 1946, virtually every one of the original seven chief designers logged time in Germany reconstructing the v-2. And that autumn, the materiel and loose ends were quickly loaded onto rail trains and shipped to Moscow, along with thousands of German engineers and their families who had not exactly been asked.

Riding along, Korolev announced, "Just wait—we'll outrun that fop von Braun!"

Despite its production in hushed secrecy, the first released version of a Soviet v-2—titled the R-1—greatly resembled America's newly minted Redstone rocket in capability. *And* in looks. Both sported identically shaped guide fins: right triangles with the outside corners nipped off. There's a joke in engineering circles that a company's prototype device is just the competitor's product with a different nameplate. And the similarities *here*? Nothing shocking: both were just modified v-2s, right down to the fin design. The original shape was like that only to permit their transport through the existing German railway tunnels. The major differences between the V-2 and the Redstone were lengthened fuel tanks and the name.

From Russian soil, the R-1 finally launched in mid-October of 1947. "R" stands for *raketa* ("rocket"), and it was almost immediately embraced by the nation's scientific community. Their quick recognition of it as a viable platform for high-altitude atmospheric study curiously mirrored the action back in America.

Over in southwestern Moscow at the Lebedev Institute, nuclear physicist Sergei Vernov had been entranced by cosmic rays since the early thirties. After hearing of the new toy, he eagerly initiated a round of introductions and discussion with the responsible parties about marrying science with technology. Remarkably, Vernov gained access to Korolev, who loved the idea. The main use of heavy-lift rockets was supposedly for carrying nuclear weapons, and the chief designer much preferred science projects.

The president of the Soviet Academy of Sciences caught wind of Vernov's scheme and met with Korolev to ratchet things up a notch. With R-1 tests nearly finished, the two lobbed options back and forth. What kinds of experiments? Where to put them? The interior was already pretty crowded. Korolev studied the blueprints. By nudging different parts this way and that, enough space could probably be freed up to squeeze in the electronics. The cosmic ray detectors themselves could be tacked on down at the base, nestled between the fins. How would that do? Vernov and his colleagues slapped their hands with glee. This magical new creation could *revolutionize* their work of examining and measuring cosmic ray activity, the origins of which bewildered physicists on every side of the globe.

All the launches—tests or otherwise—were going to be held at a new facility Korolev had coming to life. Out in the middle of nowhere, north of the Caspian Sea and immediately west of today's Kazakhstan, field scouts had located a decent spot near a bend in the Volga River, just down a ways from the town of Kapustin Yar (translated "cabbage patch"). Arguably the remotest section of real estate in all the Soviet Union, its netherworld landscape would have been strangely familiar to anyone having seen White Sands.

Korolev's R-1 became the R-2: picture a stretched V-2 with upgraded engine components and twice the range. Visually, it appeared to be mere baby steps away from the v-2. But under the hood, things had taken quite a surprising leap, including a lighter frame and refined guidance system. Such rapid advancement was especially impressive considering the daunting work environment. Things were fairly rough in Soviet Russia at the time: a thousand cities had been destroyed there during World War II, and tens of millions were dead. The nation's infrastructure was much like wet tissue paper—you push it and it breaks. So for any worker on a project as intangible as high-powered rockets, things were tight. Many an engineer went home at day's end and scoured through their own personal possessions looking for a specific fastener, spring, or fitting. A few grandfather clocks became organ donors.

Over five years' time the R-2 would evolve further still, maturing into what's known as the R-7—a brawny launch vehicle amazingly still in use today. The engineers called it "Semyorka," or "Old Number Seven." A silver torpedo ninety-one feet long, adorned by four strap-on boosters ringing a core that was half again as tall. A tapering obelisk, it stood as the Soviet Union's triumph of worker solidarity. Eventual success demanded years of hell working that rocket, fueled by Korolev's one-track gumption and salary of six thousand rubles a month.

The R-7 tallied an abysmal record even by the time it was decided to bolt *Sputnik* on top. Funny how things change: today, Semyorka's descendant is the most reliable heavy-launch booster currently available within excess of a *thousand* flawless launches. NASA actually has a contract with Russia to lease crew space on them. However, in 1954 the rocket was decidedly in bad shape, still failing left and right.

That same year, Ernst Stuhlinger paid a little visit to James Van Allen. Stuhlinger had been on von Braun's v-2 team, and he could be regarded as one of the few true scientists in the whole Peenemünde clan. Wistfully he had longed to place cosmic ray experiments of his own aboard early v-2 test flights, but Allied bombing raids ended up destroying much of the gear he built.

Stuhlinger had the knack for showing up at especially critical times. After the war, he was seen escorting visiting scientists out at White Sands, playing ambassador of science and whispering in the army's ear about the romantic pursuit of knowledge. But what all the army people really wanted was a nifty way to bomb the hell out of someone. And there, amid all that swirling dust and heat and politics and chaos, Stuhlinger had luckily found a friend in James Van Allen.

They reconvened on the East Coast, where Van Allen was on leave from Iowa to work at Princeton University on an emerging technology for nuclear fusion. Months earlier, Wernher von Braun had been discussing the flowering Redstone program with Stuhlinger and mentioned how nice it would be for "a real, honest-to-goodness scientist" to provide a worthwhile and suitable orbiting experiment—and therefore legitimacy.

"I'm sure you know a scientist somewhere who would fit the bill," von Braun coyly teased, "possibly in the Nobel Prize class, willing to work with us and to put some instruments on our satellite?" Von Braun was the kind of guy who preferred having the answers before he asked, and he obviously already knew of the friendship. So Stuhlinger came a'visiting Princeton with what amounted to an invitation.

Reclined at Van Allen's loaner house, he led into it slowly: things had been progressing well on their army rocketry experiments. They were in the process of test-flying Redstone boosters that had clusters of small solid-fuel upper stages added on at the top. The stacked configuration had been christened Jupiter, and all of it working together could conceivably loft something quite high. Really, really high.

Van Allen's ears pricked up. He leaned forward in his chair and drew a breath. "I would love," he began, "nothing more than to do a survey of cosmic ray intensities *above* the atmosphere." An atypical statement like that would appeal only to a very minor slice of the population—but Van Allen knew his audience. Stuhlinger nodded in understanding and continued on. In a few years, he explained, Jupiter might be suitable for carrying a nuclear weapon. Sooner than that—in the next year or so—it could probably orbit some kind of experiment package.

Both men smiled. The amount of data from just a few *days* of satellite orbits would laughingly supersede that from many *years* of work firing sounding rockets like Aerobees or rockoons. If the setup was put together well enough, researchers could just tune in every morning to see what their little black box was finding out up there. The coolest radio program in the world. And their data would be so much better to start with: higher, purer.

That sounded wonderful. But James Van Allen was a man to exercise restraint and therefore offered Stuhlinger little more than a lukewarm, non-committal response: "Thanks for telling me all this," he responded to his guest. "Keep me posted on your progress, will you?" The German blinked at his host and realized it was going to be a short meeting.

Years later Stuhlinger offered, "I was disappointed by this apparent lack of interest, but then I remembered from our meetings at White Sands that Dr. Van Allen was a very cautious scientist, far too careful to jump to any conclusions."

In spite of this, the Iowa man gently engaged his gears. He didn't yet have anyone to pitch it to, but Van Allen wasted little time in constructing a proposal for an Earth-orbiting satellite experiment. Its explicit purpose was "to measure total cosmic ray intensity above the atmosphere as a function of geomagnetic latitude." In English, that meant extending his rockoon-based program into Earth orbit. Over the plan's four typewritten pages, additional sections guesstimated weight and major components. Today Van Allen's treatise comes off as a bit primitive, as exemplified in the choice of power supply for the equipment: a gas turbine with rotating electrical generator.

Outer space contains no fences or border guards; the delicate issue of *airspace* had yet to be resolved or even satisfactorily discussed. If Van Allen got his way, a rocket would someday lift scientific experiments up above a hundred miles or so into orbit. And there they would fly, blissfully crossing over whatever nation happens to be underneath.

Orbiting spacecraft opened up a new and somewhat bottomless can of worms. The United States wasn't supposed to fly aircraft over Soviet territory, but what about orbiting a satellite over it? How far up did a nation's airspace go anyway?

So along came the U.S. Government's National Security Council Memorandum No. 5520, dated May 20, 1955, which was at the time very much classified. It specifically recommended that the United States "launch a small scientific satellite under international auspices, such as the International Geophysical Year, in order to emphasize its peaceful purposes."

Nothing's better than some multinational program of noble scientific pursuit to piggyback on for self-serving interests. A Trojan horse. Without question, the intent was to establish a precedent of "peaceful" satellite flyovers so that when the nasty spy satellites came along later, it would be no big deal. Only six days elapsed between introduction and adoption, although a public announcement indicating governmental sponsorship of the International Geophysical Year (IGY) didn't materialize until later that July. There was more: any scientific satellite program would be overseen by the Defense Department and couldn't interfere with any existing ballistic missile program.

In short, the IGY would have its satellite.

The Pentagon's next major consideration entailed deciding *who* got to launch. Van Allen knew from Stuhlinger that von Braun's Jupiter looked promising enough. But that was an army division, and a couple of the other defense branches wanted a shot. In response, yet another committee formed. Eight men united under the leadership of Homer Joe Stewart, a respected professor at the California Institute of Technology. His cohorts arrived at three feasible options for getting the job done. They could use the army rocket, which was basically ready to go. On the other hand, the Naval Research Laboratory promised success with a completely brand-spanking new rocket, to be developed and built only *after* being given the go-ahead. Or there was the air force and their newfangled Atlas missile—the machine to supposedly become America's great ICBM.

All showed promise. They would be compared and contrasted, argued and vilified, condemned and championed, thumbed-up and thumbed-down, in a ruthless winner-take-all cage match known as the Stewart Committee.

One approach could be dropped fairly easily, and that was the air force plan. The Atlas was practically still in the wrapper; it hadn't been officially qualified for flight or even completely tested all the way through. The fact that it was originally meant for a less-than-peaceful ICBM program violated a major tenet of the secret government resolution. Another nail in the coffin was pounded home by Major General Bernard Schriever of the U.S. Air Force, in charge of the branch's ICBM department. He remained patently uninterested in allowing his shiny new plaything to be appropriated for a series of frivolous science fair projects. Such intrusions could radically postpone his ability to dust Moscow at will.

That left the army and navy options. If the football teams weren't going after each other, it was the rocket teams. Both approaches combined existing and new technology to reach orbit. The army's scheme utilized exactly what

Stuhlinger had told Van Allen about that day in Princeton: a Jupiter booster with multiple solid-fuel upper stages. They were called scale Sergeants and came from that Jet Propulsion Laboratory place. Each kicked out a little over a thousand pounds of thrust from a tube four feet long and six inches in diameter. About a dozen of them would ring the Jupiter's nose and make stage two, encircling a trio nested in the middle, which comprised the third stage. Stage four took the shape of one elongated Sergeant—the bottom half was the rocket itself, with the payload chamber above. Just enough alley-oop to kick a modest amount of weight into orbit. And for simplicity's sake, the two would be firmly bolted together: payload on top, empty Sergeant dragging along in orbit. The clustered arrangement was a bit odd, but everyone on the project had complete faith in the scale Sergeant. More than six hundred static tests had gone down without any problems.

Working from scratch, the Naval Research Lab submitted two possible arrangements. The one that could conceivably launch sooner involved something called a Viking rocket getting things off the ground first, then handing off the work to a pair of as of yet nonexistent solid-fuel stages, to be developed by the Martin Company. On paper, this project, called Vanguard, was supposed to be one heck of a rig—but that's all they had to back it up—paperwork.

In hindsight the strength of the army's configuration is obvious. Already they had years of flight testing and real-world use under their belt. But consider the roots. "The President was very reluctant to use the Redstone," Van Allen suggested, "because of its sort of feeling of threatening character, that we were taunting the Russians with it or something." So the Vanguard was less threatening? He smiled and shook his head. "It was said to have no military capability—well that wasn't true at all of course. It had just as much as the other!"

Then in late July, Eisenhower's press secretary James Hagerty made a few small ripply waves during a press conference. That national security Trojan horse memo was still eyes only, and the Stewart Committee was not exactly public knowledge, either. But Hagerty stood at the White House in front of hovering reporters and did an "oops." He indicated the United States' intentions to launch a "small Earth-circling satellite" during the IGY.

The Sixth International Astronomical Congress was concurrently underway in Copenhagen, Denmark, and the speed of communications was quite different back then. Four days later, Hagerty's comments drifted in across the

Atlantic and smacked the Soviet delegation right square in the face. Hagerty probably wasn't trying to start a satellite war, but that's exactly what happened. One of the Copenhagen attendees taking notice was Leonid Sedov, a highly respected Soviet gas dynamics expert and head of his nation's contingent in attendance. Being nobody's fool, he wrangled up a press conference *that afternoon* at the Soviet embassy there in Copenhagen. For all the world it resembled a farmhouse living room. Wearing a suit and tie, he sat in delightfully comfortable chairs with his associate Kirill Ogorodnikov, on either side of a large window amid heavy drapery. About fifty reporters crowded the small space.

"From a technical point of view," Sedov opined from behind thin glasses, "it is possible to create a satellite of larger dimensions than that reported in the newspapers, which we had the opportunity of scanning today." A remarkable offering. "In my opinion," he calmly prognosticated, "it will be possible to launch an artificial Earth satellite within the next two years. The realization of the Soviet project can be expected in the near future."

"I knew Sedov," Van Allen said. "He was not bluffing."

The next day's *Los Angeles Examiner* carried a story called "We'll Launch 1st Moon, and Bigger, Says Russ." Sedov had lit a fire. Almost immediately, the scientific-satellite proposals that Korolev and his associates had been pushing and jabbering about to their superiors took on new significance. If the men could somehow obtain suitable permission from the special committee, their next critical steps could be taken.

Only a few months old, the Russian Special Committee for Armaments for the Army and Navy united several legacy departments under one gigantic umbrella of organization and control that supposedly oversaw any implementation of Soviet weapons technology. Therefore, seeing as how rockets were thought of primarily as delivery devices for nuclear weapons, they fell under the special committee's oversight. And that August 30, at a hastily convened meeting, Sergei Korolev quietly walked up front to speak of his dreams.

Assertively commanding the podium, he launched, grandstanding, preaching the inherent and *obvious* political significance of being the first to orbit a satellite. *You sinners! Didn't you people understand?!* And he wanted to go even farther than a measly hundred miles up: at Korolev's direction, one of his sector chiefs had already roughed out a plan to reach the Moon. The Moon! His performance went on and on as he spoke of these plans—

Korolev's emotional, convincing orations touching on his love of space. The passion. The fervor. He knew it was a tough sell. Many of those watching wanted to know when in the heck that ballyhooed R-7 would finally be operational. You know, the one they kept kissing all the rubles goodbye for? Shouldn't they talk about satellites *then*?

Committee Chairman Vasiliy Ryabikov finally halted the sermon. Korolev stood there sweating, motionless, ready to dive in again. He did not like being interrupted. His socks were hot. With finality Ryabikov extended to Korolev a very narrow range of permissions: when completely tested and flight worthy, the R-7 could be *applied* toward a modest Earth satellite— no crazy aiming for the Moon. *Prozteetyeh*—sorry.

After the meeting broke, Korolev hurried to another at the Academy of Sciences where his poignant orations continued. With people like Valentin Glushko gawking, Korolev—fueled by Chairman Ryabikov's consent—proclaimed the ongoing and detailed R-7 research as having led to major improvements. Reliability. Capability. Richness of service. "As for the booster rocket," Korolev proclaimed, "we hope to begin the first launches in April– July 1957." Confidently he declared its ability to loft nearly any cargo. Even passengers. He wanted to launch scientific satellites. He wanted to launch military satellites. He wanted to send up some dogs. Emotion built up on the podium. It was all laid out on the calendar in his head. All planned. Korolev wrapped the sermon with a fervent rallying call to action: *Launch before the IGY even starts!* he cried, fists clenched.

After such an exhilarating one-man pep rally, the impressed attendees could only nod in agreement. The effort called for a bit of structure. Academician Mstislav Keldysh would head up the undertaking with Korolev directly beneath. And the scientific community would have to be entrusted with developing a proper course of spaceborne research. No problem; that relationship was flourishing: many involved had already been flying experiments on various *raketa* iteratives, ballistically arcing up over a hundred miles and then plummeting back down. They'd been waiting for someone like Korolev, someone capable of wheedling their government into the next big step.

Now he was here.

By October of 1955 the time had arrived to make some binding decisions about what all the American IGY satellite would be. In charge of these preparations, a man named Joseph Kaplan employed a noticeably different strategy than Korolev's grandstanding. He did a very American thing, and that was to organize a committee to serve as a technical panel on the subject. The resultant list of names happened to include Van Allen's. That following January, the panel's chair, Richard Porter, approached Van Allen with the idea of having him serve as chairman of a subgroup. Their name would come to be known as the Working Group on Internal Instrumentation. Formation of the group was predicated on the conclusion that, since the business of rocketry proper seemed viable and afloat, now it was time to pick out a few goodies to stick in the nose.

They met again two months later, having made some fundamental decisions. Chosen experiments—and experimenters—had to meet four major criteria:

> *Scientific Importance*: Would your research contribute to the existing base of knowledge? Or was it going to be the same-old same-old?
>
> *Technical Feasibility*: Can your doodad actually be built? Has anything like it been done before?
>
> *Competence*: Who are you, and what makes you think you have the chops to pull this off?
>
> *Importance of a Satellite Vehicle to Proposed Work*: Is a satellite the best way to do what you propose? How else could the work be done?

On June 1, 1956, there was yet another meeting. Somebody had paid attention to the marketing; twenty-five proposals loaded the group's in-box. But Van Allen's associates already knew that not every gizmo could fly, even if they had all been perfectly suited for the program. They would never have enough rockets. So the pool got whittled to four, one of which was Van Allen's proposal, "Cosmic Ray Observations in Earth Satellites." It happened to be the most expensive. Each contender was assigned "Flight Priority A," meaning that whenever the glacial Stewart Committee finally got around to choosing a booster—and if it ever worked—then at least one of them would get to fly.

Van Allen continued sputtering to work each day in his Volkswagen. He juggled classes, faculty meetings, drop-in students with quick questions. The students were definitely his weakness: faculty in general suffered much longer lead times when making an appointment. He taught General Astronomy to undeclared freshmen. He sifted through day-to-day issues within the physics department, all the while somehow keeping tabs on the accreting satellite program. As it slowly grew toward reality, a small cadre of devoted grad students gravitated around him. Ernie Ray. George Ludwig. Carl McIlwain. Larry Cahill. They formed something of a close-knit pack of Cub Scouts. Each in a sense was untested yet exercised a sharp, analytical mind and the ability to clearly focus on the uncharted waters ahead.

Possessing a great interest in most anything electronic, Ludwig perhaps best fit the critical role of translating Van Allen's proposal into real hardware. If your stereo was broken, you'd visit Ludwig. Larry Cahill had already known him a couple of years. "He had considerable experience in electronics," Cahill explained of his associate's prowess even as an undergrad. And they were friends. "He and his family lived a couple of blocks from us in east Iowa City. We met socially quite frequently."

Ludwig rated as perhaps a nonconforming physicist. "I've always been most interested in the technology," he explained of his passions. Ludwig's bachelor's degree, also from Iowa, was in physics. Then he'd begun a master's program in the same field. But after taking virtually every physics course in the catalog—even the weird ones—Ludwig realized the multitude of knowledge to be gained from engineering studies and began snatching up those classes too. "It turns out that that combination was made to order for me," he said.

In the spring of '56 the group dived headfirst into the details. Issues of weight and power were becoming painfully obvious. These were new problems—who cared what your stereo's turntable weighed? Electronics builders normally didn't have to worry about things like shake testing either, but during its ascent, that rocket would shake like Elvis. How the heck do you make delicate electronics survive four minutes of *that*? What about the huge temperature variations during orbit, as it went in and out of direct sunlight? *Grrr...*

One thing was sure: the heart of the system would be Geiger tubes. Named after inventor Hans Geiger, they are also known as Geiger-Müller tubes, depending on how much importance is being ascribed to Geiger's assistant. A Geiger tube senses radiation. Typically they're located at the core of a Geiger counter—a handheld device used to measure radiation in any environment, from a nuclear-weapons plant to an X-ray facility. Geiger tubes had often found use in many of Van Allen's rockoon experiments, and now they were folded into Ludwig's design.

You can't just stick a tube in a box and send it up there. Different flavors of Geiger tubes measure different levels of activity. Depending on what you guesstimate the radiation level to be, complementary tubes are utilized in one single instrument. They would need batteries, since Van Allen's gas turbine idea would never have made the weight requirements. Solar power? It was not yet well-developed enough. The tubes would also need a way to report their measurements, meaning a radio transmitter—plus *its* electronics. Finally, all of this would have to be secured on some kind of mounting chassis. They only had a few pounds to work with.

To accomplish all this, Ludwig entered the picture, and the guy didn't waste much time. Only three weeks before, he had cranked out a rough diagram showing the basic components and signal flow. It demonstrated enough to meet Van Allen's proposal requirements, but now they needed a total design that would lay out every single piece and wiring connection in detailed schematics. A drawing to actually *build* from.

Within fifteen supersonic days Ludwig had it all on paper. As a bonus, the complete experiment could serve as the foundation of his PhD thesis. He tacked on a small magnetic tape recorder, reel-to-reel style, able to keep a record of the collected data and then radio it down on demand. Brilliantly, this enabled global coverage, avoiding the wholly undesirable situation of a continually broadcasting satellite whose signal could only be received while in range of the United States. Ludwig's addition overcame this weakness. Once over an American receiving station, a simple radio command could be sent up to play the tape, and a contiguous stretch of data would burst from the transmitter in only five seconds. Afterward, the machine's circuitry would reset itself back at the beginning for another run. It would weigh half a pound and nicely fit in the palm of a hand. At the time, young Ludwig had no way of appreciating how the recorder alone would become his own private hell.

For a good twelve hours a day the place buzzed with activity. Now known as McLean Hall, the physics building rests at the southwest corner of Iowa City's square-shaped arrangement of buildings more familiarly called the Pentacrest. Four squared, immense, weathered gray buildings mark the outside perimeter, accenting Iowa's original capitol building at the center of the square. The grassy areas in between have, over time, played host to inaugurations, rallies, sermons, protests, and innumerable Frisbee games. Just up the hill from the Pentacrest, past Clinton Street, an unimposing downtown area begins, featuring sporting goods stores, banks, a post office, a luncheonette. Occasionally Ludwig's wife, Rosalie, would drop into the lab, lugging along the couple's two young daughters to peel George away for a family lunch.

Every single part of the Iowa instrument, every diode and fastener, had to be painstakingly researched and hand selected from a nationwide bevy of vendors. All from scratch. There were no guidelines or templates. Each element had to pass demanding tests for an environment the original manufacturers neither anticipated nor designed for. Every wire, fitting, every blob of solder, all for something easily cradled in one arm. The timing circuit involved a tuning fork. Van Allen heard about the Hamilton Watch Company's new design for an electrically driven wristwatch supposedly coming out in the next few months. It might work better than the tuning fork; it had to be investigated. Memos, meetings, coffee. Their scavenger hunt went on.

Not until months later—at the beginning of November 1956—was the tape recorder's design finalized and its mechanical parameters solidified. Ludwig was going to go with tape heads from a company named Brush, but he later found this other great company in Minnesota called Dynamu, whose heads gave a more localized pulse and allowed smaller movements of the tape. Dynamu tossed their doors wide open for Ludwig, whisking him special tape heads and any other assistance they could. Before settling on Dynamu, Van Allen and Ludwig discussed subbing out the entire tape recorder to someplace else. General Mills trotted down from Minneapolis packing a lengthy, detailed proposal down to the man-hours: 336 hours of straight engineering at $4.45 an hour and 2,586 hours of model-shop machining at $2.45 an hour. It went on, stacking up nearly $45,000 in expenses to produce six flight-ready tape recorders. Strong concerns over weight led General Mills to a number of unordinary suggestions, like eliminating brass in favor of exotically coated stainless steel shafts. Their proposal was on the table for a while, but Ludwig ultimately passed.

He would build it himself.

Sergei Korolev felt momentum. He now commanded his own design bureau—Central Design Bureau *No. 1*, even. His satellite plans were so close to breaking out of the shell and being *real*. Colleagues understood his vision. Many shared it. Even the R-7 was coming along.

Pressing toward a common goal, Soviet scientists enjoyed wide-open, professional collaboration with rocket engineers on the most suitable payloads. The process closely paralleled the operations of Van Allen's group and the experiments they had chosen: define a program of investigations, then agree on flight priority. Back in January 1956, a group of scientists rallying behind Korolev's ideas had convinced the Presidium of the Academy of Sciences—that is, the board of directors—to contact many hundreds of Soviet scientists with a to-the-point interrogatory: "Please comment on the use of artificial Earth satellites. What do you think they could carry? What experiments do you think could be conducted?" All manner of replies came back. One of the more interesting ones read, "Fantasy: I visualize a space shot in the year 2000."

Was Korolev getting ahead of himself? The Soviet Union's inefficient and somewhat paranoid bureaucracy meant a considerable amount of decision making would operate far above the man's head—despite the levels he'd risen to. Korolev's emotional performance before Ryabikov and the special committee had led to a very helpful decree permitting the "creation of an unoriented Earth satellite," as the formal document stated. But in the old-school Soviet hierarchy, mere words on paper frequently weren't enough to get the wheels rolling. *Hundreds* of such decrees came pouring forth from committees every month like wastewater.

Prevalent in Korolev's day was an atmosphere of very informal, highly political culture. The unofficial yet critically *essential* go-ahead from some high-level bureaucrat was often what separated realized projects from those that were stillborn. What Korolev really needed was a personal nod from the Man himself, Premier Nikita Khrushchev. The opportunity came later that February. A former member of the politburo, Khrushchev had recently taken over from the now on-ice Josef Stalin and would come to be something of a breath of fresh air for Soviet technical endeavors and the citizenry at large. He resembled the stereotypical crazy uncle: rotund, bald headed, aloof. Stalin had kept a fairly heavy lid on general military matters; the ballistic missile effort in particular was airtight.

It happened when Sergei Pavlovich was down in Kapustin Yar, readying things for another test of the new R-5M short-range missile. Khrushchev materialized with an entourage of hotshot party ministers to watch and maybe draw a finer bead on how the expensive R-7 was coming along. The Communist Party at large was bad enough to deal with, but its inner circle comprised a load of humorless slugs. This could be rough.

Strategically, the facility tour had been carefully orchestrated ahead of time. A real Broadway performance that went off exactly as planned. Blindly the group rounded a sharp corner only to be confronted by the presence of a full-scale R-7 model. None of the officials had seen one in person. Korolev let everybody take it in for a bit. Later, Khrushchev would say, "We gawked at what he showed us as if we were a bunch of sheep seeing a gate for the first time. We walked round the rocket, touching it, tapping it . . ." And Korolev's window had snapped open.

He planned to deliver a short executive summary on what all this wonder machine would be able to do. That's when Valentin Glushko stepped up front and began filling the air with obscure technical details more suited for an engineering conference. Korolev cut him off and tried to slide back into his own talk again, but the group's interest was fading. It had been a long shuffle through the place anyway. They were ready to move on. Korolev's opening narrowed.

He squarely faced his premier. "Nikita Sergeyevich," Korolev began, desperately trying to redirect, "we want to introduce you to an application of our rockets for research into the upper layers of the atmosphere and for experiments *outside* the atmosphere."

Khrushchev was polite enough but not overly intrigued. He didn't seem to really get what was being said. Korolev's one chance was playing out in real time, right in front of his eyes, and it was failing. There was a rope in his hands—a lifeline to space—and it was slipping out no matter how hard he clung on. Pivoting, Korolev whipped out gigantic photographs of missiles used for suborbital research. Big yawn. Then he tried drawing everyone's attention to a satellite mock-up over in the corner. The ministers looked at their feet.

The R-7 is so much larger than any American rocket as to be almost laughable, Korolev started babbling. *It will be cheap, too—the budget is already in place.* Silence. His mind roamed. Audience attention nil. Heart beat like a trip-hammer. Hanging on by fingertips, palms red, the rope almost out of his hands. Crickets chirping. There was one trick left up his sleeve, so

Sergei Pavlovich Korolev played the last and perhaps best card in his worn, ragged hand.

"*It's possible to realize the dreams of Tsiolkovskiy,*" he emotionally blurted from out of nowhere. Tsiolkovskiy's name was never one to be used lightly by any Soviet of any stature. The great pioneer enjoyed a context of healthy respect, right up there with God Almighty; misuse could prove to be very bad indeed. Korolev brought him up out of simple desperation.

Khrushchev's eyebrows caught on. "Won't that harm the R-7 weapons program?" he wanted to know, about-facing onto Korolev's comment.

The designer told him, "No." All they really had to do was change out the nose cone up on top. Remove the nuclear bomb and drop in a satellite. Piece of cake.

Khrushchev kind of paused for a second and then mumbled words to transform history. "If the main task doesn't suffer, do it."

He'd got it. Firm grip. Rope back in his hands now with length to spare, Korolev wrapped it securely around and pulled tight. He was in. An immense unweighting greeted the man's tired shoulders. He'd finally played the right card.

5. New Moon

My feeling was one of great congratulations to them.
They were just smart as hell.

James Van Allen, regarding the Sputnik team

A man on the Stewart Committee was a friend of Van Allen's and called him right away with the sour news: the committee had finally gotten around to the up-down vote, and it was 5–2 in favor of the navy. Its Vanguard rocket would carry the first American satellite—just as soon as the navy could figure out how to make one.

All Van Allen could say was a very despondent "Oh." He got off the phone and recorded the event in his 1956 notebook without publicly commenting all that much on the decision.

"He was I think disappointed," George Ludwig explained of his associate's internalized feelings.

Even people from the Naval Research Laboratory didn't believe they had beaten the army. "I thought they'd win too," said Milt Rosen, who had done the original NRL pitch to the Stewart Committee. "They had a rocket and we didn't."

"Actually, Stewart voted in favor of the army rocket himself," Van Allen later noted with deep concern. "We developed our instruments so they would fit on the Vanguard." He made it sound like a trouncing.

The losing side didn't take it lying down. In short order many prominent dissenters jumped up to remind Stewart how the army rocket had already been tested—*repeatedly*—and how any number of Jupiters could start lifting off as early as the coming 1957 New Year. How about a second chance?

Amazingly, Stewart bowed to the pressure, wiping away the earlier results and requesting that each applicant hustle back for another pitch. As soon as the NRL heard of it, they jumped to action, editing the original proposal in an attempt to boost their grades. Even their schedule went on a crash diet, trimming delivery from thirty months to eighteen. All these new figures were similar enough to the army's plan so as to be less than coincidental. The Redstone team countered with their own upgrade: they could sub in a more powerful engine and boost the payload. Wasn't it all about payload?

There was another vote—and the navy won again.

At day's end, all scientists really want is their dang data. They don't care if the launch rocket comes from the navy or the army or if it comes from little meowing kitties. The rocket just has to work, to get up there and orbit the goods; until it does, by golly, there's no data coming today. So Van Allen finally motioned Ludwig into his office to chat a little about strategy. It was early February 1956. This whole business with Stuhlinger whispering in Van Allen's ear, and then Vanguard's reselection, had given Van Allen reason to worry. Nobody fretted like Van Allen, who possessed the ability to fret all the way home from McLean Hall, completely switch it off upon walking through the front door, then click it back on the next morning to pick up exactly where he left off.

Van Allen's fretting transitioned into a bit of postulating about what might happen if two and two got put together. It gave him an idea, a simple brainstorm, the immediate consequence of which would be a steep growth of Ludwig's to-do list. Already the guy wallowed in eighteen detailed engineering drawings for the cosmic ray box and recorder, plus his normal course load of graduate studies, plus a wife and two busily toddling daughters. He was stacked.

"George," Van Allen began in a hopefully upbeat way, "I tell you what. Why don't we make it, if we possibly can, so it would fit in the Vanguard, but *also* it would fit in the army rocket."

Ludwig didn't have to noodle that around for even a second. "I immediately embraced the idea," he later said of his response. "I never had any second thoughts."

"Well, he worked for me," Van Allen pointed out, laughing. "So he didn't have any disagreement!"

A decision of paramount importance to Van Allen, Ludwig, Iowa physics, the United States, and the general history of space exploration had just been made within an informal meeting lasting ninety seconds. Ludwig headed back to his desk.

Korolev's group called their satellite *Object D*. Isn't that majestic? The first letters of the Russian alphabet are A, B, V, G, D; earlier letters had already been assigned to other payloads.

With names out of the way, the rough technical details were finalized on February 25, 1956. The latest projections for the R-7's abilities gave them a payload weight between 2,200 and 3,000 pounds—all the more startling in light of Vanguard's top-end ability to orbit only twenty-three pounds of payload. And fully 75 percent of *Object D*'s total weight was given over to experiments. Rocket noses are always cones, so *Object D* would take that shape.

Inside would be nothing short of a staggering tour de force, including eight major scientific investigations, such as the collection of atmospheric data at varying altitudes, solar radiation, ions, electrical charges, magnetic fields, and the solar spectrum. Just about everything that *could* be studied.

Mirroring the activities in Iowa City, Central Design Bureau No. 1 set about work on an apparatus with no precedent. During launch the experiments would get shaken like a martini. Then they'd spend their working lives in a total vacuum. Russian scientists wanted continuous data from all around the globe, but the country's few tracking stations were all inside the border. They'd have to record the data somehow and then be able to zap it down on command. Maybe some kind of little tape recorder . . . hmmm . . . that meant producing a new category of automated electronics out of whole cloth.

By the middle of 1956, *Object D* was completely behind schedule. Of course American technological development occasionally fell off pace, but generally not due to the apathy and carelessness that became standard fare on the road to *Object D*'s completion. In the United States it was all about the lascivious dollar. Master negotiators for enormous companies drew up elaborate sliding-scale, cost-plus-incentive contracts. Once signed, the middle managers under them juggled payment triggers based on delivery milestones, department bonuses, section bonuses, group bonuses, individual bonuses. It never ended. God bless the greenback.

In practically unfair comparison was Soviet Russia. The labor system in Korolev's native turf, archaic the day it began, offered no such wheeling-and-dealing terms or bonuses for vendors and certainly no enticements for any worker going the extra mile. It more came down to emotive appeals based on intangible concepts like the collective might of socialist labor.

Humans generally respond well to incentives. In Soviet industry at large, none were to be found. Korolev's design bureau had no choice but to rely on outside providers of electronics, batteries, metalwork. Unless somebody had the project in their heart, the work got done when it got done. Parts often showed up comprised of inferior materials failing to match their own design specs. By the time October rolled around, a single *Object D* test model had yet to be delivered or even assembled.

Sergei Korolev was about at his breaking point. He found himself constantly on the road, trying to stay abreast of all the secondary projects that multiplied like horny bunnies. The current launch facilities deeply bothered him. His out-of-state cabbage patch had been a decent place, but really, it was high time for an upgrade. For starters, it wasn't big enough any more. And it was too damn close to Turkey, where the United States had quietly installed radar posts to listen in from over the border. Most importantly, though, Kapustin Yar wasn't as close to the equator as orbital launch pads really should be. The Americans really had it made down there in Florida.

Somebody found a place even more out in the middle of nowhere than Kapustin Yar, yet unbelievably it was still on the railroad lines. The new site was at a good latitude, one that could offer Korolev and the R-7 an extra 4 percent of payload weight up in the nose. They called the place Baikonour, even though the town of Tyura-Tam was closer. Baikonour was also a town, but it lay hundreds of miles off. The odd name choice was a MacGuffin: if the United States ever woke up one day and felt like obliterating Russia's launch facilities, a name switcheroo might throw them off. The Pentagon needed less than three months to figure it out.

Baikonour would grow to be the USSR's premier space-launching facility. As a fascinating bit of ironic history, in 1830 the czar had exiled a certain Nikifor Nikitin to the same exact location for "making seditious speeches about flying to the Moon."

Kaliningrad to Kapustin Yar to Tyura-Tam. Reverse to Kaliningrad again and then back the way he came. Wearily Korolev progressed with a certain uneasiness, looking over his shoulder. Toward the end of September the

United States Army, with more than a little help from von Braun's Germans, scored a near home run during their latest Jupiter test flight. All three stages worked to perfection, boosting a sinfully empty nose more than 90 percent of the way to orbit. The fourth stage, however—that lone Sergeant up there—had involuntarily been modified. It was inert. It didn't work. The castration was a direct order to von Braun from no less than the chief of Army Ordinance. Even though the army's flight tests continued without pause, Vanguard was supposed to be first, launched as the "civilian" rocket. Von Braun was not amused at such blatant hobbling.

Sergei Korolev knew enough about the attempt to make himself leap into wild extrapolations. He surmised that the whole episode was really a secret attempt to orbit a satellite and pilfer his nation's thunder. Of course he was mistaken—and obviously knew nothing of the odd politics crippling Jupiter—but such paranoia kept him getting him out of bed each morning until he could somehow get *Object D* off the ground.

Bad parts aside, it kept getting harder to forecast a launch. The latest thorn in Korolev's side was the booster again. Something needed to be done about the R-7's engines and their laughable failure to meet Glushko's advertised specs. As things stood late in the year, its engines wouldn't be capable of pushing *Object D* into orbit, and this is why Mikhail Tikhonravov came to see his employer in the last days of November.

Anybody could approach Korolev with any kind of brainstorm; he encouraged it. Even so, Tikhonravov spoke respectfully. Innovation is historically rife with minute suggestions that magically rewrite the outcomes. And that day, Tikhonravov flipped out a simple enough idea.

"What if we make the satellite a little lighter?" he asked the boss. "Thirty kilograms or so, or even lighter?"

Korolev practically halted midbreath. It had never occurred to him. This sort of rearrangement might be just the thing. His eyes lit up. *That* was the way to work it, no doubt at all. Proceed!

A subordinate was dispatched to rough out the ballistics changes, as Korolev took pause to reflect on his creation. *Object D*. It stood there in front of him resembling a giant's ice-cream cone, looking bizarrely accessorized with handicapped-accessible grab bars. His baby, his analytical tour de force, would have to wait. Engineers are hopelessly unsentimental; *Object D* was then brutally shoved aside as Korolev faced down the abbreviation of his scientific wonder.

A few days elapsed while sizing up the tasks, and Korolev wisely focused on simplicity. The *Object D* supply chain, if it could even be called that, traversed minefields of incompetent subcontractors on top of flaccid contractors, with a pinch of uninspired outside manufacturing facilities thrown in just for complexity. That would no longer do; he felt time running out. What they really needed to do was ink a straight line through all these crumbheads and start taking names for the Communist Party hard-liners to deal with. Actually Korolev wanted to erase the stupid line altogether and build the whole thing in-house, but that was stretching things. What about firing every underperforming supplier, giving business only to the ones who produced? He could then eliminate this order-parts-and-see-if-they're-right business. He ended up having to sub out the batteries and radio, but the balance of the craft would be produced entirely within the walls of Korolev's design bureau. And so from this chaotic environment would rise what is today heralded as *Sputnik*.

That name, yes indeed. *Sputnik*'s name originates from nothing more than a poetic afterthought. All through the design and testing phases this new machine being produced was always dryly referred to as "ps-1"—a Russian abbreviation for "Simple Satellite Number One." But when everyone in the shop heard "ps," they nicknamed it "sp," in reference to Sergei Pavlovich.

Hot waves of argument crashed through the design bureau over its most optimal shape. A significant percentage of the designers felt a cone would be best. Aerodynamics already dictated the shape of the r-7's nose; wouldn't it make sense to conform? And gee, wasn't the maternal *Object D* also a lovely cone? The chief designer successfully argued for a sphere—predicated by, among other reasons, the shape's ability to measure levels of atmospheric density in its path. Additionally, a ball shape enabled simpler and more reliable methods of separating machine from booster. So a globe it would be. Each hemisphere was stamped out of an aluminum alloy, which was then machined down before final finishing.

Inside McLean Hall, George Ludwig had his fingers in knots trying to solder diodes onto ultracramped circuit boards less than six inches wide. But the ps-1's *twenty-two-inch-diameter* sphere held three batteries, two radio transmitters, a temperature control system, ventilation mechanisms, and all the electronics in between to make it work. All of this was painstakingly assembled on a thin supporting framework. Four cartoonish catfish

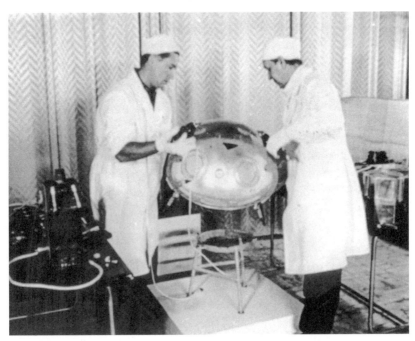

6. Technicians readying a PS-1 hemisphere. (Author's collection)

whiskers arched back off the sphere to function as radio antennae. When fully assembled, the device weighed 184 pounds—heavier than the allowable Vanguard payload by a factor of eight.

A twenty-two-inch beeping ball. The one to cause Eisenhower's frightmares—the one that Van Allen and Cahill stayed up half the night scrambling to record—was never Soviet Russia's best foot put forward or ace in the hole or even a logistical milestone on some carefully thought-out road map. Not even for a second. This first device was born of panic and reaction. It was pure compromise, a shotgun wedding.

Unbeknownst to Korolev or anyone on his team, a seemingly unimportant event was about to occur during this second half of 1956. Somewhere out in a barren part of the Asian continent, the Soviets were recovering American high-altitude Genetrix equipment. And they would look over its design with perhaps the same disbelief that anyone might feel after learning of it today. The U.S. Air Force had independently taken it upon themselves to plan, organize, construct, and then launch on flights over Soviet Russia hundreds of sophisticated camera surveillance packages—hung underneath nothing more than low-tech balloons. The cameras were some of the better long-range instruments available at the time, which were set to click every six to twelve minutes, imaging whatever terrain rolled below onto giant spools of specially formulated, large-format film. Success was easily a huge question mark. Beyond weather forecasting and the choice of launch site there was no way to otherwise control their path, or point their lenses except in a generally downward fashion. Where they went was up to the winds. More were shot down by the Red Army than were recovered by the air force. And at some point, a roll of unexposed Genetrix film began to slowly make its way toward Central Design Bureau No. 1.

Having been selected not once but twice, the Naval Research Laboratory now had to cough up their rocket—and Vanguard was of course not without growing pains. Now, this is entirely normal for a complicated, high-performance device. The V-2 had been engineered and reengineered and further tweaked for years and years before it ever worked like it was supposed to. Where Vanguard differed, however, was in the rather bright light being shone on its development by the American media. Hundreds of reporters crowded the naval test ranges, and the most insignificant of failures showed up on the next day's front page. The Soviets had copies almost as soon as the newspapers landed on American doorsteps. People like Van Allen began to think of Vanguard the way they thought about canker sores.

"Why don't we go with the army rocket?" he started bringing up at the IGY gatherings, hounding, throwing his hands wide. It was so obvious to him. "The Vanguard has all kinds of development problems." In short order, many military bureaucrats noticed their ears were hurting. It became an issue at virtually every meeting: "Why don't we just opt for the army rocket?" Van Allen would blabber on. "It's a much more tangible prospect." He kept pestering even after someone up in the higher echelons flat-out told him off. "It was contrary to national policy, so to speak," he remembered of the situation. "I never got anywhere."

Unfortunately, the American government had already spoken, and opposition to Redstone came from several angles. Look at the National Security Council: no way would the Pentagon allow something like anemic scientific research to impede the progress of studly ballistic missile development. Some of the trouble came down to perception. Right or wrong, the view persisted that von Braun's rocket was intimidating or that it would somehow taunt the Soviets. Even so, Van Allen and Ludwig continued their bilateral development. "We had a bona fide effort to make it for the Vanguard," Van Allen clarified of his intentions, puzzling over which way the loopy situation would break.

Spring of 1957 turned into summer. Out by the corroded rail tracks in Tyura-Tam, beastly heat pummeled the grounds and crew. Death row was more fun. For two hellish years the laborers had been laboring, and the place was only now beginning to come together. A rudimentary airport, plus a roofed hangar in which to assemble the rockets, had gone up on-site. The men kept working. During precious off-hours, an alluringly popular activity was catching scorpions in glass jars and then watching them fight.

The launch pad was finished. To dig its foundation, platoons of men relentlessly levered picks and shovels to open up a hole 150 feet deep and more than twice that across the length and breadth. They called it "the biggest hole in the world"—manly but unfortunately inaccurate.

Sergei Korolev stood watching it all rise before him. Engineers and technicians slowly arrived in small groups, populating the makeshift barracks; produce trucks queued up at the mess tent, and fuel tankers converged near the pad. Despite the fact that his fortunes completely depended on the rocket working, Korolev now enjoyed luxurious confidence deep inside himself. A true rarity. Korolev's entire vision slowly assembled like the facility before his eyes. Spasms of energy circled the grounds as people moved test equipment, inspected fuel, ran wires, fixed trucks. Some technicians had been out in the country away from their families for five grueling months, now. Boris Chertok, in charge of assembly and test, recalled twelve- or fourteen-hour days every day of the week, and Korolev worked himself the hardest. Beating America into space was first prize.

Far away from the biggest hole in the world, back in the main offices of Central Design Bureau No. 1, studious engineers focused themselves in ramshackle offices, trying to see how far they could get on the forthcoming satellite's trajectories. Elaborate math problems, the kind to make any tenth grader crumble, were key to accurately launching and placing their creation. The engineers ran the numbers by hand, multiplying five-digit numbers by other five-digit numbers and consulting eroded six-digit trigonometric tables in additional rounds. The formulas grew so twisty and complicated that such prehistoric stone tools couldn't handle it any more. Soon thereafter, an urgent series of phone calls gained the men access to what was perhaps the fastest computer available in the Soviet Union at the time. The room-sized machine could make a staggering *ten thousand* calculations *every second.*

Launch operations made great use of Soviet Russia's strangely dependable railway system; they still do today. Korolev assembled his R-7s on their sides in a hangar. Then, still lying down, the rockets were transferred to gigantic rail cars for pad transport and fastened onto a massive swing-arm. It was a simple operation, rough-and-ready. At Tyura-Tam's pad, the arm levered its cargo to vertical, and three enormous paddles of metal framework hinging at the launch tower base folded in to envelop the rocket from all sides. The engineers who built the pad and its attached framework nicknamed the thing *tyulpan* ("tulip"), because at launch the arms swung outward on counterweights and seemed to imitate a bizarrely oversized metal flower in bloom.

Finally the rocket was plugged into ground control and the arduous process of fueling could begin. The first R-7 finally flew—briefly. But a large red ball of flame unexpectedly grew in the sky, and hot metal bits rained down across the tundra. The second R-7 launch didn't go so well either. On June 9 of the same year, the rocket didn't blow up after liftoff, but three successive pad aborts left everyone feeling not very great about things. The last abort had also been punctuated by a heavy rainstorm that went so far as to flood several of the less-than-improved buildings.

The days that followed were low. Korolev wrote to his wife that Valentin Glushko had "arrived today and to everyone's amazement (mine included!), using the dirtiest language and the crudest phrases, began telling us all that our work was utterly worthless—and this, just half an hour after he arrived. I answered him calmly (you can imagine the nerves that that cost me!) and only criticized him for his intemperance and arrogance."

The next try came on August 21, and Sergei Korolev's lovingly prepared R-7 cruised the full intended range, all the way to Kamchatka. It ended the flight by exploding at a height of six miles, but Korolev was so jazzed that he stayed up until three in the morning just thinking about it.

Following the victory, Korolev scheduled an uncomfortable yet necessary visit with the State Commission to get a final green light for the little ball. Standing before the assembled group, he requested clearance—provided that a second triumphant R-7 flight came off—to launch his Simple Satellite Number One. Many members blankly stared at one another. Simple Satellite? Plenty were familiar enough with *Object D*, but they either knew nothing about the abrupt change in plans or, at the very least, paid scant attention to the small flurry of paperwork that had recently come through. A little ball didn't sound like the way to go.

A meeting that had already gotten off on the wrong foot went quickly downhill from there, degenerating into a handful of sniping, petty arguments. Toward day's end they adjourned with absolutely nothing accomplished. At the commission's next session, the bedraggled Korolev stood up to force a decision.

"I propose—let us put the question of national priority in launching the world's first artificial satellite to the Presidium of the Central Committee of the Communist Party." He looked around. It was a real gutsy thing to say—daring as hell. Even more powerful than the Council of Ministers, the Central Committee of the Communist Party acted in many ways as THE final word on *any* matter of policy. It was *the* committee. "Let them settle it," Korolev bluffed.

Nobody called him on it. And when September 7 saw a second R-7 fly far and true, the core engineering and design group bundled themselves off to Moscow and set about planning the actual launch effort. Among other details, they agreed on October 6 as a launch date. If everything else went according to plan, they could optimistically be ready to go inside of thirty very busy days.

Korolev arrived back in Tyura-Tam on September 29 to supervise the preparations already in full swing. One of the satellite's batteries was giving them trouble. It was swapped out before the swear words could finish crossing Korolev's tongue. And something else happened: while conspicuously omitting the particulars, a recent article in the Soviet journal *Radio* listed two common broadcast frequencies easily picked up by any amateur wishing to "tune in an orbiting satellite's signal."

In the middle of everything, the overtired chief designer abruptly made a suggestion: how about moving the launch earlier by two days? Those

Americans worried him to no end. Korolev had learned of a series of meetings to be held in Washington DC beginning on October 6—and concerned speculation arose that this was somehow going to be timed with an American satellite launch. He kept whining about history and missed opportunities. Begrudgingly the State Commission agreed to move things up.

On the morning of October 3, the R-7 destined to carry *Sputnik 1* rolled very slowly down rail tracks to the launch pad. In solemn procession Korolev walked alongside, accompanied by Vasiliy Ryabikov and several other members of the State Commission. For a mile they paced the machine, largely in silence. From Korolev came no outbursts, no swearing. Only words of caution. He constantly mumbled, "Nobody will hurry us. If you have even the tiniest doubt, we will stop the testing and make corrections on the satellite. There is still time . . ."

As the Sun began to set on the scrub and barren plains of Tyura-Tam, humongous floodlights brightened up the launch site and booster nestled into its petals. It was scheduled to fire at midnight. "Nobody back then was thinking about the magnitude of what was going on," recalled engineer Oleg Ivanovsky. "Everyone did his own job."

It was nearly time. Korolev nodded to his assistant Boris Chekunov, who was in charge of actually pushing the button. "One minute to go!" cried out the chief designer. Almost everyone was standing, Glushko near Ryabikov and the others. A senior engineer later recalled, "I kept an eye on S. P. Korolev. He seemed nervous although he tried to conceal it. He was carefully examining the readings of the various instruments without missing any nuance of our body language and tone of voice."

The clock hit zero, Chekunov thumbed his button, and Korolev's R-7— *their* R-7—sprang to life. Each strap-on booster was actually four rocket nozzles fed by a common set of tanks. It could have been one nozzle, but Glushko was never able to figure out how to make them any larger. Working together, burning liquid oxygen and kerosene, the engines blasted 876,000 pounds of thrust downward into the biggest hole in the world.

It took off like—well . . . like a rocket.

Telemetry streamed down via radio—the equivalent of EKG leads on some hapless cardiac patient. It climbed higher, black night giving way to black space. The crowd held its breath. Korolev would get his way: core stage separation happened as planned. Four boosters fell away from the R-7's central core as their PS-1 jacked into orbit.

She'd made it. Korolev surveyed the room. They had *all* made it. *Almost.* Eight hundred breezy yards from the launch pad sat a van containing one of the guys who had designed the satellite's onboard radio system. Vyecheslav Lappo hunched over a modest radio stack, eagerly waiting for signals. Downstream, Kamchatka caught them: the word was good. It bounced on back through the line, and then everybody started cheering—except Korolev. "Hold off on the celebrations," he cautioned, quieting them down. Whatever Kamchatka had was from the ascent.

"Let's judge the signals for ourselves when the satellite comes back after its first orbit around the earth."

In contrast with the frantic intensity before launch, nothing was left to do but wait. No tracking stations existed outside the Soviet Union, so their ball had to loop around past Europe before it made any sense to listen again.

"Waiting built up the stress," recalled a military colonel in attendance. "Everyone stopped talking. There was absolute silence. All that could be heard was the breathing of the people and the quiet static in the loudspeaker." And the amount of quietness rose to deafening levels, where static became the loudest thing in the crowded room. It was like being in a jam-packed bar, yet all you hear are the napkins rustling.

And then the beeps started drifting into Vyecheslav Lappo's expectant ears. Rising in strength and alternating between the two frequencies, that was it all right. A subordinate in the van snatched the phone straightaway and rang up Korolev, who was anxiously still in the bunker. The news went over well, as good news always does. Hugs and tears and claps on the back.

More details were coming down: apogee 588 miles, perigee 142. It was not as high as they wanted. But that engine had cut out too early, and without question that was the prime suspect. Ever so slowly their aluminum ball was losing altitude—not a good thing at all and definitely unfixable. Chairman Ryabikov waited until after the second orbit, pretty much following the plan, and then telephoned Khrushchev to spread a little cheer.

The Soviet leader was parked in Kiev debating policy with leaders of the Ukrainian Party. He took the call in another room then coolly returned to the discussions at hand. Naturally, the unique burden of keeping his mind on the conversation became tougher with the passing minutes. It ballooned inside him like a heavy bowl of bran flakes. Finally, he couldn't keep it quiet any longer.

"I can tell you some very pleasant and important news," Khrushchev abruptly said to change the subject. His tone then morphed into secrecy: "Korolev just called. He's one of our missile designers. Remember not to mention his name—it's classified."

People looked at one another. Um, where was this going?

"So, Korolev has just reported that today," Khrushchev continued, "a little while ago, an artificial satellite of the earth was launched." The statement drifted out over the group, settling on them, absorbed at different rates. The premier's mood decidedly perked up from then on, as he spent the rest of the evening quite jovial and bouncy, crooning ad infinitum about this grand achievement and the new technology it represented. In Khrushchev's eyes, it would "demonstrate the advantages of socialism in actual practice" to the Americans. A victory for labor.

Out at the launch site, a party exploded into full swing, prompted by the celebratory teapot of alcohol handed out to each person at the facility. Later, a commandeered movie theater in Tyura-Tam played host to speeches by Ryabikov, Keldysh, and Korolev. They went on praising everyone deep into the small hours.

Fifty-one years old, Sergei Pavlovich Korolev tossed his arms to the sky and crowed, "I've been waiting all my life for this day!"

Back in Huntsville, Alabama, Wernher von Braun was that very evening attending a party he had personally arranged. In keeping with typical von Braun policy, the whole idea was to relubricate Washington's satellite machinery now that a new secretary of defense had just come on board. So von Braun and his entourage put together a slick, expertly choreographed tour of the rocket facilities, capped off by dinner at the officer's club. There, he had the opportunity to ovate about the emerging fields of rockets and space. Nobody worked a room like Wernher von Braun.

Things were going great until Gordon Harris rushed through the door, sweating. Harris was the public relations man for the base and had some really, really bad news. "It has just been announced over the radio that the Russians have put up a successful satellite!" he blurted out to the crowd.

In near-synchronous fashion, up in Washington DC, a cocktail social picked up steam in the Soviet embassy. Barely an hour after the gates opened, American reporter Walter Sullivan barreled up the grand stairs, hustling through to the center of the ballroom where he gabbled something into the ear of Richard Porter, who swallowed and got up on a chair. Clapping his hands for attention, Porter brought forth an unprecedented announcement.

"He'd just received word from New York that the Soviets had launched a satellite," recalled George Ludwig, who had been standing not six feet away from the guy. Ludwig was in town to attend the IGY planning meetings along with other participants from around the globe. "We were all kind of taken aback. We knew that they were planning one; we didn't think they were anywhere *near* that close. And there was kind of a stunned silence for a while."

Ludwig turned his head, sighting one of the senior Russian scientists in attendance. To someone else the man coyly offered, "We don't cackle until we've laid the egg."

The gathering's flavor immediately began to change. "Then applause broke out," Ludwig continued. "There were lots of toasts . . . but then very quickly the party began to fade away as the U.S. delegates went back to their labs to try to tune in their receivers."

In New York, NBC was picking up signals. So were the BBC and the Armed Forces Radio. They were all broadcasting reports, one of which was picked up by Larry Cahill aboard the USS *Glacier*. He then ran to tell Van Allen, who was just getting ready for dinner.

Back in Huntsville, the news immediately lit von Braun's fuse. "We knew they were going to do it!" he screeched at the senior military officers. Pivoting, von Braun faced down Neil McElroy. "We have the hardware on the shelf, for God's sake!" he pleaded, in high contrast to the many congratulations offered to the Soviets at the IGY party. "Turn us loose and let us do something!" von Braun imploringly continued. "We can put up a satellite in sixty days! Just give us a green light!" He had no way of knowing the Russian machinery had been launched by a rocket that failed in five of six attempts. What was von Braun's nightmare became *the* major story throughout the civilized world by the next morning, on October 5. "Earth's Gravity Conquered," lauded France's *Le Figaro*—as just one example.

In Moscow's own *Pravda*, the term "sputnik" first saw public use. Although it's Russian for "satellite," the word can also be taken to mean "fellow traveler." It was used as a *description* of the machine versus as a *name* for it. But the Western press employed this word as a proper, capitalized title and ran with that. What didn't make it to print, conversely, were any of the human names: Glushko, Tikhonravov, Chertok, Korolev. In the Soviet press they only received oblique references, using nebulous titles quite at home in to-day's limp-wristed corporate vernacular: "Chief Designer of Rocket-Space Systems" and "Chief Theoretician of Cosmonautics." None of these could ever legibly fit on a standard-sized business card.

On the night of the fifth, Korolev and some coworkers finally straggled onto a plane bound for Moscow. They hadn't gone to bed in two days and fitfully collapsed into their seats. Once airborne, the pilot sidled out and delicately approached Korolev. "The whole world was abuzz," he updated the instant national—if undisclosed—hero.

Not everyone got as much rest that night. In James Van Allen's remote neck of the woods, so much laborious work remained to be done. Out on the steamy confines of the *Glacier*, his ragtag group absorbed what they could of the *Sputnik 1* hoopla while unfortunately in the middle of trying to execute other long-planned activities. Several difficult weeks at sea and a hold full of rockoons remained, yet this massive distraction whizzing overhead. During the launch night of October 4, Van Allen and Cahill and Gniewek persisted through the small hours, crammed into the radio shack, expending loops of audio tape while recording additional *Sputnik 1* transmissions as it passed over. Somebody suggested outputting the signal to the pen-and-ink data printer that the Iowa group had brought along. It would have been similar to the method doctors use to trace a patient's heartbeat onto a spool of paper—but no direct way existed to link the tape recorder to the printer. So Van Allen sat down and, completely off the top of his head, sketched an electronic circuit that could do the job. Then he marched downstairs to the group's equipment stash, directly above the *Glacier*'s twin screws, where he soldered it together from individual parts. Upon returning, the men quickly hooked everything together, and soon a distinct pattern began laying itself down on the paper. Van Allen also took out his weather-beaten notebook and logged the following thoughts spilling off the top of his head:

Brilliant achievement!

Tremendous propaganda coup for U.S.S.R.

Vehicle must be . . . 9 times as heavy as Vanguard

Confirms my disgust with the Stewart Committee's decision to favor N.R.L. over the Redstone proposal of Sept. 1955!

May be a genuine loss of opportunity for us at S.U.I. to assume a larger role in the future!

Van Allen's last comment summed up a great fear of his—that any work already done by his beloved physics department would now be discarded in some misguided crush to respond by any possible means. He'd later say, "Well, here I was in the South Pacific and out of touch with the situation, and we'd lose the chance to be on board with the U.S. satellites. There was going to be something big break loose up there and we were not going to be there!"

Today, disagreement persists among those who lived during this time. Certainly, no mass Western panic ensued, with people running wildly in the streets and screaming and whatnot. There did arise, however, a legitimate concern that the Big Red Bear could soar overhead at will and do whatever the heck he wanted. He could put up a Moon of his own with such audacity as to make it fly over U.S. airspace, and to be *visible* at that . . . about the luminosity of a fourth-magnitude star. If *this* could happen, what would stop such an ostentatious nation from parking nuclear warheads up there in space to lord over us all?

What about other impacts? Didn't this spotlight America's trailing scientific abilities? "In my opinion, it was not scientifically important who went first," Larry Cahill suggested, adding that the political fallout likely boded well for subsequent research endeavors. "Space scientists found it somewhat easier to get public and political support for continuing the research that they wanted to do."

Mere days after the launch, Senator Henry "Scoop" Jackson from Washington publicly regarded *Sputnik 1* as a "devastating blow"—all the more ironic in light of reactions like Van Allen's "brilliant achievement" or the applause and congratulations at the IGY party. Jackson went so far as to suggest Eisenhower officially proclaim "a week of shame and danger." Senator Syles Bridges from New Hampshire joined in, commenting on how Americans should be "prepared to shed blood, sweat, and tears if this country and the free world are to survive."

When looked at even *just below* the surface, the Soviets almost made it seem like they never had anything to hide in the first place. *Sputnik 1*'s frequencies were deliberately chosen because they were compatible with any amateur radio setup. They were published far in advance, in foreign-language radio-enthusiast magazines. The craft broadcast on five watts of power. All this meant that its beeping tallyho could be instantly received by thousands on the ground to provide quick, free verification that the thing was up there and working. In comparison, Vanguard's payload was to run on a special frequency set aside just for the IGY—and only on a few milliwatts of juice, thereby greatly increasing the difficulty of monitoring the craft.

In the double-time atmosphere created by *Sputnik 1*, something, *anything*, had to fly and it had to go *right now*. America probably would've launched a urinal cake into orbit that day as long as it fit the nose cone. President Eisenhower tried calming the waters at a news conference on October 9. He said, "From what they say they have put one small ball in the air." The remark prompted Van Allen to record another thought in his logbook: "The pompous character of the White House announcements!"

In simultaneous occurrence many time zones over, Sergei Korolev sent a bunch of his closest *Sputnik 1* teammates to sack out at a dacha in Sochi near the Black Sea for a mellow five days. Then, after deplaning in Moscow he finally managed to drive home and steal a night of rest. The next morning, his phone rang. It was Khrushchev over at the Kremlin, and he got right to the point.

"We never thought that you would launch a sputnik before the Americans," the premier casually applauded. "But you did it. Now please launch something new in space for the next anniversary of our revolution."

Korolev gripped the receiver. His head must have been spinning . . . *the anniversary of the revolution?* Why, it wasn't more than a month away—coming up like a freight train! The time interval was more suited to planning a dinner party than a satellite launch. *And something "new" in space?* The only "new" thing the Soviets had lying around was *Object D*, and enough booster power still wasn't coming from Glushko's wimpy engines to shove it all into orbit.

The inability to send *Object D* up, then, still left an unfulfilled "new" criterion—by necessity different from *Sputnik 1* yet familiar enough to the engineers to meet the deadline. Korolev kept his premier waiting as ideas silently clicked by. *Something different, although familiar. A step up, yet within reach.* And then a natty thought swam into his mind.

There should be a live creature on this next one, he thought. Not a handful of crickets or spiders or mice or some weak overture such as that, but a true and noble animal like a *dog*... something capable of affection, happiness, and eliciting a positive response from the world. There's no way that an adorable, furry critter for people to fall in love with could be even remotely construed as menacing.

Steeling himself, Korolev made this suggestion to the leader of the Soviet Union. It was a blindingly audacious, unthinkable proposition that could be either championed as brilliance or derided as lunacy. Khrushchev? Tickled pink. Korolev breathed relief at his idea's acceptance then hung up the phone so he could get on it. His next call was to immediately summon home all those still languishing at the dacha.

Had he not already enjoyed great success with high-altitude canine flights, Sergei Korolev may well have not offered up what he did. As early as midsummer 1951 his design bureau had been working with dogs in a string of launches through high-altitude parabolas and had safely recovered most all of them. Dogs were chosen over other creatures like apes for various hair-splitting reasons, including "less excitable" and "easier to dress." The flights incorporated tricky mixed-gas environments, about four minutes of weightlessness, and recovery by parachute. A total of nine wide-eyed canines experienced ballistic trajectories above sixty-two miles, which is generally considered the threshold of true space. Three lucky dogs flew twice; four died when their parachutes failed. One was adopted by an engineer after its flight.

When the chief designer barged back into his bureau after that decisive phone call, tall orders came pouring forth from his increasingly stout frame. Verbatim instructions from the Central Committee dictated that this launch had to happen "without fail" by the holiday, and until then the short-sheeted deadline would hang around Korolev's neck like a lead choker. There wouldn't be enough time—or *any* time, really—to come up with something from scratch or even to pursue a conservatively sound method of building the equipment.

"It was, I think, the happiest month of his life," recalled Georgi Grechko, an engineer for Central Design Bureau No. 1 who later ascended to the status of cosmonaut. "He told his staff and his workers that there would be no special drawings, no quality check; everyone would be guided by his own conscience." And that's what they did.

Working quickly, the group agreed on a Frankenstein-like mating of a canine enclosure and the reserve PS-1 globe. The engineers made the drawings and gave them directly to the fabricators, who started building the necessary parts and then handing them off down the line. An atmosphere of complete trust and confidence.

As questions about dog-handling kept arising, Korolev managed every one with ease, practically clapping his hands. He'd been through this before, every bit of it! Answers came flowing out of an open tap. He knew his favorite breeds, sizes, temperaments. He knew the best colors, the best position to set the animal in for launch. He only wanted females due to the waste management system, which was also in hand and flight-tested. He even possessed a retread environmental capsule left over from six years ago that had been flown and recovered. *Nothing like experience!*

Only six brief days after *Sputnik 1* had shocked the world, the sequel was already coming together. The kennel part resembled a sort of half cylinder with a flat bottom. Inside, the setup—liberally padded all the way around—permitted one single dog to stand or lay down as the animal desired. Any room left went to life support and scientific experiments, including a cosmic ray package. Their spare PS-1 orb would fly in an almost identical configuration as the first, right down to the four radio antennae. The animal enclosure with the ball stacked on top formed a relative cone shape, measuring seven feet across at the base and twelve feet high. It was 1,100 pounds altogether, a truly staggering amount of payload. That totaled more than six times the weight of *Sputnik 1*, easily surpassing that of all four planned Vanguard satellites added together.

They figured a seven-day mission would be entirely within their grasp. In an effort to save weight, the unified cone would remain attached to the R-7's booster core, eliminating the need for separation mechanisms (not to mention their time-consuming tests). Another bonus was that the second artificial Earth satellite contained a radio that would serve double-duty for both itself and for the booster to save even more weight.

Korolev believed that all the extra metal up in the nose would also dissipate the heat better and keep their little critter cooled off. "But," a worker later explained, "we didn't take into account that the metal of the construction could bring more heat to the animal."

Which dog should go? Six available finalists had been properly screened and trained. They all shared many of the same overall characteristics, so it

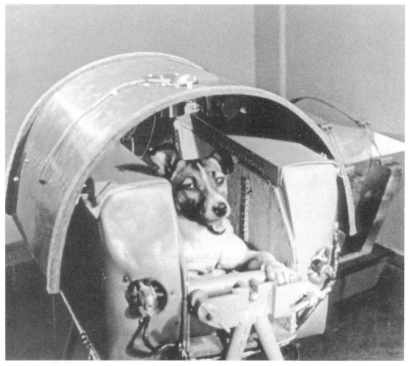

7. Laika in her recycled kennel, prior to its mating with the remainder of *Sputnik 2*. (Courtesy of Colin Burgess)

mostly came down to attitude. Biomedical specialist Vasiliy Parin picked her out. She had been a stray formerly wandering the streets of Moscow. In English the part-Samoyed's name was Barker, translating from the Russian Laika.

Another dog named Albina had already been up twice in ballistic flights and was nominated as the backup. Albina had actually emerged from all tests as the favorite, but that was exactly the problem: she was their favorite. Everyone loved Albina and didn't much care to see her go on the one-way trip. Besides, she'd just had puppies. Behind that dog a third choice waited, held in reserve in case something happened to the first two.

An unidentified member of the engineering team recalled the scene one evening. "We were in the midst of preparing the second satellite for launch," he began. "Several dogs, little mongrels dressed in cloth, were running around on the floor, and we lamented the fact that one of them would soon die a gruesome death in orbit. During an interval between tests, someone came rushing into the control room shouting: 'Why are you all sitting here? The satellite is about to pass overhead. Come on outside!'"

They all piled out, gawking up at the night sky. Korolev stood among members of the State Commission, surrounded by other personnel, out in the open, chins up, squinting and straining, waiting for the faint point of light marking *Sputnik 1* to climb over the horizon and into view.

Total silence from the group.

"When the satellite did appear," the man remembered, "it rose high in the sky as it moved from the southwest toward the northeast. It kept us spellbound for several minutes until it finally vanished." There it went. Their big shocker.

Then they went back inside and resumed working.

6. Let's Make a Deal

I feel that a certain amount of motion,
even turmoil, is necessary for my maximum enjoyment.

George Ludwig, discussing his state of mind

Three weeks after *Sputnik 1* had flown, on October 30, Van Allen was still in the thick of his rockoon flights when a radiogram arrived on the *Glacier*, addressed to his attention. And radiograms never came for stowaways like him:

TO DR. VAN ALLEN, WOULD YOU APPROVE TRANSFER OF YOUR
EXPERIMENT TO US WITH TWO COPIES IN SPRING. PLEASE ADVISE
IMMEDIATELY. PICKERING

"I didn't know what they had in mind," Van Allen recalled of his quizzical concern. "It sounded as though they were trying to take over my experiment." He had no way of knowing what all had gone down behind the scenes in the flaming wake of *Sputnik 1*—and therefore, he regarded this note from the Jet Propulsion Laboratory's director with great interest. And worry.

William Pickering had come from the fishing villages of Havelock, New Zealand, attending the same primary school as famed subatomic researcher Ernest Rutherford. Pickering embraced a natural attraction to the scientific side of things; during high school in Wellington, he assembled radio sets and chemistry experiments in the basement of a friend's house. From there, it was on to the United States and the California Institute of Technology, where the tall, slender Kiwi emerged with a PhD in physics by 1936. Work at JPL ensued, centering around telemetering equipment for instrumented balloons—and later, long-range missiles. At the time, the Lab was also heavily entrenched in rocket-assisted take off, which precariously involved strapping rocket motors onto aircraft to severely boost their initial speed. Having slowly risen through the JPL ranks, Pickering earned his directorship in 1954. On the whole, people liked the guy.

Bobbing in the middle of the ocean and therefore quite out of the loop, Van Allen hadn't been around to see the fallout: after *Sputnik 1*'s launch Pickering and Wernher von Braun endured a marathon schedule of meetings to determine America's best response. Von Braun practically lived at JPL while formulating his answer. The guy sported a well-known mug. His appearances on television and the cover of *Collier's*, going back four years, had already made him a celebrity. At one point while in the JPL cafeteria, a young part-timer approached him with a copy of von Braun's book *The Mars Project*, seeking an autograph. It happened all the time.

What the Russians did made von Braun feel like he'd gotten a spanking—a really hard and undeserved one considering Jupiter's flightworthiness. And not until after *Sputnik 1*'s launch had the White House relented—Neil McElroy finally caving, agreeing that the trouble-plagued Vanguard would realistically fail to deliver any sort of prompt response. The first American rocket in space would not be the government's civilian ideal after all, and von Braun soon had a green light.

With blistering speed a three-way pact emerged among Huntsville, JPL, and the University of Iowa. The army and von Braun would make available their heavy-lifting Redstone. In keeping with the proven Jupiter concept, JPL would then populate Redstone's upper stages with clusters of their scale Sergeants. And JPL would also handle the communications, telemetry, and a payload enclosure.

Now all they needed was something of merit to actually put in that enclosure. Another meeting convened. Pickering referred to a sheet of paper describing those four experiment finalists chosen by that nearly forgotten subcommittee. Von Braun hovered over the conversation, knowing full well the interaction between his boy Stuhlinger and that Iowa guy. Pickering continued his talk, specifically noting how the Iowa project stood out as the only one already configured to fit JPL's payload compartment.

"Isn't that interesting?" von Braun casually mentioned, and Van Allen dropped into the number one slot.

Over in Iowa City, George Ludwig got somewhat of an unexpected phone call explaining that a couple of people wanted to come speak with him about important matters. Ludwig told them sure; he well knew the classified nature of the project. Nobody was supposed to talk about it in detail on the phone or send any explicitly worded telegrams.

So two suits from JPL made their way to Iowa and met with the graduate student. Ludwig confirmed that he had indeed designed a cosmic ray experiment to fit either Vanguard or Jupiter.

However, there was a catch. The tape recorder, that little eight-ounce hockey puck of a device used to store data from an entire orbit to be played back on command, wasn't working quite right. It kept breaking down in tests replicating the stress of Jupiter's upper stages whirling at high speed—up to 750 rotations a minute. Unfortunately, Jupiter *had* to spin like that in order to even out any thrust imbalances from the multitudinous Sergeants all linked together. It was the only way von Braun's crackerjacks in Huntsville could get the upper stages to fly true. That year, millions of Americans spun Peggy Lee's "Fever" on the turntable at 33 rotations a minute—which by comparison was standing utterly still.

No way could JPL just throw something else in the nose—Iowa's box *needed* to fly with them; it meant credibility. But the United States had only a couple of tracking stations, so without a recorder their harvest would have untold gaps. "The experiment we wanted to do depended so heavily on getting data throughout the whole orbit," Ludwig pointed out. He really wanted the onboard storage. Well, was there some other possibility? Could it still fly and provide some measure of results even without the recorder?

Ludwig said, "So the agreement we came up with right away was that there would be an all-out crash effort to put the abbreviated version on." The "abbreviated version" he referred to would be an otherwise complete cosmic ray experiment—sans onboard recording. For the time being they'd go with a method of transmission that continuously broadcast its measurements. Later on when the kinks were out of it, they could add the recorder back in for the next flight.

That arrangement sounded great. Thoroughly pleased, the men returned to Pasadena. Five days later, JPL determined it was time to lay their hands on the actual cookie jar; so one of the suits from that original visit, Eberhardt Rechtin, picked up the phone and called Ludwig at five-thirty in the afternoon on Monday, October 28.

"Okay, now we're all set to go," Rechtin told him. "Do you have authority to switch the experiment over?"

A hiss of static in response.

"That kind of took me aback," Ludwig candidly recalled of the situation. He had this vision of giving Iowa's instrument to JPL and then being totally cut out of the action. "I thought for a moment, and then I said, 'No, really it's Van Allen's experiment. He's the principal investigator. He's the one who has the final authority to switch it over.'" JPL would have to wait for a thumbs-up from the infected grubby guy in the middle of the ocean.

Still on the phone, Ludwig promised to find some way of communicating with Van Allen . . . just as soon as he found the guy.

"I had no doubt *at all* that he would want it done," Ludwig emphatically stated, expressing his belief that Van Allen definitely preferred Jupiter over Vanguard. "He thought it was more assured."

Wasn't Ludwig worried about JPL pulling the plug on them?

"No, no," he insisted. "We were the only ones really under consideration."

Two days later Van Allen received that first radiogram. Pickering's message didn't quite make sense; its deliberate vagueness, combined with Van Allen's disconnect from the headlines, required a bit of clarification:

DR. W. H. PICKERING

UNABLE INTERPRET YOUR WORD TRANSFER DUE IGNORANCE RECENT DEVELOPMENTS. OUR APPARATUS FOR ORIGINAL VEHICLE NEARLY FINISHED. DELIGHTED PREPARE THREE SETS NON-STORAGE TYPE FOR JPL PROGRAM.

VAN ALLEN

Deep concern rang through the man's head as he prepared this reply. Van Allen possessed something an already powerful and influential organization wanted. They wanted it badly and in no easy way could reproduce the same thing by their own hands. It wasn't that farfetched to suggest that JPL could lean on the University of Iowa physics department, maybe pull rank somehow, or have the Defense Department strong-arm Ludwig into coughing up the gizmo. Heck, it was sitting right there in the unguarded physics building. A confederate probably could have grabbed the thing in the dead of night, and later on at some watery closed-door inquest JPL would have pleaded national security.

Conversely, the Lab could ostensibly have "borrowed" or "leased" the device from Iowa. Such a move would have forever relegated the apparatus—and the harvest of data from it—to JPL's domain, thereby reducing Iowa to a mere subcontractor.

Van Allen later commented, "The main apprehension was, you know, that JPL was sort of proposing to take over our experiment, and I wasn't very happy to contemplate that. So I gave an equivocal answer."

Just one day following Van Allen's equivocal answer, way the heck out in Tyura-Tam, Laika received a few final pets, then was sealed into her flight container and installed on the top of a hastily modified R-7. She could stand, sit, or lie down on the cork floor. It was quite cold on the ground, so a few men were put in charge of keeping Laika warm. Quickly they converted an air-conditioning system into a heat source by rigging it to a hose and then trundled it up to the top of the rocket to make sure little Barker stayed comfortable. Three whole days stood between her incarceration and the actual launch.

Despite having had minor surgery to implant biomedical leads, and that irritating fecal collection system she never quite got used to, this dog had it as good as any canine ever did. Multiple technicians swooned over her with the utmost regard for the dog's safety and comfort. She enjoyed a palette of specially prepared foods, served on demand in the most exotic of kennels designed with her in mind. Clean air and film cameras rolling.

About the only thing the second artificial satellite *didn't* have was a way to bring Laika back. Nobody felt good about such a harsh decision. There just wasn't enough time to figure it out.

"Laika was a wonderful dog," remembered a Korolev teammate. "I once brought her home and showed her to the children. They played with her. I wanted to do something nice for the dog. She had only a short time to live, you see."

With only one day left before launch the medical staff raised a commotion with Korolev about the capsule pressure. It should be at normal Earth loads during launch, they contended, permitting an observation of how things changed while ascending—an argument totally out of the blue.

No matter how bizarre the idea seemed—and especially so close to launch—Korolev acquiesced and waved it through. Shortly thereafter, en-

8. Fitting the insulating nose cone to *Sputnik 2*, late 1957. To the technician's immediate left is the spare *Sputnik 1*, with animal compartment to the left of that. Laika's view would be out the circular porthole. (Author's collection)

gineers by the pad noticed a horde of doctors approaching them with syringes. Hello? What do needles have to do with cabin pressure? Well, they weren't really interested in changing the pressure, the men explained. They just wanted to give Laika a drink. Everybody looked at the syringes. They were filled with water. The engineers nodded, then climbed up the rocket gantry and squirted water into the dog's food trough through a series of air holes, which were still open. Humanity had struck again.

Another radiogram showed up on the *Glacier* for Van Allen on November 2, again from Pickering:

PRESENT PLANNING SUGGESTS MOST OF EXISTING EQUIPMENT BUT TRANSFERRING RESPONSIBILITY TO US INSTEAD OF CARRYING TWO PROGRAMS. FIRST EXPERIMENT WOULD BE CONTINUOUS. SECOND WOULD BE STORAGE TYPE. SUGGEST YOU PHONE ME ON ARRIVAL IN NEW ZEALAND.

If only Pickering had known what was afoot at the end of the Tyura-Tam railway line. Laika and her electronic entourage permanently left Earth the next morning at five thirty Moscow time. Launch came right on schedule, and the dog's heartbeat fully tripled during the climb to orbit. Every experiment worked like it was supposed to. The *Sputnik 2* array contained a pair of Geiger tubes mounted just above the ball, measuring the same radioactive phenomena that Van Allen's and Ludwig's project eventually would.

"Alive! Victory!" the men cried after their ship entered orbit.

Sputnik 2's radiation detectors worked for a week—six days longer than Laika stayed alive.

The official plan relied on a small device in her living compartment. About ten days in, it would have automatically injected a drug to put her to sleep. The simple contrivance had unfortunately been easier to design than a method of safe recovery.

But it wouldn't be needed. By her second orbit the temperature in Laika's kennel had soared above 104 degrees, and flight controllers on the ground had no way to control it. All they could do was study the problem. The booster's nose shroud had popped off like it was supposed to, but in the process a wad of thermal insulation was torn loose. There would be no way to control the heat buildup, and within seven hours of launch she was dead.

Although it invoked the ire of many animal-rights groups, Laika's demise did nothing to lessen the mission's impact. Clearly, the Soviet Union was no longer some agrarian swath of peasantry. They had climbed into the twentieth century, dazzling humankind with heretofore unthinkable achievements like sending living beings into space. Three days after launch, at the fortieth anniversary of the Russian Revolution, Nikita Khrushchev spoke publicly: "In orbiting our Earth, the Soviet Sputniks proclaim the heights of the development of science and technology, and of the entire economy of the Soviet Union, whose people are building a new life under the banner of Marxism-Leninism." Everyone beyond the flight controllers figured Laika was still kicking. Not until nine anxious days after launch did Radio Moscow finally confirm her demise.

After *Sputnik 2*, Korolev's name went unmentioned in *Pravda*, or in any other publication. Sergey Pavlovich Korolev, the individual most responsible for his country's greatest space exploits, would never live to see his name celebrated in print.

The *Glacier* finally bobbed to port in New Zealand. Van Allen gathered up his precious suitcases of paper tape rolls and raced for the phone to call home. "I'm back," he swooned to an excited Abigail through what was surely a delicate connection. "I'm back on dry land and hope to be home in a few days." He hung up just in time to get the mail: another radiogram from his newest best friend Bill Pickering:

URGENTLY NEED YOUR APPROVAL ON PROPOSED CHANGE OF LUDWIG EXPERIMENT. PORTER COMMITTEE HAS GIVEN THEIR APPROVAL TO PROPOSAL TO CHANGE EXPERIMENT TO JPL AND TO MODIFY EXPERIMENTS AS AGREED UPON BETWEEN LUDWIG AND JPL.

Van Allen wasn't sure what to do. Details were agonizingly slow in coming his way. Still reacquiring his land legs a few days later, on November 13 he cabled Ludwig directly:

QUESTION. IS PICKERING PLAN FOR OUR EXPERIMENT AGREEABLE WITH YOU? PLEASE CABLE ANSWER IGY REP CHRISTCHURCH.

The fact that James Van Allen held up the career decision of a lifetime to consult with a mere graduate student of his, from no less than the other side of the globe, is perhaps the finest available example of this man's character.

Ludwig elaborated, "He wanted to be assured that this was being done in such a way that Iowa would remain the experimenter. And that we would still be in control of the experiment. And he wanted my assurance that it was being set up that way." Such was the relationship and level of trust between the scoutmaster and his cub.

But they'd known each other for a long time already. Way back in December of 1952, a twenty-five-year-old Ludwig, fresh from the air force and B-29 flights, returned to his parents' farm. He was the boy whose father interviewed Van Allen about rockoon launching on *Fresh from the Country*, the radio program broadcast from their farmhouse breakfast table. At the time, Ludwig had never heard of the guy. "I was a captain in the air force," he said, "and had done an awful lot of things, and I had a strong tendency not to be overly awed by anybody. I was pretty mature by that time."

Down in New Zealand five years later, Van Allen now greeted a new morning cable from the pipe-smoking Ernie Ray, who was busy handling Ludwig's mail so the latter could quickly pack his family for an extended trip to Pasadena:

AFTER HIGH LEVEL APPROVAL AND OBVIOUS REARRANGEMENT OF OLD
PROGRAM, GEORGE LEFT TOWN FOR EXTENDED STAY. GEORGE QUITE HAPPY
PICKERING PLANS. HOPE YOU SAY YES.

That was all Van Allen needed to know. Immediately after receiving Ray's cable, Van Allen sent a definitive response:

APPROVE TRANSFER OUR EXPERIMENT ACCORDANCE JPL PLAN.

Why all the hassle? How come JPL didn't just take the thing outright? "Well, I think that basically it was that Pickering's an honest man," Van Allen suggested. "And I had significant stature in the field, so it was not exactly easy to take it over." Van Allen never asked the director point-blank if the idea had entered his mind. "But I said that we expected to continue having primary scientific cognizance for the experiment," the midwestern physicist summarized. "And he did not disagree with that."

At the time, George and Rosalie Ludwig counted on a black-and-cream 1956 Mercury to fulfill their transportation needs. Even before Van Allen's final approval, Ludwig had returned home to their little rental house on Rochester Avenue on the east side of Iowa City. "JPL and I came to agreement very early that this thing was really only gonna work the way we wanted it to if I went out there," he said of the impending exodus. "I loaded what drawings and notebooks and things I had with my prototype engineering model, and parts that I had accumulated for flight units—and some clothing—and a few odds into the trunk of our Mercury; and we proceeded out to the West Coast with our two children."

The family drove four days straight, finally arriving late at night at a motel in La Cañada—virtually outside the JPL gates. Later on they'd rent a house in Pasadena, but this would have to do for the time being. The four of them crowded the small room and tried to catch up on sleep. As Ludwig told it, "Lo and behold, that night our number two daughter, Sharon, fell out of bed, and she started crying and whimpering." Groggily Ludwig arose to see what in the world was going on. "She wouldn't quiet till finally I started feeling around and found that she'd broken her collarbone. And so our first order of business the next morning was to scurry around and find a doctor, about the same time that I was also scurrying around to go in the gate at JPL and get started." The next morning, Ludwig entered the JPL gates carrying a small wooden box holding the coveted experiment. And with that, the major pieces of America's first satellite finally met.

Somewhere along the way, it acquired a name: the Deal. Van Allen thought a JPL man named Jack Froehlich may well have come up with it, but he wasn't sure. In Van Allen's opinion, what it meant was more or less what it sounded like: *Here's the Deal. The Deal is what is going to happen between Huntsville, JPL, and the University of Iowa. We've made a deal that JPL is going to use the Iowa instrument in JPL's payload shell, on Huntsville's rocket.*

Other interpretations persist. George Ludwig asserts it to be more of a playing-card analogy: the Soviets took the opening hand, and now it's America's "deal." This latter explanation seems closest to what most old-school JPL workers recall. Apparently, the Lab employed a quintet of engineers who habitually played gin rummy with so much enthusiasm and focus that even launches became irritating distractions. Froehlich was one of these card players, along with Henry Richter. "We played in airplanes, motels, conference rooms, and wherever we could," Richter explained. One day in 1956 another round was just beginning when news came through of a successful launch. One of the men jokingly shouted "Deal!" and simple as that it entered the lexicon. "Froehlich had decks of Deal cards made for us," Richter said. "I don't have a full deck any more (several have told me that!)."

Either way, the name stuck, finding its way into almost all official paperwork about the project. A launch date was planned and kept confidential. And to Van Allen, "the Deal," no matter what its origin, struck him as very appropriate. The history books identify it as *Explorer 1*.

George Ludwig was always a man to write things down. "I've maintained a journal on and off all my life," he said. And as the decidedly historical events began to unfold around him, Ludwig continually made time to record his thoughts. Usually they found a home in one of many hardbound computation books filled with graph paper:

January 20th, 1958

My wife says this is the early writings of George Harry Ludwig and that when I become famous they will become very valuable. I rather doubt that anyone will ever want to read them but me, but upon reflection, the past several days I decided that I am forgetting interesting happenings from my earlier days and that unless I keep such a journal as this, these very interesting days that are occurring now will also fade in my memory. Therefore, my journal. At the time of this writing I am working at the Jet Propulsion Lab, 4800 Oak Grove, Pasadena California, on a temporary basis, to adapt the cosmic ray instrumentation which I developed at State University of Iowa.

"My first task was actually to draw out the schematic diagrams from my notebooks," Ludwig described of the hectic environment greeting him at the Lab. "Because I'd never really up to that point put together a final set of schematics and wiring diagrams. So the engineers at JPL needed that *immediately*, and we went from there."

The mild-voiced farm boy from Tiffin, Iowa, born on November 13, 1927, had always known a preoccupation with things electrical and mechanical. Perhaps it's because so little of it surrounded him as a youngster. "We didn't have electricity until 1941," Ludwig said of his childhood, which included five siblings. "I grew up in a very rural setting with outdoor outhouse and carrying water up a hill to the house from the well."

Ludwig enjoyed taking things apart to study how they worked. Overall, his mother encouraged the interest, giving him an old clock to dismantle when he was a youngster. As Ludwig grew, he got more into radios. His enjoyment of them was definitely facilitated by the addition of electricity to the house. Allied Radio in Chicago had a nifty catalog filled with lots of magical bits; when he could gather the funds, Ludwig ordered parts to build his own creations. Usually though, the bulk of the pieces had to come from old radios or other salvaged parts. His mother cleared off one end of the kitchen counter so as to provide her busy son with a work space. When Ludwig needed a hot soldering iron, Mom invited him to heat it up in the coals of the wood-fired stove.

In contrast to his mother, the hobby caused Ludwig's father heaps of concern. "He didn't want any of this stuff to get in the way of my responsibilities for chores and helping on the farm," Ludwig explained of his father's attitude. As a teenager, George Ludwig sat gazing one day over a half-finished radio project; with flying magazines strewn about, he overheard a spirited conversation between his parents.

Emphatically Ludwig's dad pleaded, "These interests could ruin his life!"

Perhaps the elder man only wanted his son to walk a similar road. The life of Ludwig's father had greatly focused on education—studying philosophy and religion in college, then briefly serving as a full-time minister. Finally, he concluded that teaching could serve double-duty as a most enjoyable way of ministering. In fact, Ludwig's father taught all his son's high school science classes. "I always liked to get into arguments with my dad when he was teaching science or math," the younger Ludwig remembered, blatantly amused by the situation. His classmates—all four of them— always got a kick out of the fireworks. "They could just sit back and relax while he and I argued." Occasionally, George even won.

Within two weeks of graduation Ludwig had joined the air force with his eyes on its great choice of specialties. He went through basic training in San Antonio, and by the end of the first year of his service, he had worked his way up to corporal. The force accepted him into flight training in July of 1947. Ludwig graduated a year later as a pilot and second lieutenant and received orders to McCord Field in Tacoma, Washington.

Ludwig flew c-82s, and the assignment appealed to him on a variety of levels. He went to airborne electronics school there and snuggled up to radar —taking apart state-of-the-art radar equipment was heaven compared to old clocks and salvaged radios. A nearby church fulfilled his worship needs, and after one of the services, Ludwig noticed an attractive young woman named Rosalie. "We hit it off right from the very beginning," he recollected. She was still in high school. The pair commenced a heavy and uninterrupted schedule of dates that effloresced into love.

After choosing a ring, an enchanted Ludwig picked her up in his '47 Buick Roadmaster and proposed on one of their evenings out. They'd been together less than a year; Rosalie was just about to graduate. "I mean, if I'd found the right woman, I wasn't going to wait around!" Ludwig gaily recalled of the moment. Before the I dos could happen, he underwent a transfer to Keesler Air Force Base in Biloxi, Mississippi. Ludwig attended radar school there and corresponded with his sweetie pie almost daily. Rosalie's parents brought her down that July for the pair to join in matrimony.

"The Korean War started," as Ludwig continued the story, "and as she's said many times since then, she wasn't going to let me go off to war without having married me." Not a whole lot of time was wasted in starting a family as daughters Barbara and Sharon were born eleven months apart while Ludwig was still in the service—now stationed at Mountain Home, Idaho.

By 1952 George Ludwig had frustratingly explored several methods of gaining college admittance while remaining in the air force. He found no luck with any of them. A different story waited for those in the air force *with* prior college—the branch frequently sent people back to finish it up. Such was not the case for Ludwig; he'd gone into the service straightaway. So late that year with his military obligation fulfilled, Ludwig hung up the uniform as he and Rosalie and the two wee ones moved into his parents' Tiffin farmhouse. "I went back with the single-minded goal of going to the University of Iowa," he said, "and just naturally I was going to go into physics. Nothing else even entered my mind." That chance meeting with Van Allen led to a job in the department's research lab, where he earned seventy-five cents an hour to maintain and repair equipment. In short order the physics faculty realized how unbelievably overqualified Ludwig was and promoted the hell out of him. "Very quickly I started building equipment for the balloons," he explained. "They'd give me the wiring diagram, and I would lay it out and build the instruments."

Now, five years later, grad-student Ludwig readied a machine to make history.

The basic arrangement was effectively finished as far as Ludwig's original, more-expansive design was concerned. But that one included the tape recorder. On the other hand, JPL's barebones setup wasn't totally planned out in the drawings. As Ludwig clarified, "There were a few final details of the design that I was still working out. But outside of that, it was complete. At JPL they took the simpler version and ran with it immediately." By mid-January, Ludwig had completed his work on the initial flight package, "Deal 1," and was neck-deep in preparing its follow-up, "Deal 2," to finally marry detectors with recorder. He faced an unrelenting schedule of work. At the end of the month he was supposed to get on a plane for Florida to be present at Deal 1's launch. Yet all was not well in Pasadena as is evidenced by his journal entry:

26 January

The conclusion of a rather discouraging day. The past two days I've been working hard to get the recorders and GM *tube calibrations in order so I could leave for Florida today. But at 2am this morning I called to cancel my reservations because the* GM *tube calibrations have not worked out. Today I brought the equip-*

ment home to see if the electromagnetic fields at the lab have been disturbing. No better luck today. Will have to get something done tomorrow for sure . . .

But it was time to leave. Ludwig scooted down southeast to the cape. His priceless experiment traveled separately. They were going to try to shoot it off on the last day of January. The Jupiter was already down there on the pad.

Ludwig and Van Allen never drafted any kind of formal agreement that one would be chief experimenter and one would be chief instrumentor. It just sort of evolved that way. As an undergrad, Ludwig built more and more of the rockoon payloads; by 1955 he was building the lion's share of them. All ten of Iowa's groundbreaking Deacon-based rocket experiments that flew from the Davis Strait in 1955 came to life in Ludwig's hands. So as IGY activities picked up and the idea of a satellite became more of a reality, the two men talked more and more about how a cosmic ray experiment would actually be pieced together. Relationships and roles naturally grew and matured from that point. "There was never any moment," Ludwig explained, "when I went to Van Allen, said, 'I wanna do it,' or when he came to me and said, 'Would you like to do it?' Or when we sat down together at any moment in time where we said, 'Ah ha! I'm going to do it!' It was just a tacit agreement that grew by small steps."

Friday January 31st 1958 A motel on the surf at Melbourne, Florida
I just got up a few minutes ago. Stood out at the surf for awhile and brought my expense record up to date. Am getting somewhat fed up with JPL, especially some of the supervisors. They have treated me like a piece of excess baggage. A necessary evil for the duration of my stay there. They're a bunch of local politicians trying to feather their own nests at SUI expense.

Years later in 1981 Ludwig annotated his own journal entry in the margins: "JPL wouldn't let me in their trailer. Right or wrong, I always felt it was because they wanted all the glory for the first satellite."

Van Allen wasn't in Florida for the launch. He'd been rounded up along with von Braun, Pickering, and a bunch of military gooney birds and stuffed into the war room of the Pentagon. News came through about the launch being successful, but there weren't enough details to know for sure that every stage had cooperated to send *Explorer 1* the full distance. So in the stuffy war room, everyone took chairs and made small talk. They sat there amid stained coffee urns and hanging drapes of cigarette smoke for an hour and a half, at which point everyone had pretty much stopped talking.

"Gosh, I guess it didn't make it," was finally offered up by someone in the room.

Another half hour passed before a phone rang. It was for Pickering. Everybody stayed dead quiet, looking right at him. Pickering took the headset and began to softly converse with the other end.

Somebody finally asked, "What's the word, Bill?"

He held the phone away from his ear. "Earthquake Valley says they have a signal." The entire room leapt to its collective feet and began a celebratory dance of hand shaking and back thumping; within seconds von Braun, Pickering, and Van Allen were extracted from the scrum and rushed into a car idling outside. It screamed twenty minutes over to the back entrance at the National Academy of Sciences Building. The trio bumped out, knowing a press conference was planned if everything went well . . . but were slightly unprepared for the crushing melee of jostling reporters who'd rushed the room in festival-seating fashion and unpacked their cameras. It was two in the morning, and Van Allen was astonished at the number of people waiting for them. Each of the men had a chance to deliver remarks on the mission, and then an assistant brought out a full-scale mockup of the Explorer satellite, antennae and all, attached to a stage-four scale Sergeant. The bulbs popped as all three smiled over the long tube.

One of the reporters tossed out, "Why don't you hold it up there so we can see it!"

The three explorers grasped their creation and in unison held it high, held it like a trophy, like the underdog football team that just took state. They didn't talk to each other, didn't even have to *look* at each other before it went up. Three minds worked as one: the rocket man, the satellite man, the experiment man. Take one out and it all topples like a house of cards. They held it for five full minutes as the image inhaled rolls of film, killed hundreds of flashbulbs—three movie-stars on the red carpet. Two o'clock in the morning, and every one of the three still had his tie securely knotted. Two o'clock in the morning, and they were fresh as daisies.

Walking away from the meeting, Van Allen looked at Pickering and halfway joked, "Well, I hope it's still working." Then he tucked himself into a Washington hotel bed and caught an early plane to Iowa the next day. It'd

9. The Big Three. *From left*: Pickering, Van Allen, and von Braun at the press conference, February 1, 1958. (Courtesy of NASA)

been a short night's sleep, but the gleam in his eyes the next morning didn't give it away. Walking into the arms of his Abigail, James Van Allen said, "Hi, honey, I'm home. You probably heard the news!"

At 3:30 in the morning on February 1, Ludwig took a minute to grab his nearby journal and quickly note the success. He managed a few hours of sleep before heading back out to collect his stuff and face the press. Out on the Cocoa Beach Strip, a party built to full rockin' swing at the Bahama

Beach Club. Ludwig dropped in for a while but couldn't stay long; he faced a six o'clock flight to begin journeying home to Iowa City. Reporters milled around waiting for him to get off the plane. As Ludwig did so, he noticed his dad, like a guiding force, emerge from the cloud of media people and escort him to the car. Late that afternoon Ludwig finally alighted at the Tiffin farm and immediately collapsed for a nap.

On the same day, Van Allen had a minute between endless phone calls to jot down with satisfaction the following notes about his cosmic ray experiment:

Temp 20 degrees C
Holding well
C. R. working like a bomb

And the telegrams started pouring in. Most of them gushed with congratulations, but Ludwig got this one on February 4 at 11:30 in the morning:

TO: GEORGE LUDWIG

FROM: HERBERT KALISMAN ANTON ELECTRONIC LABORATORIES

CONFIRMING OUR TELEPHONE CONVERSATION CONGRATULATIONS AGAIN COULD YOU PLEASE WIRE AUTHORIZATION FOR US TO PUBLICIZE THE FACT THAT WE SUPPLIED OUR TYPE-314 GEIGER-MUELLER TUBE FOR THE UNIVERSITY'S COSMIC RAY MEASURING INSTRUMENTATION IN THE EXPLORER. PLEASE CONFIRM BY RETURN WIRE. WE SHALL GIVE UNIVERSITY FULL CREDIT. THANK YOU.

Ludwig was already making his way back to California and the JPL workbenches to ready Deal 2 for the next launch. At the airport he even crossed paths with James and Abigail Van Allen, who were returning from dinner at the White House.

Before leaving Iowa City, Ludwig gave several interviews and appeared on his dad's radio program. And his journal now reflected a more upbeat attitude.

Wednesday 12th February 1958
Explorer certainly has taken the public's fancy. I've received more than my share of publicity. It has been a very gratifying project in all respects.

And the intrepid *Explorer 1* continued on, her paternal creators watching from the ground. She would go, in fact, for many years and over fifty-eight thousand orbits, long after men had visited the Moon. But even before *Explorer 1*'s third orbit finished, while Van Allen smiled for pictures and Ludwig updated his journal, the world's superpowers planned and plotted. Both parties in this most peculiar, expansive, and expensive contest were already thinking about their next target.

It looked down on all of them most every night.

7. The Creators and the Makers

Machinists, instrument makers, students wiring boards,
reading the data, and so on . . . these all tend to get left behind.

George Ludwig

The tape recorder, egad. Without its conceptually simple yet fundamental role, Van Allen remained short of contiguous, round-the-world cosmic ray data. And without *that*, atmospheric physicists lacked a complete picture of what all was going on up there above our sunny skies. Ludwig chewed his lip. He just *had* to conquer the dang-blasted spinning and give his recorder a chance:

4th of March 1958
The last few days I have been in the "dreaming at night" stage. I've worked so long to perfect this system. There are so many things to go wrong. I sometimes wonder how it can possibly work properly. This is a grueling business. One can become a hero or a heel depending on the turn of events. Which will it be for me?

Luckily, *Explorer 2* offered him a chance to send the whole casserole. By late February Ludwig confidently felt he'd nailed the problem and returned to Florida for his prearranged date with a rocket. He came in via Washington DC on National Airlines with one stopover en route. Nobody met him at the airport, so Ludwig caught a ride with a few of the regulars from Naval Research and checked in at the Sea Missile Motel in Cocoa Beach. Through large picture windows on the second floor he could look down over the lovely central courtyard and brick-edged swimming pool, but there was no time for lounging.

The equipment he built came down separately with JPL: a primary instrument and two full backups. One day they were luggage; the next, they were headlines.

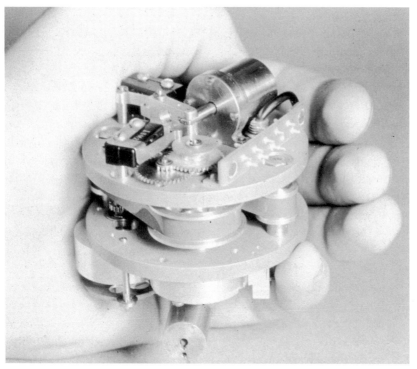
10. Ludwig's amazing hand-built tape recorder. (Courtesy of George Ludwig)

5th of March 1958 L-day, 4:20am
This sky is light cloudy and broken rather high. Payload activation in 40 minutes. This is the day for which I have been working since January 1956. If successful, this is to provide my PhD thesis. I'll have to give that payload a goodbye pat.

Launch minus three hundred minutes: the recorder double-advanced in a test. Befuddled, Ludwig asked for a repeat, and there were no problems.

Launch minus eleven minutes: with upper stages spinning at 550 rotations per minute, the recorder wasn't playing back. They stopped the spin altogether then brought it back up gradually. It worked at 450 rotations per minute but not at 500. The final approval for a multimillion-dollar rocket launch involving thousands of people now rested in the hands of one single graduate student. Ludwig contemplated the anomaly for a bit, then said go ahead.

Jupiter's early stages performed beautifully. With everyone loaded on its shoulders, Redstone erupted from the blocks—pushing, straining up there, giving all. It handed off the chores to the second-stage cluster of eleven Sergeants showing good burn and high spin; another handoff pushed the workload onto the third-stage cluster of three Sergeants. Magnificent work.

The launch baton was then supposed to pass into the hands of that lone remaining Sergeant permanently fastened to *Explorer 2*. All they needed was that anchor leg to just hold on and win gold. But after six hundred perfect test firings, this one didn't light. *Explorer 2* tumbled back toward Earth, helpless, carrying with it Ludwig's carefully designed, excruciatingly tested cosmic ray detector and tape recorder, screaming down for an unceremonious smack into the planet. Ludwig swallowed hard, then called his boss to explain what happened.

Van Allen did not directly admit to swearing. "Oh . . . *shucks*," he reported his response to be and then laughingly admitted, "Maybe a little more profane than that!"

But George Ludwig cared enough to make sure another one of his instruments again operated flawlessly when it was installed on *Explorer 3* less than a month later. While readying things, both Ludwig and Van Allen managed to keep one ear on the *Explorer 1* telemetry. Up there on the circling bird something seemed amiss. *Explorer*'s signal was perplexingly ramping in and out and getting noisy, too. Both men took pause, speculating on what combination of behaviors might lead to the odd reception. As Ludwig put it, "Of course, at first we scurried around trying to figure out whether there was some problem with the instrument."

Although riddled with gaps, the received data seemed clean—and that's what was so bothersome. At times, the level of cosmic radiation hung right in a constant zone, around twelve to eighty counts per second. But for other periods it would drop out completely. What the heck? Put together with a discontinuous ground track, all they really had was a block of swiss cheese. Maybe there *was* some problem up there beyond reach. Van Allen puzzled over the situation while trying to readjust to university life and class lectures to undergrads. Students didn't stop asking questions. His phone wouldn't stop ringing. *Time* magazine put his face on the cover.

"We made every assumption that we could think of," Ludwig remembered of their frustrations. Nearly every waking hour went to sustaining his mini assembly line of instruments. All the while, different *Explorer 1* scenarios ran through his mind. Bad Geiger tubes? Well, gee willikers . . . they'd been tested about as well as anything could.

"In late February," Ludwig said of the frenetic times, "we were coming to the conclusion that there must be something real there. We didn't think the instruments were misbehaving."

Data is pure narcotic for these guys, so right after *Explorer 3* made orbit Van Allen hustled out to the Naval Research Laboratory to collect its first tape dump. Twenty-one brisk days had passed since *Explorer 2*'s big downer of a truncated flight.

By this time, Van Allen was able to fly on commercial airplanes, requiring considerably less transit time than the old method of spending days in the swaying belly of a train. From the airport, he cabbed out to Pennsylvania Avenue and the one of the Vanguard offices, collecting a printout of data.

While these accordioned reams of paper appeared unimpressive to the lay, Van Allen's mitts now held the very first stored data from any orbiting spacecraft. Ludwig's tape recorder was chugging like a dream. On the way to his hotel, Van Allen stopped at a nearby drug store to load up on pencils and graph paper. Did the cashier realize they'd be used to make history? Flicking on the light in his meager room, Van Allen emptied his purchases from the sack, unsheathed his slide rule, then sat down to see just exactly what the data told him.

Lay out the X and Y lines . . . Each orbit took 102 minutes. Ludwig's equipment could measure from 0 to 128 counts per second. The raw data was more boring than income tax forms—just infinite streams of numbers at first blush, over and over again through seemingly endless pages upon pages. Every single data point had a home somewhere on that graph paper, and only after being plotted correctly would the embedded codes unlock any secrets contained within. Such is the process of reducing data. Around him the daylight began to fade. Van Allen was in for a protracted session of dot placement.

The afternoon became evening and then nighttime. His perceptible world shrank to a tiny sphere of light and activity, and beyond it was the void—no Iowa City, no lectures, no George or Abigail. Van Allen never budged, remaining hunched over the papers in front of him. Life reduced further still, focusing down to a small cocoon enveloping his upper body and work table awash in light and activity. Everything else blurry and faded. Then again it shrank, down to hands and paper. Nothing more existed—just black. The dots in front of him morphed into lines, which then grew, bent, arced . . . *revealed.*

The data lines told a similar story to those from *Explorer 1*—count rates that hovered low, then jumped up to a second drop, and plummeted down to zero. The Iowa boys, they'd seen a few steps of this dance before, but there were always such pesky gaps in the data. Thanks to an elegantly constructed and rather bulletproof tape recorder, the complete picture was now in front of one man wise enough to understand.

Van Allen clicked off the light at three in the morning and finally turned in, face smiling all the way down to the pillow. He knew the answer. He *knew* it. He also had a plane to catch the next day to Iowa City, where he spread out his results on a huge table. Carl McIlwain and Ernie Ray came running. Everybody leaned over and sucked in the answer for themselves.

The counting rate wasn't falling to zero, it was zooming up high and going completely off the scale. Radiation levels above Earth measured exponentially higher than anyone imagined. The detectors were telling them something all along, but not in an obvious way.

Carl McIlwain tested the theory after running to get the portable X-ray machine they used for calibrating instruments. He blasted a matching *Explorer 1* Geiger tube with spasms of energy. And much to everyone's general oohing and aahing, the equipment's counting rate dropped to zero, precisely mimicking the behavior from their orbiting hardware. Ernie Ray looked up and announced, "My God, space is radioactive!"

George and Rosalie Ludwig finally packed up their Pasadena rental house in April for the trip back home. Enough new acquisitions prompted the couple to rent a U-Haul trailer to hook on the back of the Mercury. One thing Ludwig trundled out of the house was the spare *Explorer 1* flight article. He paused, and posed, as Rosalie took a picture.

Over twenty years later, while casually rereading his journals, Ludwig felt the need to grab a pen. In the margins he confided a couple of small thoughts: "I have only two very mild regrets from that period: 1) That I didn't stay on the SUI payroll while at JPL, and 2) That I was so immersed in building and launching the satellites that I couldn't participate more in 'The Discovery.'"

Two weeks before *Sputnik 3* went up, Laika came back. Since her departure from terra firma, 162 days had elapsed. The night of her return, a smattering

11. George Ludwig with the flight spare *Explorer 1* on the back porch of his Pasadena rental house, April 1958. (Courtesy of George Ludwig)

of UFO reports came in along the eastern coast of the United States. Most reported pretty much the same thing: a distinct, bluish-white light flying across the sky, going red, shedding blobs.

"The more time passes, the more I am sorry about it," explained a member of the *Sputnik 2* team, many years on. "We did not learn enough from the mission to justify the death of the dog."

Attention now unceasingly focused on a subject of much longing. The D machine, what *Sputnik 1* was intended to be all along. Complexity and weight kept pushing it aside—oh yes, that staggering weight. When contemplating *Object D*, never think of it as an instrument. Think of it more as a facility—it had so much weight that the mighty Soviet R-7, in spite of its gargantuan size and twenty booming exhaust nozzles and nose-thumbing victory for socialist labor, couldn't lift this very payload Korolev and his band of scientific followers had wanted to fly all this time.

Engineers, however, tend to be a sturdy bunch—used to persevering through setbacks and initial failures. And these Soviets had been clocking overtime to tweak the capabilities of their monster rocket. After much nudging and fudging, they had it. Supposedly. There was, at the very least, enough confidence to strap the three-thousand-pound *Object D* cone into the rocket's nose to see what happened. That payload weight was impressive enough all by itself. But what really guaranteed to amaze was how so much of it had been dedicated to such a broad spectrum of research above the atmosphere—a dozen or so experiments covering primary cosmic radiation, micrometeoroids, electronic charges, solar radiation, ions. No American booster at the time would have come anywhere close to lifting all that. And it got better: aboard would also be a recording system, christened Tral, capable of storing data for later transmission to the ground on demand.

"Tral was much bigger than mine," conceded George Ludwig, "and would have recorded much more data." Otherwise Tral ran very similar to Ludwig's device in terms of operation, but it could not match the reliability.

Once *Object D* was finalized, technicians made two of them. It turned out to be a very smart move because the particular R-7 cradling their first copy blew up shortly after launch on April 27; *Object D*, along with its intricately designed experiments and recording system, came tumbling back into the hostile netherworld of Russia. Two years of work lay scattered in the tundra. Oleg Ivanovsky described what happened next: "There was a search for the satellite. I remember that the pilots conducting the search were not allowed to know what they were looking for. 'Just search the area for anything unusual,' they were told, 'and don't attract the camels.'"

On May 1 James Van Allen sat before a Washington DC press conference at the National Academy of Sciences. Patiently he articulated what had been found, which amounted to no less than a significant natural phenomenon. The Iowa physics department had even resolved its shape. Picture a glazed Krispy-Kreme doughnut. Suspend Earth in the center of the hole. Add a larger Krispy-Kreme around the first, in concentric fashion. Both doughnuts represent groupings of radiation that surround Earth, swarming in trapped orbits. Van Allen opined that any planet with a magnetic field would probably have one, although the shape might vary.

A reporter there didn't get it. "Do you mean like a belt?" he asked.

Van Allen said, "Yes, like a belt." Later that summer, a man from the NRL named Bob Jastrow coined the phrase "Van Allen Radiation Belt" while attending a scientific conference in Europe. No official naming process or method of recording a title of this sort exists, but it stuck all the same.

Van Allen heard the term and said, "Ah! Oh, that's nice!"

Over the years, he gave innumerable talks to various high schools. "I've had kids say, what is the purpose of a radiation belt?" he recollected. "Well, you know in a way it's a silly question, and I always give a courteous answer: 'It doesn't have to have a purpose, it's a natural phenomenon, and so we're investigating what nature's up to!'"

The very real possibility exists they may have been called the Vernov Belts. At the time, accomplished Soviet physicist Sergei Vernov was in the process of lending his experience to the design of many *Object D* instruments. Iowa freely shared all their scientific results as they became available, and Vernov had been as flummoxed as Van Allen about all this radiation that seemed to be dillydallying high above Earth. Tantalizing glimpses of a possible radiation layer had already been discovered on Laika's flight. But was it localized—an isolated spot above the tracking station? Vernov knew the dilemma as well as Van Allen or Ludwig: to get a complete picture, the spaceship needed some way to remember things when it wasn't in range of a ground station. Hence the equipment on *Object D*.

However, George Ludwig hadn't manufactured *Object D*'s tape recorder. Rather, it was designed and produced by an Experimental Design Bureau team, working in service to the uneven mind of forty-five-year-old Aleksei Bogomolov, whose failures would begin to stack up more and more as time went on. He displayed exceptional skill at delivering on few of his promises.

Chief Designer Bogomolov's bureau—a wing of the Moscow Power Institute—had supplied the telemetry system for use on *Sputnik 1*. His fingerprints were also on the gear that recorded Laika's heartbeat during *Sputnik 2*. Both bits of machinery more or less worked up to their expectations. This time around Bogomolov had to deliver a tape recorder able to store up the blips from *Object D*'s instruments and send them down on command. The recorder was supposed to be a cinch. But right in crunch time the equipment rose up and bit Korolev on the nose, leaving Bogomolov to wish somebody like Ludwig had been around to offer up a few pointers.

Matters came to a head during the final hardware check before launch.

Every science experiment tested positively, but not the stupid recorder. It was having a problem announcing itself to ground stations, which had to happen before off-loading the data. Most of those kibitzing around the spacecraft were scientists; they looked at each other for several minutes in grave, wordless concern. Korolev stood there not entirely sure what to do. Somebody then logically suggested that the issue be properly resolved on the ground before anything flew. Soon all eyes rested on Bogomolov, who drew himself up and, with a completely straight face, postulated that the test failures were due to the large amount of electronic interference in the room. His conclusion? Good to go. Suddenly an urgent phone call came through for Korolev. Shortly, he returned with his face tightly screwed into purple.

"I don't want to have anything more to do with you," he screeched at Bogomolov. "Go away—I can't even be in the same room with you!" And much to the shock of his colleagues, Korolev waved the recorder through.

On the same day as his press conference announcing radiation belts, Van Allen received notice of unofficial backing for his next Explorer crafts. Although the new roster of flights themselves were certainly unclassified and public, a curious twist involved half of their true purpose being uncharacteristically hushed . . . at least for a while. Van Allen and anyone else working on the program would have to keep it quiet. Said Ludwig, "We were able to work on it openly because of the first reason, but never mentioned the second one outside our group."

Things got rolling in mid-April. Van Allen maintained regular contact with Bill Pickering at JPL, keeping him updated on the Explorer results. The Iowa physics department had been loading spare classrooms with volunteers to process and reduce *Explorer 1*'s hundreds of thousands of data points by hand. Too many people volunteered to help, and they spilled over into McLean's basement hallways. Naked light bulbs hung down from the ceilings.

On the same subject, Van Allen sporadically corresponded with Wolfgang Panofsky, a respected Stanford physicist who occasionally consulted with the U.S. government. Both remained patently intrigued by the Explorer findings, which summarily raised a heretofore unanticipated question within the minds of the U.S. Department of Defense: *How do we know this radiation is natural?* This became a diplomatic way of suggesting, *Isn't it possible*

12. State University of Iowa volunteers reduce *Explorer 1* data, April 1958.
(Courtesy of the University of Iowa)

that the Soviets have conducted high-altitude nuclear tests? Maybe that's what Explorers 1 *and* 3 *had been measuring all this time?* Until that point, Van Allen had no reason to think the belts were anything *other* than natural activity. The suspicion had led to calls for more research, more flights.

A secret name was then floated to Van Allen's attention—Argus. The U.S. Atomic Energy Commission (AEC), he learned, wanted to perform some atmospheric tests of its own—the Argus Tests. Plans involved staging high-altitude nuclear bomb explosions while orbiting radiation detectors recorded the event. If it worked, the scheme might evolve into an ongoing presence from hundreds of miles up. Properly equipped satellites could detect and report the testing of nuclear bombs by any nation. The AEC was quite serious about the whole idea and wanted Iowa's proven equipment onboard. In a cold war context, their concerns seemed entirely realistic.

Van Allen called a pack meeting and put the challenge to his Scouts. Ludwig, Ray, McIlwain, and company quickly recommended an appropriate menu of detectors. "Officially" the work became a modest extension of Iowa's IGY activities. But not even graduate students can work for free all the time, so the entire tab went to the army.

For blanket coverage the army wanted two satellite payloads, which Iowa would provide, and three rocket-launched bomb explosions, which Iowa would not. Arguably the University of Iowa physics department had the easy end of the deal. They also had a bit more wiggle room—von Braun's tweaked rocket could now lift a bit more—and that meant some nice heavy batteries to provide a couple months of run time.

By June, George Ludwig's journal entries showed him to be back to the old routine:

Having to work long hours on the project, 12–14–16 hours, up 'til 2am, back at 9 to 9:30, sure hope it's worth it when it's all over.

The world's fair came to Brussels, Belgium, in 1958. And along with it came hundreds of nations. "FIRST WORLD'S FAIR OF THE ATOMIC AGE," proclaimed the billing, which no doubt made for an irresistible selling point. Belgium went all out. Fifteen thousand workers had spent three years building the site. Without question, its signature landmark was the huge building shaped like an iron crystal, which remains standing even today, the Atomium. Forty million people showed up between the April and October run dates. Many Belgians drank their first Coca-Cola there.

The site resembled a campus. Larger nations got an entire building to themselves. The Americans yanked their doors open to show off grandiose exhibits on such topics as progress toward overcoming segregation issues and the latest hot fashion trends. There were American cars to sit in. There was a snack bar. And literally right next door, the Soviets filled *their* majestic aluminum-and-glass pavilion with, among other goodies, full-scale replicas of *Sputniks 1* and *2*.

A month into the fair, the Soviets added an upgrade. *Sputnik 3*—the renamed *Object D*—was unveiled after its refreshingly successful May 15 launch.

In keeping with preflight speculation, Bogomolov's tape recorder had indeed failed—leaving the experiments only able to report a broken and spotty trail of cosmic ray activity. The problems went unmentioned in Brussels; far too much other cool stuff lay spread about under the Soviet roof. They had farming machinery, heavy presses, trucks, sporting equipment. And one night they even had some unannounced visitors.

A group of operatives from a variety of U.S. agencies—including the CIA

13. Korolev's original satellite finally nears launch. It's *Sputnik 3*, 1958. (Author's collection)

and the air force—broke into the Soviet building and ripped off the *Sputnik 1* display. Realize it hung by wires thirty-five feet off the floor and was accordingly something of a challenge to remove. The ball was examined and photographed while one guy kept an eye on the time. Three hours later the Sputnik craft was reassembled and hung up in its exact original place. The Soviets never caught on.

Getting up-close looks at a Sputnik satellite didn't have to be so risky. Around the same time *Sputnik 2* blasted off to orbit Laika, an American radio astronomer named Herb Friedman happened to be at a scientific conference in Leningrad. His curiosity about its onboard program of study led Friedman's Russian hosts to open up. Giddily, they trotted out a detailed cutaway exhibit of the craft, discussing every component inside. The men dripped satisfaction, proud papas, displaying the same attitude James Van Allen would later note in his own experiences with Soviet scientists. Friedman jotted down some of the part numbers and the next day found a little equipment shop with a few of the doodads in stock. He bought two Geiger tubes that were model-number specific to the *Sputnik 2* flight hardware, which he brought home wrapped in a copy of the previous day's *Pravda*. No questions asked.

Not long after returning, an officer from Naval Intelligence visited

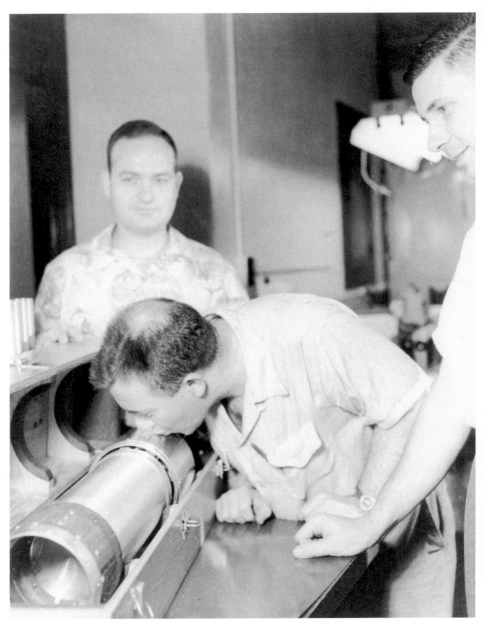

14. Van Allen kisses *Explorer 4* goodbye, 1958. Carl McIlwain stands behind him, and to the right is an amused George Ludwig. (Courtesy of the University of Iowa)

Friedman on a routine follow-up and inquired whether he had learned anything about *Sputnik 2*. From his coat pocket Friedman pulled out the Geiger tubes, still wrapped in *Pravda*, and handed them over to the speechless officer.

Iowa's first delivery to support Argus received the title *Explorer 4*. It went up on July 29, working beautifully. Of course George Ludwig noted a few thoughts in his journal:

It was a real personal thrill to take part in the blockhouse operation. For one thing, it is immensely interesting and thrilling. For another it represents some kind of SUI victory. For this time we were not regarded as mere outsiders. JPL's part in the operation was small so they were much easier to get along with. This is not a pleasant thing to say about anyone, but ABMA agrees with me. The Explorer 4 orbit is a good one.

Ludwig's ongoing comments about JPL reinforced a peculiarly evolving tension noted by both himself and his scoutmaster. "On the one hand we were very fond of JPL," Van Allen suggested. "But then when we went down to the subordinate levels of JPL, we usually found an arrogant attitude there, that we were just a kind of a minor subcontractor." Unaccustomed to such behavior, Van Allen's recollection was marked by a slowed tempo of his words, a change in tone to one of confusion and disappointment. "We were just supplying a part, you know? Like ones who furnish a bag of bolts for the spacecraft. And it was *their* box then, and *they* could do *anything* they wanted to with it. That was totally in violent disagreement with our point of view, which was that we'd sealed it up, we did everything we could think of to do to assure its reliability!" Occasionally Van Allen had to get Bill Pickering on the line and straighten things out at that level. "We've met the specs, and we have been through all the environmental tests," Van Allen said to Pickering in one particularly stressful phone call. "We don't want any low-level technicians mucking around with it! It's too delicate!" And Pickering sent the word on down.

"There were three groups that I was working with," George Ludwig began, illustrating the contrast in attitudes. First came von Braun's rocket collective down in Huntsville. "I have nothing but admiration for that group," Ludwig said. "They always included me as a colleague and treated me *won-*

derfully." His voice lilted up as if describing some amazing movie he'd just come out of. "Likewise for the Naval Research Group, the people who were working on the Vanguard. In spite of the fact that we switched our instrument over, we did so very openly with them. There was never any rancor at all, and in fact we continued to benefit a lot from the knowledge of the people there at the Naval Research Lab after we made the switchover to the army vehicle."

The Jet Propulsion Laboratory, however, did not rate equal marks. "JPL had what I referred to at the time as a fairly heavy dose of institutional arrogance," he said. "And that, I think, maybe got in their way just a little bit from time to time."

Ludwig was quiet for a minute.

"There were a few people in the project who, from time to time, I thought were more interested in getting something up there than they were of doing a meaningful task with it. And I did have to do a little hand wrestling with them." The Iowa physicists were not the only ones unimpressed by JPL's attitude. Within a couple of years it would vividly fissure between the Lab and the National Aeronautics and Space Administration (NASA).

Shortly after *Explorer 4* orbit confirmation, technicians detonated two ten-megaton bombs at high altitude above the middle of the Pacific Ocean. *Explorer 4* dutifully noted the upper-atmosphere changes and passed them down to ground track in real time. "We had no recorder on *4* and *5*," Ludwig noted. "We depended on direct reception." This success was justification enough for James and Abigail to load their four smaller Van Allens into the station wagon and drive all the way to Long Island to kick off a much-deferred vacation. Even so, Van Allen tucked his field notebook into his bags and kept tabs on this newest member of the family.

During a press conference following *Explorer 4*'s launch, Wernher von Braun received most of the attention while Ludwig and Carl McIlwain stood idly by. Afterward, the trio stopped for lunch at a roadside café. In the middle of their meal, von Braun impulsively turned to McIlwain.

"You are the important ones," he said. "I'm just the trucker."

Remaining in Florida for *Explorer 5*, Ludwig prepared for the second and final Argus launch. On August 15, he jotted down some feelings relating his work to these tests:

Perhaps these devices which we scientists develop for our own purposes, but whose control is taken by the politicians, who control the means for making them—the money—may do great harm in our future. But I would rather believe that they will do great good both in their scientific endeavors and for the politicians.

Certainly this work has given SUI—*and ourselves—much prestige. In fact, I am in the middle of another dilemma. Should I continue in this work now, while so much is happening, or return to school?*

Explorer 5 left the pad on August 24 and immediately fell victim to the one variable for which Ludwig and Van Allen possessed absolutely no control: booster success. The fourth stage didn't ignite. And their fantastic cosmic ray package would spend the last ten minutes of its life operating precisely as designed until greeting the ocean at terminal velocity.

Three more bomb detonations followed. World media knew enough of the blasts outright—and, of course, these twin Explorer firings went off in full public view. Yet the reporters didn't manage to link them together in print until six months later.

"The Americans are at our heels," Sergei Korolev warned Oleg Ivanovsky. Later Ivanovsky clarified, "He didn't mean this as an insult but as a show of respect for the competition."

That December in Moscow, Korolev weighed in on the subject of this developing space race and where the next heat's finish line might be. In one of his occasional *Pravda* articles—always done up using a pen name— Korolev claimed, "Reaching the Moon by a rocket launched from the Earth is technically possible even at the present time."

Back in Iowa City, the physics department had brought in a part-time translator, who sat unscrambling *Pravda* into English for just such a purpose as reading the above tidbit. James Van Allen wasn't familiar with this "Professor K. Sergeev" who authored the article, but he agreed with the sentiment and even held a few reasons of his own for embarking on such a lunar trip. Reasons beyond good ol' national pride.

Ever determined to huff the delectable scent of raw data, Van Allen had recently dispatched a memo to the IGY committee. In graspable terms his prose laid out the gains to be made from additional broad brushstrokes of information on Earth's marginally understood radiation belts. The best way to do that? Gosh—in his opinion, it would be by taking measurements with an experiment *on its way to the Moon.* In this fashion a complete three-dimensional picture could be formed of what, exactly, surrounded us. Soon Van Allen's memo fell into the dangerously deep shuffle of IGY paperwork.

William Pickering had been thinking along those same lines—ever since *Sputnik 1* really. In the wake of that historic launch, Pickering had boldly trumpeted the need for lunar flights, although Van Allen's aforementioned rationale made the JPL director's scheme ring hollow. Barely three weeks after *Sputnik 1,* while America sat chewing its fingernails with itchy dismay, Pickering finished revising a tidy proposal calling for what he termed Project Red Socks. In spite of such an idea getting lost in the trampling melee up to *Explorer 1,* he'd never quite let the thing die.

Red Socks's stated objective had *nothing* to do with science: "regain stature in the eyes of the world," it demanded, "by producing a significant technological advance over the Soviet Union." The path to such grandeur? Moon flights—celestial mechanics enable about one trip a month. Pickering wanted nine missions.

His Red Socks baby sat collecting dust for a couple of months, until some greenbacks came filtering in by way of the Department of Defense's spanking-new Advanced Research Projects Agency (ARPA). ARPA's birthright and mission statement were to temporarily oversee all American space programs. History would ultimately categorize the outfit as a stopgap governmental agency bridging both the hodgepodge of wheel spinning brought about by *Sputnik* and the eventual creation of NASA.

ARPA's first director was a former General Electric executive named Roy Johnson, who lengthily orated on how excited he was to "surpass the Soviet Union in any way possible." That was it. That was his guiding thought. The word "research" apparently did not cross the man's lips.

Before the nameplate was even installed, they were receiving more mail and more proposals than Sophia Loren. A deputy of ARPA, John Clark, later refocused the outfit's early notoriety into sharper perspective. "After we had been in business a short time, it seemed to me that everybody in the country had come in with a proposal except Fanny Farmer Candy, and I expected them at any minute." In spite of all the competition, Red Socks sifted to the top. And despite a name change, the spirit of the Socks ebbed through ARPA's first program.

On March 27, 1958—nearly two months downstream of *Explorer 1* and ten days after *Vanguard* finally orbited—the Defense Department tenderly sashayed onto the IGY's dance floor, beginning what they later christened the Pioneer series of lunar missions. It picked up a name only after the first flight. "This program," announced Defense Secretary Neil McElroy in a news release, aimed to "determine our capability of exploring space in the vicinity of the Moon, to obtain useful data concerning the Moon, and to provide a close look at the Moon." All anybody had to do was figure out *how*. Responsibility for achieving these would first be delegated to one branch of the military, then it would switch over to another—presumably to keep things fair.

The U.S. National Committee for the IGY simply loved the idea and pledged support to shore up Pioneer with a desperately needed scientific angle. A follow-up announcement identified Pioneer flights as an ongoing U.S. contribution to the IGY. Some of them were intended for lunar orbit and photography; some would just wing past the Moon. In fact, the Pioneer name would find itself affixed to a colorful smorgasbord of projects resembling one another in no fashion whatsoever.

ARPA first gave money to the air force, which saw great promise in creatively stacking together *Vanguard*'s second and third stages on top of their new Thor booster, which had come from Douglas Aircraft. Thor was pure missile—150,000 pounds of thrust sporting a range of 1,500 miles, all designed to neatly carry thermonuclear weapons from midtown Europe toward the general direction of Nikita Khrushchev's privates. Thor had survived more than a year of development, and about sixty of them would find homes in England until a true intercontinental missile could be produced. The official paper from ARPA to the air force instructed them to launch "as soon as possible consistent with the requirement that a minimal amount of useful data concerning the moon be obtained."

The air force directly unburdened themselves with procreating this new moon ship. Nobody there knew how to concoct such a beast. Wisely, they subcontracted out the dirty work, to a relatively new yet whip-smart outfit on the West Coast that had more or less sweet-talked themselves into a high-buck tryst. The name of the new place was Space Technology Laboratories (STL). Intelligently managed, staffed with top-shelf brainiacs, the company was in the abbreviated process of refining itself into a highly technical, rock-solid outlet for nearly anything that needed to operate above the atmosphere.

With unreal alacrity STL was becoming the Jet Propulsion Laboratory's new rival, and JPL wasn't used to competition.

Who knew the Cleveland Cap Screw Company would go so far. This Ohio firm had been making screws and bolts since 1901. Following a merger with Thompson Products, the business expanded into car parts. Thompson made the kinds of things most drivers rarely consider—things like gear assemblies and pushrods. Thompson built the wooden wheels for the Model T. It evolved from there into crafting more intricate parts like engine valves, and then airplane power plants. When Lindbergh crossed the Atlantic, Thompson's sodium-cooled valves thumped away inside his engine.

Their stuff was good enough that by the early 1950s the ballooning company looked to diversify. Specialty electronics were definitely becoming a hot item, so Thompson's executives got themselves into bed with a sweaty-hot LA firm known as Ramo-Wooldridge. The namesakes of R-W were two exceedingly bright refugees from Hughes Aircraft. They'd fled in September 1953, pursuing the business of missile development for the air force.

Simon Ramo and Dean Wooldridge had already established themselves as bona fide geniuses in the arena of grand-scale program execution. Innately and confidently they broached extraordinarily complicated problems—like building a high-powered missile—and sectioned them down into manageable components, keeping tight rein on how the innumerable pieces fit together. They didn't do the actual building, just masterminded the logistics of it all—"just."

Hold on. *Two guys* came along trying to steal the whole show? Where was General Electric? What about RCA (the Radio Corporation of America)? Weren't they players in this game?

Ramo explained it like this: "The established industry that should have had the strength to do it, didn't do it, and this is because of their preoccupation with the civilian things." In other words, despite their technical proficiencies and vast infrastructures, companies like RCA and GE were more interested in building television sets. "They didn't even appreciate adequately the A-bomb," Ramo quipped.

Managing large technology projects is kind of like running a war. Building great tanks and airplanes can definitely get your side off on the right foot. And great benefit is certainly to be reaped from quality soldier training. But these alone don't guarantee a win. The soldiers have to be fed and supplied with clean, drinkable water, and they must be sent fuel for the tanks and cars and trucks. The correct ammo has to be shipped to the right airfields to be put in the correct planes. Victory results only through *superior application* and *integration* of all these elements. It's the same deal when putting a missile on the streets.

The air force wanted keen minds to oversee America's first ICBM, the Atlas. Si and Dean understood the challenges, were interested, and presented their strategy to the air force. They landed the project barely a year after incorporation and didn't even have to bid for it. The force gave it to them on merit alone. "We were drafted for the job," Ramo summarized. "It was not a competitive program—you couldn't do that today."

The pair had met at Caltech and immediately hit it off. "We were exactly alike in our way of thinking," Ramo said. "We could finish sentences for each other." Even their builds matched. Ramo ended up at Hughes first then enticed his old chum to move on out and join him in the fun. One of their more colorful successes at Hughes was development of the Falcon. This electronically guided missile superseded such aging technology as the radio proximity fuze, on which James Van Allen had labored so hard.

Ramo and Wooldridge carried their legacy of excellence forward to the new company, converting an LA barbershop as their first office—with a grand total of four employees. From this highly unlikely platform they tackled the Atlas.

Methodically perfecting the then-abstract concept of systems engineering, the duo oversaw the divvying of Atlas components to a widespread assortment of the most suited contractors. They then brilliantly orchestrated the vendors as a conductor would a symphony. Convair prepared the structure, North American worked on propulsion, Burroughs handled the onboard computer. "A lot of separate pieces of equipment have to be designed," Simon Ramo generalized, "but it doesn't make sense to design them in isolation from the fact that they have to work together, of course." *That* is the essence of systems engineering. As Ramo tersely put it, "Equipment didn't just fall from heaven and happen to work together well."

Within six months the mushrooming Ramo-Wooldridge added over 20 bodies; it then swelled to 170. During this boom time the dynamic duo also managed to hire Louis G. Dunn, none other than the JPL director himself—and the guy even combed out a slew of additional worker bees from the Lab to cart along with him. Dunn's hastily scrounged replacement was some benchwarmer named William Pickering.

Ramo and Wooldridge supervised details to an inhuman degree of precision. On the Atlas job, calculations indicated that if even 1 percent of the missile's fuel went unused, the extra weight would cost six hundred miles in range. Though it seemed an unascertainable conclusion, the guys just shrugged and said it was all there in the numbers. As word of the company's proficiencies spread, hordes of employee candidates queued up at the door. Then more joined the line right behind. Many university researchers —full professors and department heads among them—took leaves of absence in order to work at Ramo-Wooldridge. The 170 new hires bulged to 2,000. Their business plan was working.

By 1958 Si and Dean decided to formally comingle with Thompson. Perhaps in a hurry, they chose the name Thompson-Ramo-Wooldridge Inc. Regardless, TRW emerged as nothing less than a giant, joining ranks with the likes of Raytheon, Grumman, and North American Aviation as suppliers of government solutions. It offered complete, operational, mostly dependable hardware and no-nonsense systems integration. Nobody there suffered any fools.

Atlas merely opened their floodgates. Open? Heck, the gates washed away in a torrent of projects. TRW earned a second contract for the Titan booster, then another fat one for the Minuteman. No client ever wore out a welcome. At one point, air force personnel were at the Los Angeles company offices so much that TRW bought a vacant lot down Bellanca Avenue, around the corner there on Arbor Vitae Street. There, they built completely new offices and moved all their employees over, leaving the air force back in the old space.

In the wake of Sputnik and Explorer, Si Ramo sensed what was coming and decided there was enough brainpower lying around the shop to warrant carving out an entirely separate enterprise. It would be wholly owned by TRW yet focused on the particularly extreme needs of equipment designed to operate in space. Within a week of *Sputnik 1* going up, Ramo had already filed all the papers to trademark the name Space Technology Laboratories. It was that simple.

The first chairman of STL was none other than Tokyo raider Jimmy Doolittle. Ramo stood in as pro forma president. Louis Dunn was tapped for general manager. First item on the docket? Greasing a few palms. Ever proactive, the company had boldly sketched out their own lunar program, independent of Pickering's or anyone else's. STL campaigned for this, their Project Baker, during a January 1958 visit by James Killian, who was then on leave from MIT as one of President Eisenhower's science advisory reps. The only thing STL heard as he left that day was, "Send in a proposal, and we'll go ahead." Within no time a dense thirty-page report thudded onto Killian's desk outlining Baker. It advanced the novel idea of stacking various military rockets on top of one another, thrusting beyond Earth's gravity in order to land a modest experiment package on the Moon. They actually thought a landing to be possible.

Baker joined those hundreds of other proposals that ARPA's John Clark had seen fit to complain about. But STL's lusty liaisons with the air force clinched easy victory. They would supply ARPA's first space probe. So a guy named Dolph Thiel quickly found himself project manager, reporting directly to Dunn, and rather in over his head.

Now Thiel took his instructions and scurried off into the Inglewood catacombs, desperate to realize a plan that could transform Red Socks rhetoric into reality. He faced a hard road. Thirty pages of dog-eared marketing certainly *help* define scope on a program, but really, they only scratch the surface. Thiel and his team faced less than six months to design and instrument the spacecraft, get it on top of a booster that was still in flight testing, and then get to the Moon utilizing a tracking system that hadn't even been designed yet. Thus began what could only be described as an intense period of effort. The company was already stacked up with the Thor, Atlas, Titan, and Minuteman missiles—all for the air force. Louis Dunn felt the increasing, impinging weight on his back. It was coming at him from all sides. Officially—and unofficially—probes were in no way supposed to interfere with the missile projects. Dunn needed to set a mood for the work ahead, so he gathered his flock for a little sermon on the mount.

"Ze forty hour veek is out ze vindow," Dunn's thick Afrikaner accent informed the assembled team. "And eets impossible zu pay overtime, zo yeel 'ave to do most of dees on yeer uwn time." Patriotism and the Red Scare drove them all to 120-hour work weeks and to so little sleep that it was almost easier just to stay up. "Like being in a war," said one engineer.

The mission parameters were cut back almost immediately. Actually landing something—even the scabby twenty pounds called for in the pro-posal—was admitted to be completely unrealistic. They settled for lunar or-bit, naively concluding the ease of doing so. On August 15, 1958, it stood on the pad, and two days later they sent the command to fire. George Ludwig was there:

17th August—Moon Attempt

Well, it was a failure. After 77 seconds, it blew. I saw it from Hangar S five miles away. Bill Whelpley said after it climbed 300–400 feet a small fireball ap-peared for 3 seconds about the diameter of the second stage, at the point where stages 1 and 2 join. I didn't see this from Hangar S and to me it appeared to climb normally until 77 seconds when two white puffs of smoke were seen in quick succession. Following the first one, several small pieces were thrown off. It looked like it might have been an engine explosion. The beaches were lined for the shot. It is surely a shame it failed because now USSR will have another chance to beat us.

Lovingly tucked inside the probe rode an experiment put together not by Ludwig but by Carl McIlwain, who was in the process of assuming his share of the mad workload. Debilitating strain had worn heavily on McIlwain's classmate and friend. It soaked into his pores. As of late Ludwig straddled a bottomless crevasse with one foot in his work and one in his family. Fatigued like crumpled metal, he was starting to break:

My family is another consideration. I have been away from them so much the past year, and Rosalie complains bitterly. I cannot continue to spend one hour per day or less with my family and expect to have a livable home relationship and happy, well-adjusted children.

Despite this inglorious launch failure, the probe itself wasn't an entire washout. *Able 1*, as it was known then, cleanly separated from its booster in one piece. It telemetered out over two minutes of useable data on south-ern Florida's balmy atmosphere during this unplanned trip down to the Atlantic. Then *ploink*.

Two months later an air force man suggested that his branch of service "show who the real pioneers in space were." He proposed that STL's next ship carry the name—easy guess—"Pioneer." But he'd gone on to suggest that the new launch be referred to as *Pioneer 1*, which meant that the previous ship would either remain *Able* or be called something else. Humorously enough, this first attempt has come to be known as *Pioneer 0*.

Ludwig took the agonizing failure as hard as McIlwain, and he hadn't even built the payload. Ludwig's attendance related more to the impending launch of his *Explorer 5* instrument. In his journal he had this to say about *Explorer 5*'s own failure—while returning to Iowa from the cape via Huntsville:

All that work to prepare this experiment and the damned rocket failed to put it in orbit.

The future is confusing and a big question mark. At least four projects can be seen in the near future, with no sign of letup. This in view of the fact that I plan to return to school this fall. I can hardly bear to think of not obtaining my PhD. Yet how can I escape from this other work.

The time had finally come. Shortly after his return to Iowa City, Ludwig commandeered a bit of time in Van Allen's disheveled office. "You know," Ludwig said with a touch of finality, "this is getting beyond what I can handle as a graduate student." The men spoke for a while, Van Allen in complete understanding yet naturally concerned about the continuation of Iowa's space involvement. Really, they were only getting started with the exploration. But more than ever Ludwig felt a need to complete his studies and get on with life.

They reached an agreement. Building on the exceptional work done so far, Carl McIlwain would no doubt serve as a dandy replacement, freeing his chum from the certain turmoil of lunar instruments. Ludwig happily finished out his commitment to deliver the remainder of Iowa's satellite payloads, and then that was it for him. No moon probes.

Over the difficult course of his graduate activities, Ludwig acquired at least one unanticipated cheerleader. "My dad turned out to be one of my greatest boosters with the radio program," he said, clearly surprised by the outcome. "He was always talking about the work I was doing."

Now, having freed up some real estate in his brain, George Ludwig could finally turn his full attention on finishing up his doctorate. Despite the failed launch of *Explorer 5*, he completed writing his thesis and endured the oral exams, in which Van Allen participated.

"Then finally," Ludwig said, "I remember going home to Rosalie and saying, 'Oh, I think I'm gonna make it.'"

That was the end of his early satellite-building career. "I've never made a complete tally of all of the instruments I've developed in my lifetime," Ludwig explained of his efforts. In recent cooperation with the Smithsonian Air and Space Museum, he endeavors to reconcile the production figures. "But to my knowledge, not a one of them has ever failed in flight."

Some years later Konstantin Gringauz was out for a walk in Gorky Park and happened to bump into Sergei Korolev. A contributor of science instruments to *Object D*, Gringauz had long felt an uncomfortable question eating at his brain. He apparently figured the chance encounter that afternoon was as good a time as any to get an answer.

While the two strolled together, Gringauz politely asked Korolev how he possibly could have trusted Bogomolov's idiotic rationale for why the *Object D* tape recorder wasn't passing any tests. They could have claimed discovery of the radiation belts! *Why did he let it fly?*

"You think that I don't feel guilty about that whole story?" Korolev snapped. Gringauz didn't know what to say. Korolev kept talking. "You think I was simply a fool that day, issuing instructions to launch when the tape recorder had failed to work? If you want to know what happened, I can tell you."

Gringauz could only listen.

"On that very unfortunate morning, Nikita Sergeyevich called me on the phone at the Baikonur Cosmodrome," Korolev explained, using Khrushchev's patronymic. "He said that the Italian Communists had urged him to do something spectacular, like sending something into space."

Gringauz was starting to get it.

"The next day, Italy was going to have parliamentary elections. 'If you Soviets would make one more show in space,' Khrushchev had been told, 'it will bring our Communist Party a few more million votes.'" And when Korolev finished talking, it was hard to look at his friend.

There was the whole story. An order from Khrushchev himself. No concern about science, only stupid politics. S. P. Korolev launched a defective probe because he had to.

8. Storming the Sea of Dreams

Korolev dreamed of the cosmos, and he clung to that dream.

Andrei Sakharov

In Russia tea is served after meals and as sort of a break in the midafternoon. It's an important part of the culture—every Russian family owns at least one tea set. Always made of porcelain. Oftentimes particularly valuable collections, such as those from the Lomonosov Porcelain Factory in St. Petersburg, are ceremoniously passed from one generation to the next. Lomonosov, as a point of curiosity, was founded in 1744 via royal decree to "serve the cause of national industry and art." Indeed, Russians value the idea of tea more seriously than most.

Never make the mistake of showing up with prepackaged bags. Only by placing tea flakes directly in the tea pot, and then delicately submerging them under boiling water from a samovar, is the preparation considered acceptable. Russian samovars are special, beautiful kettles with an enclosed heating chamber. Proper water application ranks as a fundamental skill; kids learn it almost from infancy. The moment of perfection is, as a Russian would term it, "sparkling boiling"—when air bubbles first appear in the water. Not until precisely then has mixing time arrived. And you don't just dump it in; relatively miniscule amounts of water are used to first craft a "mother" batch. A little bit of this concentrate is then transferred into a small glass, hot water diluting to taste.

Rounding out a proper Russian tea is the generously presented assortment of cakes, cookies, jams, and honey. Honey *never* goes directly in the teacup. To properly enjoy it, the wise participant spoons a discreet amount onto a saucer for parallel consumption with the tea itself. That's all there is to it. Tea is a simple little bit of culture, so reverently executed.

Many Russians use teatime to prop their feet up and decompress. People like Sergei Korolev used it to noodle over the daily dilemmas continually hampering their plans. Lately most of Korolev's post-Sputnik desires centered on lunar probe flights. And not entirely by chance the requisite technology was now right around the corner. The tantrum-prone chief designer had been seriously chewing on the Moon problem for about three years already and thought he might nearly have it cornered.

Of course, Korolev never operated solo. A lot of solid groundwork was laid by his pal Mikhail Tikhonravov, who was as close to a right-hand man as S.P. would ever get. They shared the same vision, as if both dreamed the same little movies in their heads every night. Born right at the turn of the century, young Mikhail inhaled the writings of Tsiolkovsky and very soon afterward tumbled into an incestuous relationship with rockets. He was nuts about them. After graduating from the Zhukovsky Air Force Engineering Academy in 1925, Tikhonravov set about finding some good company. His long, lean face quickly crossed paths with Korolev at the GIRD meetings. While other men their age indulged in things like attending ballet and cinema programs, *these* two began ironing out what problems stood between them and other simple pleasures—like a liquid-fuel rocket engine.

A year after they met, anyone looking for eager Misha—a form of Mikhail used by friends and family—had only to track down his improvised launch facilities out on the remote tundra. Throughout this early period, he spent great volumes of time nurturing capable sounding rockets; repeatedly he'd launch them in ballistic arcs and then try to improve on whatever just happened. He couldn't get enough. When Stalin's dark purges came through, Tikhonravov completely avoided the chaos and remained at work. He just walked between the hailstones. Even today the reasons aren't completely understood.

Knowledge of his talent quickly spread. Midway through 1944, when the Soviets learned about the existence of German Vengeance Weapon debris only thirty miles beyond the front lines, Tikhonravov was one of the men clandestinely going in to look for it. His capacity had been appreciated by more than kindred countrymen: During the Soviet recovery efforts inside Germany, somebody made quite a find. Pawing through a stack of abandoned papers uncovered some recorded work with Cyrillic writing. They were a bunch of Tikhonravov's rocket and missile blueprints—deeply classified work. How did the Nazis get their hands on them?

15. Good buddies: Tikhonravov at left with Korolev on the right. This picture looks half-painted because it likely *was*; extensive (and awful) Soviet touch-ups were common at the time. (Courtesy of Colin Burgess)

By April 1957, six months before *Sputnik 1* motivated American classrooms to hold nuclear attack drills (remember: just "duck and cover"), Korolev and Tikhonravov had formalized their exploratory plans in one mammoth document. It was called "A Project Research Plan for the Creation of Manned Satellites and Automatic Spacecraft for Lunar Exploration," the filing of which probably required an extra-jumbo manila folder tab. As it was laid out, this game plan entertained no mere fantasy or half-baked rubber-band engineering. It resulted from soul-searchingly intense analysis and discussion among learned professionals regarding two logically distinct branches of their work. One branch was piloted flight; the other, robotic lunar probes.

With such extraordinarily magnificent planning under their belts, what actually had followed was off the map, in the rough, and entirely reactionary. Whiskered balls and kenneled puppies? Each Sputnik satellite exemplified a stunning technical achievement, but exactly what it was and did had unquestioningly boiled down to politics and the juvenile whims of Nikita Khrushchev. Yeah they shocked the world, but even shock was only worth so much.

"These are approaches with no future!" Korolev sparked at one particular meeting. "Although gradual visits to space are effective, they are of no significance for science—and for spaceflight!" The most painful dissonance raged on inside his own head. Quit stunting, Korolev told himself. Stop it with these asinine gambits.

And so in mid-February 1958, with only two meek Earth orbiters behind them, Mikhail Tikhonravov stood in Korolev's office one morning to learn that he was supposed to take point on one of those new branches, to start figuring out how to shoot real live human beings into space. Over the next three years a clay wad of enthusiasm, design, manpower, and engineering would become molded and shaped into Vostok—an orbiting spacecraft piloted by the likes of Yuri Gagarin and Valentina Tereshkova. The term "Vostok" is Russian for "East."

Now the other branch needed attention. Receiving managerial control of this new unmanned division was the dark-haired, slender, bespectacled Gleb Maksimov. A structures man, Maksimov saw to the creation of the probe itself, while people like Tikhonravov worried more about the booster.

Maksimov's face, which often bore the expression of someone just returning from a funeral, wasn't at all new around the bureau. Already, he'd spent years roaming in Tikhonravov's circles and well understood many of the problems facing success. At the first meeting between Maksimov and Korolev, their budding relationship stumbled on rocky footing. As the story goes, one day back in the early fifties, Tikhonravov's people received Korolev, who met with the engineers for a scheduled conference on ballistic missiles. While there, the chief designer caught wind of some ongoing side work by a sprightly young engineer named Maksimov. Something to do with artificial satellites. It sounded worth a look and he went to check it out.

Unfortunately, Maksimov had no idea at all who this imposing, arrogant, and unbelievably blunt man coming to see him was. The two sat alone for a while, side by side as Maksimov lectured a grade-school primer on satellite fundamentals. Ever helpful, he intermittently punctuated the talk with "Do you understand? Do you understand?"

Over the next many minutes, these innocent questions piled up to critical mass. Finally, Korolev rose over him "like a snorting bull," remembered young Gleb, hollering, "I do understand—everything!"

With *that* to look back on, perhaps Maksimov could appreciate these new marching orders as the closing of some awkward circle. "He was actually very warmhearted," Maksimov later recalled of his boss—*much* later. During this lunar quest the good times and warm fuzzies between the men occupied an easily measurable amount of time. Don't blink. "He had a penchant for punishing by shouting, swearing, and this was not always justified," Maksimov elaborated. "The next day, he would see you and be embarrassed. He wouldn't apologize directly, but would wordlessly convey his feelings with maybe a pat on the back." In just a few years Maksimov's deepening criticisms of Korolev would ultimately become too much for the chief designer. Amazingly, S.P. coordinated a demotive reassignment in order to shove Maksimov out of the picture.

Gleb Maksimov never really came to like Korolev per se, but he acquired a healthy respect for the man's abilities. "He was very careful in choosing the technical means to solve problems," Maksimov suggested. "Korolev 'sucked out the problem'; looked at it from various sides. But after choosing the technical means he became very decisive."

Maksimov continued, "Everyone in the design bureau was afraid of him, although he never fired anybody."

To assist with the growing workload, Maksimov appropriated several key people from the Ministry of Aviation. They fell into solid rhythm: nine- to eleven-hour days. A real sense of urgency befell the group, in part because a March 20, 1958, governmental decree had actually mandated the launch of a lunar probe within one year. Slowly the days grew longer.

Harassing pressure like this from up top was nothing less than a slap in the face to Dmitri Ustinov, acting minister of armaments and officially Korolev's boss. Ustinov liked to shoot things. He was supposed to be overseeing artillery development, which, by bureaucratic extension, included missiles. Dapper, smiling, and coldly manipulative, he gamely watched as the R-7 matured to ICBM status. Ustinov hovered over it like a kid about to get his first BB gun.

16. Tikhonravov's core engineering team, whose work contributed to so many early Soviet spacecraft. Seated individuals are, *from left*: Vladimir Galkovskiy, Gleb Maksimov, Lidiya Soldatova, Mikhail Tikhonravov, and Igor Yatsunskiy. Standing in the back row, *from left to right*: Grigoriy Moskalenko, Oleg Gurko, and Igor Bazhinov. (Author's collection)

Alas, Ustinov did not have the same little movies running in his brain as Tikhonravov and Korolev. What was all the space hoopla about? Moon probes? Ustinov just couldn't wrap himself around their importance. With great enthusiasm he ignored, hampered, or subverted most any lunar proposal that came down the line, no matter what any silly governmental decree told him to do.

Dmitri Ustinov had not been matched with his position via career counseling. He'd fallen into it more due to Stalin's purges clearing out all the qualified people than due to any sort of innate skill. And Korolev didn't need a boss—he needed a handler. "Ustinov would complain that it was difficult to work with Korolev," explained Boris Chertok, "because he liked working with people who were obedient and Korolev wasn't. All the other general designers followed orders."

Ustinov liked pretending to be in charge. He liked playing bad cop. He liked to nitpick. A couple years back, he'd been visiting Korolev's design bureau to meticulously inspect a new building—right down to the washrooms. What kind of problem was he expecting to find in the can? The toilet partitions went all the way down to the floor, which didn't seem to bother anyone but him. Ustinov flipped and immediately ordered a trimming down of the stall doors, reportedly to catch loiterers.

But a man like Ustinov wore two different faces, and the one that showed depended on who was around. In the final years of Stalin's life, for example, Ustinov had personally shuffled around Jewish engineers whom he felt to be at risk from bigoted superiors. His methods ranged from indirect to confounding—whatever got the job done. One night Chief Designer Mikhail Ryazanskiy got a call from Ustinov, who said he felt like walking in a nearby park and yearned for company. Intrigued by the bizarre invitation, Ryazanskiy showed up and began peppering the minister with questions. With great economy of speech, Ustinov apprised him to immediately dispatch one particular employee on an important trip "to anywhere."

Being a detail-oriented sort of fellow, Riazanskii wanted the whole story. Ustinov seemingly had no time or patience left to mince words. Abruptly he instructed the designer to, without delay, transfer the named individual to some other location—anyplace really, as long as it was far away. Later, Ryazanskiy would realize this action probably saved the man's life.

Sergei Pavlovich was never one to let any superior change his plans, even a Good Samaritan. Sideswiping Ustinov called for more troops. Again Korolev teamed up with Mstislav Keldysh, now the VP of the Soviet Academy of Sciences. Like golden trophies, they doled out experiment slots for the upcoming lunar probes, and in no time at all Keldysh and Korolev had the scientific community eating out of their hands. Then, like good government employees, they went over Ustinov's head—only now with the scientists in tow. A few conversations with Premier Khrushchev were all that remained to set the wheels turning like they needed. See how easy that was?

He had the scientists, the structures men, the rocket men. Korolev, like Louis Dunn, needed them all to be thinking the same thoughts. "Comrades!" Korolev preached to the congealed mass at his kick-off meeting. "We've received an order from the government to deliver the Soviet coat of arms to the Moon. We have two years to do this." An overview of the program followed, explaining the various milestones and predicted launch configurations. The rules were discussed. A couple of them seemed odd at first—like the one about how any travel between Moscow and the Tyura-Tam launch pads would only be permitted at night. "Korolev could not imagine wasting a day on travel," one of his colleagues later clarified. "He considered that a nearly sleepless night in an airliner's seat would give sufficient rest."

Internal paperwork described their goal as "the Automatic Interplanetary Station." Today the program is remembered as "Luna," and fulfilling its promise required solutions to an exceptionally large number of headaches. No single biggest hurdle existed. Where to start? The Moon-bound probe would fail outright without little maneuvering rockets called verniers to adjust its flight path along the way. But there had to be a method of commanding those rockets from the ground, which meant staying in contact with the probe during critical periods. *And* they needed a way of confirming such maneuvers had been properly executed.

Most of these abilities didn't yet exist behind the iron curtain. But Korolev figured the problems could be bulldozed over like any other. Gleb Maksimov said, "Most people use the simplest path in difficult situations—it's usually the optimal way—but Korolev used *all* the paths, even the most difficult. . . . For example, he ordered a number of people to solve the same problem of getting the right trajectory for the *Lunas* to hit the Moon." This theme—secretly directing multiple entities to work on the same problem, then choosing his favorite solution—became a Korolev trademark. Some knew about it; many didn't.

On average, the distance between Earth and the Moon is roughly 239,000 miles. For Korolev's group, getting anywhere close would rely heavily on a fashionably accessorized R-7 booster, sporting a debutante third stage up on top. That was just what Korolev had brazenly predicted almost two years before, and now struggled to realize.

Even his nemesis was lending a hand; Valentin Glushko's energies as of late had been channeled toward his compactly buxom new RD-109 engine to power this up-and-coming third stage. To put in the gas tank, Glushko found much promise in an emerging offshoot of hydrazine, which ignited on contact with liquid oxygen. It didn't need a pilot light or a spark or anything. Fuels that do so are known as hypergolic, favored when the bottom line is fail-safe simplicity and reliability.

In any kind of fuel there are trade-offs. Your misguided neighbor can put 110 octane in his Dodge Diplomat and get a little more zip, but the trade-offs are cost and availability and probably increased transmission wear. Rocket engineers like to talk about something called specific impulse—sort of a ratio describing how much thrust you get versus the type of fuel being used. Apply that concept to solid-fuel rockets, which is kind of like making a roman candle. The propellant can be whipped up and packed inside, where it sits quietly until the moment of use. Incredibly convenient. But it doesn't offer much bang for the buck, delivering a low specific impulse. Great stuff for submarine torpedoes! Not so hot for lunar rockets.

Sending anything to the Moon requires high specific-impulse fuel and all the massively complex baggage that comes along with it. The bottom stage is the most fussy and demanding, because it has the most to lift. But the rocket's *upper* stages tend to permit more wiggle room. There isn't as much weight to lift and not as much gravity to fight. So they can run on fuel with a lower specific impulse, which in general is easier to handle.

Glushko felt his latest model balanced these tradeoffs well—that is, when it was working. His expanding lists of bugs were growing into colossal show-stoppers, and by midyear 1958 the RD-109's construction and temperature-control foibles seemed largely insurmountable. Fuel tank welds weren't even staying together.

S. P. Korolev blackened the air with concern over endless burned months of Glushko's men laboring to debug the new engine. After pondering a whole slew of what-ifs, the chief designer informed Glushko that the clock had run out; he was going to look for something else. Korolev probably wasn't excited about whom he'd have to call for tech support, either.

These instincts proved correct. Glushko's heralded RD-109 didn't pass muster in 1958, or for that matter in '59, either. Virtually abandoned on the road-side, Glushko and his empty pockets left Korolev relieved on one hand but burdened to locate a suitable replacement on the other. He was in a pickle.

Then a light went on in the back of his head: what about Kosberg?

Back in the depths of World War II a middle-aged marvel named Semyon Kosberg turned the tide for Soviet aviation when he developed fuel-injected aircraft engines. The guy became a war hero. Since then, he had set up quarters down in southwestern Russia to focus on aviation design and had even garnered a chief designer title. Perhaps as a little back-burner project, Kosberg tinkered with prototype rocket engines. Some months back he had called on Sergei Pavlovich to introduce himself and talk a little shop; now it all flooded back into Korolev's mind.

Kosberg's stocky, upbeat presence was received with great excitement and many a samovar of tea. A platoon of his own men trailed in behind. Plugging away toward the first Soviet engine for use in true space, both groups intertwined resources to practice laudable cooperation. Where one of them lacked a solid turbopump, the other proffered a reliably sound one. Such well-spirited teamwork enabled the finished product to go from drawings to flight hardware in nine rapid months. Their product wasn't as mighty as Glushko's fantasy engine; it would easily lose in a drag race. But it *was* going on the rocket.

Korolev's first crack at hitting the Moon went up—and blew up—on September 23, less than a month after the Iowa physics department saw their *Pioneer 0* experiment unexpectedly double back after launch. In eight days, Eisenhower's new NASA would officially cut the ribbon and open for business.

Misfortune also descended on the second Soviet attempt of October 12, mere hours after *Pioneer 1*'s booster launched and started malfunctioning. "Don't worry that the American rocket is flying to the Moon," Korolev reassured his flock. "We will reach the Moon several hours before the Americans!" In spite of their tardy liftoff, Tikhonravov's team had shrewdly chosen a corner-cutting flight path that would have reached the Moon significantly ahead of *Pioneer 1*. This time the R-7 ran one hundred seconds. And then it had kittens, rattling to pieces. The *exact* same problem as before. Tikhonravov frowned. Somewhere deep inside the maze of plumbing and wires they had a very big glitch.

All work ground to a screeching halt while recovery teams marched out with brooms and dustpans, collecting what shards could be identified littering the Kazakhstan steppes. Vicious blood-and-guts shouting began, in parallel, over at the bureau offices, rising to painful thresholds not encountered since Sputnik. The whole program hung on a string. Korolev singled out a worker named Svet Lavrov and ordered him to identify the goddamn problem right there and then. Lavrov was chief of ballistics, quite high in the food chain, and patently interested in restoring a neighborly feel to the situation. He gulped.

Before Lavrov could respond, a stream of recovered hardware pieces began trickling in. They spread it all out on the big work tables amid rolls of telemetry printouts and began sleuthing. Their machine, so sleek on paper, became unforgiving in flight. And this detritus was all that remained.

Seldom are there easy finds while troubleshooting complicated equipment. Slowly, in a methodical fashion so brutally drawn out, the group uncovered a thought-to-be-dormant vibration bugaboo in their main stage. Like a drunk in rehab, it got the shakes; they'd never been completely resolved, although endless design iterations minimized the evil behavior. At that point, everybody had filed their notes and gotten back to work. Development continued along: people's short-term memories filled with trajectory calculations, experiment testing, final payload configuration. And that's when something got overlooked. Adding in their new third stage made the shuddering come back: the rocket moved in such a way as to invite the center of gravity outward, allowing ugly tremors to go racing up the side of the booster and tear it to shreds. It had done the same thing on both tries. Tikhonravov's faction modified some of the engine supply piping and readied another gleaming booster. Surely nothing would squelch this one . . . right?

Their third attempt—December 4—did not go well either. Normal booster separation came and went with no vibration problems whatsoever. The probe reached high in the atmosphere, riding on the core engine. Solid burn. That core, however, depended on the faultless operation of two unexciting yet critical hydrogen peroxide pumps. And when their lubrication system failed, 70 percent of the thrust vanished. Then the core engine quit. Four minutes after launch, the group's beloved creation agonizingly started whistling backward into its own contrails. And the timing of this attempt? Less than a month after *Pioneer 2*.

Determined to rise from the ashes of three successive failures, Sergei Korolev christened the next payload *Mechta*. Russian for "dream."

Come 1959 the men stood eager to have another go. Fade-in on New Year's Day. The mercury in Tyura-Tam read twenty below. Many on the launch team were staying in a nearby hostel, but the heat pipes had broken. In desperation, the men tried using portable stoves indoors, but they produced so much smoke it was getting hard to breathe with the windows closed. The launch team was icing up, their eyeballs practically freezing. They wanted time off. The radios weren't working right. The wind was too whippy. They wanted booze. Korolev waved his hands like a referee and said no way—a gaggle of Communist Party high rollers were due any minute.

Wails of discord thickened.

Something finally penetrated the chill—a procession materialized, carloads of party loyalists straggling out in their formal suits, walking stiffly through the operation to nod approval. They tried to make a show of importance, but didn't last long in the cold.

After the last car withdrew, one of the logistics officers stepped in front of Korolev to command his attention. No words were exchanged; the officer's eyes did all the talking. After a pause, Korolev begrudgingly nodded, saying, "To hell with it. Give it away." Within milliseconds an orderly line formed around the supply shed. Everyone knew where the liquor was.

The probe looked as otherworldly as the place it was being sent to: a deliciously polished globe with several antennae on its top half, sticking up like weekend hair; sensors freckling both hemispheres; micrometeorite detectors for cosmic rubbish; Geiger counters for cosmic rays. She tipped the scales at *eight hundred pounds'* worth of science and equipment, when a U.S. Pioneer spacecraft weighed less than fifteen. The world would come to know her as *Luna 1*, and she flew on January 2, albeit in a slightly different path than intended.

Sergei Pavlovich had been his usual uppity self heading right straight toward launch day. He liked to personally oversee every aspect of the preparations. At one point during a visit to the assembly shop, Korolev had dashed pell-mell screaming into one area where a round of hammering noises had kicked up. This was the spot where *Luna 1* was being mated to its third-stage adapter, and the epicenter of clanging metal-on-metal tones echoed through the plant like some bizarre, thumping heartbeat. Korolev's bellowing instantly eclipsed it.

"What on earth are you doing!" he angrily demanded, arms wide, virtually shaking the walls. The misfortunate tech in front of him straightened up. Words drizzled slowly out of his mouth. He had some stuck bolts on his hands and was trying to free them with a mallet and a huge wrench.

"Why are you pounding it?" Korolev boiled, hurricane red, eyes popping. "This is a spacecraft!"

Somehow keeping his cool, the worker patiently explained that he was pounding on the mounting frame, the work stand. Most assuredly, it was *not* the spacecraft itself. Despite this innocent explanation, the employee swore to never pound on anything again—at least, not while Korolev was around.

They had a good telemetry lock on *Luna 1*—the radio problems were gone. Things were looking rosy at seventy thousand miles, when an onboard experiment jetted out a cloud of sodium vapor. It was visible to anyone in the vicinity of the Indian Ocean and lit up to about the brightness of a sixth-magnitude star. Years later Gleb Maksimov recalled how Korolev signed up a contractor for that experiment "literally in five or ten minutes," after Maksimov himself had been unable to locate a vendor who would take it on.

"And you spent a month on *this*!" S.P. scolded.

Besides giving scientists an idea of how gases behaved in space, the sodium cloud allowed flight controllers to assertively pinpoint their craft's position, which wasn't where it was supposed to be. A few rounds of pointed questions and math and head scratching and duck-duck-goose uncovered a booster programming error that, owing to the technology of the day, couldn't be fixed.

Today, if the backyard satellite dish for a television starts making fuzzy pictures, the guy on the tech support line can regimbal it from over the phone and make everything right again. Gleb Maksimov was not so lucky. No provision existed for radioing the interplanetary station to adjust course. There would be no lunar impact. However, out of dozens of launch failures, the team had flung a payload barely four thousand miles shy of the Moon.

Luna 1's batteries gave out after nearly three days. Contact faded, and the probe began drawing a long, looping solar orbit somewhere between Earth and Mars, where it still is today. The Soviet government would hold high the success of its station and claim for years that the path she went down was what they'd intended all along.

Arvid Pallo probably would not agree. After involuntarily donating much of his free time to Stalin's forced-labor camps, this gifted engineer arrived to commune with Maksimov and Korolev on the *Lunas*. He acted as nothing less than a central figure in the probes' existence—sort of a godparent, immersing himself in most every phase of design and testing.

Pallo also budgeted time to prepare a unique set of commemorative gestures intended to be left in moondust for all time. There is the matter of a tiny sphere aboard *Luna 1* that took inspiration from Tsiolkovsky, containing different kinds of substances all held in a fluid. Deep in the center rested a small strip of metal decorated "USSR" in handsome script, featuring the launch month and year on the reverse.

Another sphere rode along with it. Borne of titanium, *this* one resembled a polished metal soccer ball and was tantalizingly designed to fracture itself into seventy-two pentagonal segments. Each one carefully imprinted with the launch information on one side and a Soviet hammer and sickle on the other. Upon impact an explosive charge deep inside the sphere would blast it open, flinging the pieces apart to the radius of a few miles. These inclusions had no place on a craft intended for solar orbit.

In spite of the failure, Nikita Khrushchev got up in public on January 27 and spoke in confident terms: "Soviet scientists, designers, engineers, and workers achieved a new exploit of worldwide importance, having successfully launched a multistage cosmic rocket in the direction of the Moon. . . . Even the enemies of socialism have been forced, in the face of incontrovertible facts, to admit that this is one of the greatest achievements of the cosmic era." And thirty years whisked by until the country fessed up.

The premier delivered encouraging words, but Gleb Maksimov had future Luna upgrades taking shape in his head even as Khrushchev droned on to the press. Maksimov worked his tail feathers off in the effort to design a suitable encore. Some nights he took a quick breather to stare longingly at the gray, cratered surface hanging in his night sky. Somehow, they had to get one there.

Maksimov also found himself immersed in much more good company. The whole crazy business of solar wind intrigued the participating scientists almost to no end. See, *Luna 1* accomplished much more than failing to hit the Moon. It had happened upon a constant stream of *really* high-energy protons and electrons erupting from the Sun in all directions, spewing forth for no apparent purpose other than to cause naughty time wasters like the disruption of power grids. The way a comet tail always points away from the Sun? That's the push of solar wind.

Hypothesized in 1916, the notion grew legs by virtue of British mathematician Sydney Chapman, who sat on Van Allen's living room couch that night in 1950 giving rise to the IGY. Over the next few years, Chapman stirred the idea on low heat then published something of a hypothesis. He argued that in light of how intensely these alleged solar discharges scooted along, they ought to run billions of miles out into space.

With another opportunity on the horizon, scientist Konstantin Gringauz set out to beef up his unassuming Luna experiment that had apparently proved solar wind's existence. One of Gringauz's last responsibilities had been the part of *Sputnik 1* that produced the beeping Van Allen, Cahill, and the rest of the world heard on October 4, 1957. So when the time came to swing for the Moon, Gringauz was a natural go-to for a goody to send along.

Forty years old at the time, Gringauz was the son of a pharmacist and grew up on the banks of the Volga River. Early on he'd been stricken with a bug for radio. It became his overriding hobby during adolescence and consequently prompted his 1935 enrollment in the Leningrad Electrotechnical Institute. At the time, frequency modulation (FM) was a fairly new concept. Seemingly intent on living on the cutting edge, Gringauz made it the topic of his diploma work.

Amid the foggy blackness of World War II, Siberian tank makers depended on Gringauz and his like-minded colleagues for the development of radio transceivers. "He was obviously a true Russian patriot," opined Stanford physicist Don Carpenter, who pursued similar avenues of research as Gringauz in later years and got to know the man. "A true Russian patriot in the sense of being proud of the cultural traditions of Russia; perhaps somewhat more skeptical of the socialist realism of the Communist era."

Gringauz and *Luna 1* hadn't exactly gone looking for solar wind, because nobody knew for certain if it could even be found—solar wind was only baseline theory. The Luna spacecraft's instrumentation, therefore, was not specifically tuned. But with this new knowledge, the stout, bald-headed, and somber-looking Gringauz reconsidered his half-spherical ion traps, which had first pointed the accusatory finger. Their configuration or operation or something else inside would need to change. What made sense? How do you build an instrument to study something that hasn't ever been studied? No generally accepted model existed for the state of matter between Earth and Moon. Often Gringauz frowned in the effort; it was tough and nerve-rackingly delicate. "He looked like a man preoccupied with serious work," Carpenter remembered of his friend's most commonly displayed emotion. "And I don't recall him laughing or smiling hardly at all. I think he tended to have a kind of a serious demeanor."

Gringauz had a brainstorm: instead of laying them out flat in the equipment bay, his traps could be arranged more three-dimensionally, like a pyramid. This alteration might better describe to him the particles' direction and density because not every sensor would catch sunlight at the

17. *Luna 2* nestled inside its spacecraft adapter. (Author's collection)

same time. "It was fairly crude but I think rather effective," Carpenter said of the design.

As the *Luna 2* odometer flipped over forty-seven thousand miles, Radio Moscow personality Yuri Levitan was finally handed an announcement to read. Soviet citizenry then learned of the attempt, after it was already 20 percent of the way there. Levitan's grandiose voice would, over the course of his career, broadcast some sixty thousand programs. He'd been such an ingrained facet of Soviet life for so many years that during World War II Hitler had once personally ordered him killed.

Luna 2 flew on course, trajectory holding. A matching sodium cloud experiment traveled aboard. When it operated perfectly, Gringauz pulled up a chair and held his breath. The ship lacked any active controls. One shot only. Telemetry messengered in the vital signs, along with experiment progress. And with much interest the sober Gringauz noted how his ion detector's activity rose and fell as the spacecraft rotated—almost in perfect sync, like two embraced lovers panting in rhythm. Weeding out the definitive answers would require time and effort, but up there it sure looked like a blast of ions rushed in whenever the detectors were facing the Sun.

Gringauz's relationship with Korolev went back over a decade. Early launchings at Kapustin Yar carried Gringauz-hewn radio sounders, which probed the ionosphere. The results would've made him a hit at scientific conferences, except the Soviet government shushed any work done on rocket flights prior to 1957. Glory would have to wait. *Sputnik 1*'s final transmitter design had been his own. At the time, debates had raged through the bureau like brush fires as competing radio systems were going head-to-head for the honor. Gringauz won out. Really, it became an easy decision when Korolev realized how the signals could effortlessly be heard around the globe by any radio hobbyist—simple validation of Gringauz's favorite pastime. Then, the day before *Sputnik 1*'s launch, Konstantin Gringauz had scaled the tyulpan for a final, paternally reassuring check of his transmitters. He was one of the last to touch the ship. For a man resigned to the shadows of space fame, it became a proud memory to share with his daughter, Tatiana.

This time they were going for the Moon. There was no way to speed it up—no fast-forward if Gringauz didn't like the scary part of the movie. And as the Moon drew closer and closer, the probe went faster and faster still, tugged by gravity. Onboard instruments reported no detection of magnetic fields or radiation belts, like Earth possessed. Telemetry still rolling back from nearly a quarter-million miles away, almost a second and a half delayed from real time because of the distance. *Would* it connect? *Did* they have it? Nobody knew for sure. Lengthy strings of numbers on paper, hand-calculated, told them it might be. But it wasn't over until it was over.

The news came two minutes later than planned. On that Sunday, September 13, 1959, the Cubs beat the Cardinals 8–0 at Wrigley Field. And in frigid Asia, Soviet telemetry abruptly cut off because *Luna 2* had slammed into Earth's moon near the crater Archimedes. An on-site witness standing nearby would have observed *Luna 2* silently roar from the black of space, hitting and instantly detonating as Arvid Pallo's two spheres ruptured like bar glasses on a tile floor, dozens of titanium pentagons scattering themselves about in a jarring, heinous bubble of entropy to wake the Moon from deep slumber, a concussion of waves trembling the surface and rattling dust.

Then the Moon was still again.

Two days later NASA deputy administrator Hugh Dryden output a remarkably humble press release, mentioning how "NASA joins with the world's scientific community in acclaiming this as a truly great engineering achievement. It is further evidence of the excellence of the Soviet capability in the propulsion field." A *New York Times* editorial spoke of how "all humanity joins today in congratulations to those responsible for this achievement."

Ever desirous of impeccable timing, Nikita Khrushchev landed in New York City the very next day for a previously scheduled visit. His appearance marked the first by a Soviet leader and, as such, the first in-person opportunity to have a few digs at the competition. Gifting Pallo's replica spheres to Eisenhower, the Soviet premier commented, "We have no doubt that the excellent scientists, engineers, and workers of the United States of America . . . will also carry their pennant over to the Moon. The Soviet pennant, as an old resident, will then welcome your pennant, and they will live there together in peace and friendship."

Konstantin Gringauz had found his elusive, invisible solar wind. He spent the rest of his days analyzing data and writing about it, along with a thin region around Earth christened the "plasmasphere"—a completely unexpected discovery he shared with Don Carpenter, who'd been probing the same region with ground-based equipment.

Gringauz didn't stop there. He would even go on to directly investigate Halley's comet in 1986. "People were impressed by Gringauz," Carpenter fondly suggested in regard to the man's body of work. "He seemed to have acquired a certain, more-prestigious position in the hierarchy." By 1964 the Lenin Prize came Gringauz's way, and with it the privilege of cutting to the front of the line for everything from basic necessities to cinema tickets. So much for a classless society.

Today Carpenter holds the reputation of Konstantin Gringauz in very high regard. "I considered him to have had such a fine, distinguished scientific career." This diminutive Russian had nattily proved the existence of solar wind. "His instrument may have been one of the first to make what you might call definitive measurements of the flow."

At the moment of *Luna 2*'s impact, America's space effort had nothing to offer. Nobody at JPL or Huntsville or over at the newly established NASA would in turn respond with any lunar launch for almost an entire year. Little did they grasp how things behind the iron curtain were just heating up. Coinciding with *Sputnik 1*'s second anniversary, Maksimov and his band of explorers prepared their follow-on *Mechta*. *Luna 3*. Forget about

18. "He was very modest and self-effacing in his manner." Konstantin Gringauz, 1993, the year he died. (Photo by and courtesy of Brian Harvey)

hitting a target; they wanted to dance with it. And, therefore, *Luna 3* was not under any circumstances to be some kind of milquetoast rerun. They built to deliver, and it did just that—stealing the world's collective breath in a skilled demonstration of thoroughbred design and world-class orienteering, not to mention the onboard gadget whose innards were as shockingly amazing as what it accomplished.

To pull off the mission, what started as *Luna 2* was souped-up with enough additional gear to make any hot-rodder jealous and the original *Luna 2* unrecognizable. At its equator the craft's spherical body was sectioned, and a squat cylinder, to be *pressurized* no less, was fixed in between. It compared nicely in size with that of a Radio Flyer spring-loaded riding horse. Solar panels were then hung all the way around, funneling lifeblood electrical juice to rechargeable batteries inside. And what if it got too hot in there? The panels cohabited with a low-tech arrangement of glorified venetian blinds, designed to passively open or close for heat regulation—a spacecraft equivalent of rolling down the windows.

Now known as *Luna 3*, the automatic interplanetary station (AIS) would be spin-stabilized for almost its entire trip. But Maksimov's showstopping trick at the end didn't permit any spinning. It needed a bona fide attitude control system capable of reorienting the vessel during flight to make it look in a certain direction on cue. It needed magic. They weren't sure how to invent magic.

Creating the solution fell on an engineer named Boris Raushenbach. He tossed a thousand rubles at one of his engineers, who promptly took a walk through turgid city streets. Midmorning the guy discovered a jumble shop loaded with secondhand electronics. He bought the place out and hauled it all back to the design bureau. The cache would be a jumping-off point for the eventual complement of sensors and controls as they sweated out *Luna 3*'s unprecedented accessory.

At the time, Rauschenbach was in his early forties and probably surprised to be experiencing them. Like many, he'd been carted off in 1942 for unspeakably horrific crimes, and in Rauschenbach's situation the charge was possession of a German-sounding last name. No, *really*. Six harrowing years vanished in futile attempts to clear himself while poor Raushenbach clung to life in a gulag averaging ten fatalities a day. Without his determination to live, learn, excel, and then make his way to Korolev and Maksimov, Soviet Russia may well have conceded a fair percentage of their historic firsts in space to the Americans. One of these firsts made use of Raushenbach's control system known as Chaika, meaning "seagull." To any lay observer the finished handiwork presented itself as just a bunch of little openings on the bottom half of *Luna 3*. But this technological stunner boasted a design mature enough so as to be later used in Yuri Gagarin's capsule.

Tyura-Tam operations ran at a fairly intensive pace 24-7; some people didn't leave their assigned work areas for days. *Mechta* itself arrived by August, though not all of the particulars were resolved. Once there, it was swept into a vortex of final preparations rivaling the backstage of any Broadway show. A moody electrical subsystem continued giving the engineers fits and had to be fine-tuned. Korolev remained upbeat regardless, actually contemplating the idea of letting Soviet radio announce their groundbreaking plans in advance.

19. If not the real deal, this is at least a model of *Luna 3*. Camera port at the top with hemispherical door hinged open. (Author's collection)

After a pleasantly uneventful launch on October 4, *Luna 3* began tracing what would hopefully be a slow, lazy figure eight around the Moon. It would take days. Everything looked great, so the principals, along with S.P., all dragged their overtired selves aboard a plane headed back into Moscow. Forty-eight hours later *Luna 3*'s critical radio connection had all but died.

The ship was also about to hit a dead-air gap between ground stations, killing any immediate shots at isolating the problem. By now they had additional uplink facilities spread out, plus a handful of tracking ships. But their coverage still fell far short of uninterrupted. Not even the immense British radio telescope at Jodrell Bank could get a lock, and their gear was ten times more sensitive than anything the Soviets had.

Korolev looked at his watch. There wasn't a blasted thing he could do about celestial mechanics, and it wouldn't be long before the probe came in range of the Crimean ground station at Ai-Petri. *Crimea*—hell and gone from where he stood right at that second. Sergei Pavlovich loved Moscow, but suddenly it became the wrong place to be. Four hours remained to save his *Mechta*.

Right on the spot, Korolev decided that matters would be best taken into his own hands. And the unchoreographed cavalcade of events that happened next simply never would have occurred within the architecture of America's space organization. It was as if Wernher von Braun himself were to have gone traipsing cross-country through Cape Canaveral, Iowa City, and the Jet Propulsion Laboratory in search of a blown fuse.

Boris Chertok was also back in Moscow—at a Venus study meeting— when word came that Korolev wanted him immediately. Chertok showed up at the chief designer's office a short time later, finding Korolev on the phone making arrangements for a plane.

Chertok watched him hang up. "Radio communications with the spacecraft are very poor," Korolev said flatly, as a way of bringing his man up to speed. "The telemetry is not coming through; radio commands are not getting on board."

Chertok frowned.

"We will fly to the Crimea," Korolev instructed, "and have to be in place before the communications session starting at 4:00 p.m.—that's the time of radio coverage from the Crimea." By this time senior engineer Arkady Ostashev had joined the group and also turned his full attention on S. P. Korolev. He and Chertok were informed that a pair of cars had already been summoned and were waiting down in front of the building. The chief designer ordered, "Stop by your homes, grab what you need, and drive to Vnukovo." A specially chartered plane would be sitting there ready for takeoff.

No time remained for questions. Chertok raced out of the office to quickly complete his tasks, and then, after driving at an unhealthy rate of speed over to the Vnukovo Airport, he ran aboard a Tupolev Tu-104 aircraft with dual turbojets already spinning. In the aisles Chertok met up with Korolev and Mstislav Keldysh, who had already found places and buckled in for a stressful ride.

The plane closed up and started taxiing, but then the pilot slammed on the brakes and cranked the doors back open so Arkady Ostashev could dash up the steps and fling himself in. Soon they were airborne and arcing full throttle toward Simferopol Airport, near the tracking station. After landing, the airplane was practically still rolling down the tarmac as everybody jumped off and sprinted for a waiting helicopter, which leaped into the night sky as soon as they were all on with chuck-chuck-chucking rotor blades making it hard to talk. Past the Crimean Mountains the pilot stuck his head in the cabin explaining how the weather in Ai-Petri was nothing less than wall-to-wall sleet. They were about to fly right into it. What should he do?

Everybody has choices to make in life, and some of them suck. Visibility at the landing site was absolute zero. Korolev thought hard for a moment about having the pilot fly there anyway. He looked at the people shoehorned in with him. Had that copter gone down, it would have torn the head off the entire Soviet space program.

Then Korolev had a better idea. From the helicopter a barrage of urgent requests shot out over the radio to area officials of the Communist Party, who instantly commandeered a pristine six-seater ZIM V-8 limousine and had it fueled and waiting at the Yalta Airport when the men landed. They all squished in the back like a bunch of hamsters. It was ten miles to the station. The driver had no idea who any of them were at all. Korolev told him where they were going and challenged, "Well dear boy, show us what you and the ZIM can do. We are really in a great hurry." Their eyes locked for a moment; then the driver stomped on the gas pedal and covered those ten miles like a Formula 1 champion, never once slowing for any of the icy hairpin switchbacks peppered along the way. He practically dodged the sleet itself.

The ZIM rolled up to Ai-Petri in one piece. They'd made it with ninety minutes to spare before *Luna 3* came in range.

The local tracking station resembled more of a camping trailer than a facility, sporting wheels underneath and its radar dish Scotch-taped onto the side of a nearby rock face. Few people typically remained on-site to man the cramped equipment room. Korolev huddled everyone together, launching into a stand-up roundtable discussion of the symptoms. Could it be the power supply? Batteries not charging? What about the antennae? Sixty minutes to go.

Perhaps cognizant of the need for some other approach, Korolev retreated to a quieter spot and took turns with every radio station controller individually, speaking one-on-one. The technicians came and went in an orderly, impromptu confessional; as they did, a picture of the situation meticulously assembled inside Korolev's head—building truth and illuminating conflict and loopholes. Now he understood much. For starters, he understood how the spacecraft control process itself—as written and practiced—sheltered a dreadful oversight: no formal procedure verified the correctness of any one step before moving on to the next. It more came down to operational diligence, which varied with the wind. What kind of mood were the workers in? Had they gotten enough sleep? Any marital strife on the way out the door that morning? In the wrong frame of mind, anyone could just blow down the list.

Before S.P. could process all this, a litany of repressed, snippy complaints began gushing forth, practically soiling his boots. People out here were seriously unhappy. The work flow between controllers lacked clarity and harmony. The job was boring. Nameless military types drifted in sometimes and told them to do things differently.

An unprepared Korolev had no place to shovel all this muck. It only worked to smother any forward progress on resolving *Luna 3*'s fate. Ah, couldn't they talk about this some other time? Twenty minutes left.

Tugging hard on the reins, Korolev spat new orders that from now on, by golly, only Yevgeniy Boguslavskiy would tell a radio operator what to do. Nobody else, no matter *what* medals hung on their uniforms. Then he motioned to Boguslavskiy himself, and the two men spoke privately for a few minutes. Without mincing words Korolev admonished him for losing control of the situation and then impressed upon him the significance of asserting his authority. Now, what of the problem with the interplanetary station? Boguslavskiy twisted his mouth in befuddlement. He had put *Luna 3*'s radio system together in the first place and had overseen construction of the ground stations. If *he* was tongue-tied, then anybody would be. Complicated equipment made for complicated problems.

Finally, he offered a thought: the antenna setup contained a few design flaws they knew about in development but hadn't been able to overcome. It wasn't perfect, so running into a few signal dropouts wouldn't be out of the question. Korolev pursed his lips. *A few signal dropouts?* Boguslavskiy continued speaking, but as he did so, *Luna 3* shuffled into range of the station. One of the controllers called for everyone's attention. Like spectators at a tennis match, they synchronously pivoted toward the radio stack.

It was the moment of truth.

Suddenly, the telemetry came streaming in just like it was supposed to. The radio was fine. Paper printouts of *Luna 3*'s vital signs began to unspool.

Tension in the room broke. They could get on with exploration. All systems go . . . or were they? *Schzz crackle.* Wait. As the data came in, an operator tapped his finger on one section of the telemetry printout, the section about the temperature. *Mechta* was way too hot—roasting at the hands of its own overheated instruments.

Arkady Ostashev had been glumly off to the side this entire time. Now he felt Korolev's stare level over his position. Very nearly it had the effect of roasting him from the inside out, and Ostashev knew there wasn't going to be any tea for quite a while.

"Why are we overheating?" erupted his boss.

All Ostashev could manage at the time was, "I don't know yet."

Korolev drew himself up in authority. "If you, as a test specialist, believe that questions of planning and ideology of the system are not your problem, you are deeply mistaken."

Ostashev said nothing. He was exhausted. They all were.

"Why didn't you find out the capacities of the thermoregulation system?" Korolev demanded. "This is your gross error, which cannot be allowed in the future!" He sneered, "And now I am waiting for proposals from you."

Still, Ostashev kept quiet. His brain conjured no legitimate, satisfying response. And Korolev was just rounding second base. "Involve all of us here, anyone in Moscow or in the entire Soviet Union," he snarled, "but remember—the fate of the results of the first AIS flight is in your hands."

For the rest of the day Ostashev and the others methodically untangled *Luna 3*'s ailments. Their eventual solutions were, like many things Soviet, uncomplicated and low-tech. The probe's onboard electronics generated heat just by virtue of being on. New commands instructed *Luna 3* to power off anything whenever it was not actually in use. That would get 'em started.

Boris Rauschenbach's Chaika then jumped into play, wisping to life, imparting a slow rotation to the ship. This gentle spinning promoted even, regular heat distribution. Later, Apollo Moon flights employed a similar approach called "barbecue mode."

Within a couple of days the internal temperature dipped by almost twenty degrees Fahrenheit, which was enough to get them out of the danger zone. Still in Crimea and regaining momentum, Korolev now wanted to check the flight path. He telephoned back to the Moscow command center looking for information from a trajectory specialist named Efraim Akim. A female receptionist took the call. She told Korolev, "Akim is asleep, and there is nobody available for you to talk to." S.P. banged the phone down and came completely unglued right there in the tracking station. A black hell of obscenities erupted forth.

Life offers few similarities to watching a close friend melt down. Mstislav Keldysh attempted to soothe his friend with gentle words, explaining that there was no way *everyone* in Moscow had just turned off the lights and headed out. There had to be, heh-heh, some kind of simple explanation. An hour later Akim called back with one.

The flight path guys had been killing themselves this entire time, he explained, even before the launch, right up until just a little while ago. So after calculating the last maneuver, some folding cots were brought in so everybody could steal a few winks. Their offices lacked space enough for the whole crew to stretch out in, so Akim had gone next door in search of a cozy spot.

With all of this behind them, the Crimean phalanx shuffled out into the biting wind to have a smoke. The Sun was just beginning to set on what had established itself as an altogether unpredictable day. They stood there watching a ship just off the coast. A ship whose job it was, someone mentioned, to monitor radio interference. Boguslavskiy struck a match and lit up. He was the Jewish man Dmitri Ustinov had requested be sent on a trip "to anywhere" the night of Ustinov's loony walk in the park.

A while later Korolev left Ostashev at the tracking station and went with Keldysh and the others to a nearby hotel. It was midnight when they showed up, but prior arrangements ensured a huge meal awaited. Engineer

Anatoliy Abramov recalled the odd fascination of watching Korolev dine. "He ate very quickly, paying more attention to answering questions than to the meal," Abramov described. "After finishing the food on his plate, he would wipe it clean with a piece of bread, which he subsequently put in his mouth. He even scooped up crumbs and ate them." Once, when some responded with slack-jawed silence, another man explained that their chief designer had developed this habit in the labor camps.

After dinner they got four hours of sleep and were on the road again at six.

Of course, *Luna 3* remained blissfully unaware of all the human ruminations down on Big Blue. It was just about time for the one-two grand finale, as the nimble ship arced around the back of the Moon, severing radio contact. Our only natural satellite rotates, of course, in such a fashion as to keep the same hemisphere constantly facing Earth. Humans had yet to experience the other side, so *Luna 3* made this unprecedented journey for a little sneak peek.

The reactions of a simple photocell were now the next item on the agenda, cueing an entirely automated process of machinations. All anybody back home could do now was wait to see how things unfolded.

This photocell thing had been a long haul, like nearly every other aspect of the design. It was one tiny light sensor on the spacecraft. All it had to do was wait for a caress of sunlight—any reflection from the Moon's surface would do. When it came, the sensor would initiate everything downstream; if for some reason it didn't work, there was no way to prompt it manually.

The first time they tried was during ground tests. The sensor had been installed in a prototype *Luna 3*, which was then clipped to a rope and hauled up in the rafters of a testing hangar. From the other end a searchlight washed over the probe. Nothing happened. They tried again with no luck. Somebody suggested they bring the light a little closer. It kept not working. Finally, the intense light was so close it started their mock-up smoldering. Again, nothing. The operator flipped off the searchlight and went to join the argument gaining steam down there on the floor. The sensor wasn't built right. The wiring was bad. The batteries were dead. It went on and on. Out of curiosity, one of the engineers lit a match and held it in front of the photocell. The probe sprang to life, thrusters hissing, chassis revolving slowly toward the fiery sliver of wood. What the . . . it worked after all? The problem lie in the testing. Their searchlight gave off wavelengths of light the sensor couldn't detect. But flame—or sunlight—was just the ticket.

Korolev looked over to the engineer who'd figured it out. "We would have to launch you and your matches into space," he deadpanned.

But the getup worked when it mattered. The onboard cell had indeed awoken to sunlight like it was supposed to, *Luna 3*'s nod that Earth's moon was now close enough for the big show to begin. Petite gas thrusters oriented the top end into an optimal position. Although that might sound unimpressive, this was the crown jewel of Rauschenbach's control system. The probe simply used the Moon as an absolute reference for where it was supposed to be and when it was supposed to act.

Next, a small door that had been covering the lenses popped open. After that? Get busy shooting. *Pictures.*

For forty minutes the dual-lens camera exposed image after image, capturing over 70 percent of the Moon's far side. This wasn't some wimpy television system either; *Luna 3* shot top-shelf 35 mm film—the very spool recovered from America's Genetrix spy balloon. A few years back, word had discreetly circulated that one of the design bureaus kept striking out in their search for premium-quality photographic media. They needed Hollywood stuff. It had to meet a long list of difficult criteria.

In no time at all an unmarked package skidded to a stop on Korolev's doorstep. It held a roll of the nine-by-nine-inch high-tolerance stuff the Luna team had been wistfully dreaming about all this time. A few tests proved it would do everything they needed, and without the six centimeters of lead shielding some were claiming would be necessary to prevent image fogging. Lab technicians cut the film down into 35 mm strips and punched sprocket holes. The spaceship's cramped body would only hold forty exposures—forty chances to snare the unseeable as *Luna 3* traced a single pass behind the Moon and swung back toward Earth. After twenty-nine frames clacked through, the mechanical shutter jammed and stopped advancing. It wouldn't budge.

Now *Luna 3* shunted its film through nothing less than an ingeniously hands-free onboard Fotomat, outfitted with processing tanks, fixatives, and a dryer all wedged inside. Gleb Maksimov's people called it "the Laundromat."

With the strip of images developed and fixed, it went through a scanner. Differing intensities of light and dark patches were converted into electric signals then stored until the entire job was done. The scans weren't anywhere near film quality, but one thousand–by–one thousand pixels per image were better than, say, no images at all.

The entire process took more than a day from start to finish. A pensive Korolev had retreated to a nearby resort town on the Black Sea. He tried walking some to relieve tension, but his mind was always on the flight. The chief designer's mood wasn't helped any when an astronomer visiting from the nearby observatory insisted that none of the pictures would come out, because of all the radiation in space.

Now came time to send it all home. The first frame of the roll had been taken back on Earth with a flight spare *Luna 3*—the Moon as photographed from down on the ground in Mother Russia. Then the entire length was rewound and oh so carefully loaded into their primary camera. The appearance of this image from the ether functioned as a test pattern, cleanly demonstrating the system's correct operation.

Reception would only improve with proximity. So the group waited some, until the eighteenth, when their little cylindrical explorer was much closer to home. They got impatient and began trying to make contact. Aleksandr Kashits was on the team: "We were sitting in the dark control room," he began, "staring at the screen of a monitor." Nothing came on. Of course there weren't *supposed* to be images, but Kashits and his comrades kept attempting all the same. "Again and again we were trying to see, or rather to guess, at least a hint of the picture." But nothing yet.

Local officials closed roads near the tracking station and sent word to passing ships about maintaining radio silence, lessening the contaminated airwaves. With the hardest roads behind, *Luna 3* now flew an arc to keep it in unbroken contact with Ai-Petri. Radio commands scampered up, imploring the ship to phone home and start faxing over what it had. There was a commotion of feet. Everyone crowded around a single white monitor screen as its violet cursor moved slowly, drawing one line every second and a quarter. All it made was random gobbledygook.

They tried again and got static with dark blobs. Then another attempt. All eyes were on the cursor during its repeat trips from left to right. *Luna 3* would only close the distance, and the quality could only get better. Wait. Was that their test pattern?

Finally on the fifth try, something came through worthy of a hard copy. So on October 7, 1959, the far side of the Moon printed itself on a roll of thermal paper into Sergei Korolev's hands. He blinked, asking, "Well, what do we have here?"

The image was a little light on details.

It looked like the X-ray of a fried egg.

One guy said, "At least we know the back side is round also."

Another sold it better: "Don't worry. We'll add some filters and remove distortions." Mild frustration drifted over them like gathering storm clouds. This whole mission was in service of good images—it got the best people, the best designs, best equipment, best film. And here they were, with pictures of wet newsprint?

But over the ensuing hours as more and more images came through, they could see things. They could see shapes and craters never before known. And that's when these men began to feel a little more like they'd done it. They had brought to Earth something never before seen by mankind.

A voice in the room announced *Luna 3*'s distance to be close enough that it would automatically be switching to a faster transmission mode. That meant a change in receiving equipment, so a mild jostle broke out to swap it all over. By this time people stood mostly anywhere they could fit in the overcrowded room. They were wedged in the doorways and between the equipment and the walls. The purple-cursored monitor began drawing again with increased speed. Hazy noise and static remained, but it was not so bad this time. Sharper features. Better resolution. They *had* it. More pictures came tumbling out of the printer, rapidly like a mall photo booth, data spooling onto magnetic tapes, whirling to keep up, laid down for long-term storage and to be filmed out for the world. They had the dark side of the Moon. Ai-Petri erupted; somebody cuffed, "You bet!" Hugs, wet eyes, teapots of alcohol. There would be no anticlimax—all pictures that could be sent finished rolling through as *Luna 3* charged into Earth's shadow, telemetry fading, never heard from again.

It was over.

9. Moving at the Speed of Design

The first interplanetary route is paved.

Pravda, *February 26, 1961*

Only a couple of years prior to the success of *Luna 3*—three days after the *Sputnik 1* bombshell—England's *Manchester Guardian* had published an editorial that cautioned in part, "We must be prepared to be told what the other side of the Moon looks like." Who would have thought that dungareed Soviet visionaries needed only twenty-four months to deliver?

On occasion a helping of bureaucracy really is good for business. Welcomed into the world on October 1, 1958, the National Aeronautics and Space Administration represented a gigantic model of regime change. The new legislation creating NASA effectively lumped such diverse enterprises as aeronautical research and space exploration (unmanned or otherwise) all under one agency's control. The widespread smattering of affected facilities included Lewis Field in Cleveland, the Langley Research Center in Virginia, and Ames Laboratory in California. At the stroke of midnight they bowed to a new king.

Freshly minted NASA administrator Keith Glennan called all the departmental heads together in the little auditorium at the Dolley Madison House and told them, "Gentlemen, this is no place for tired men. If you're tired, go home." Moving forward, NASA would now govern all things recognizably "civilian" and therefore "peaceful." That left barbaric nuclear missiles and outright defense to the military at large. Before NASA came along, a conspicuous amount of deliberation went into potentially changing the air force's name to the U.S. Aerospace Force.

How Sergei Korolev would likely have cherished a similar arrangement. Hearty, defiant flames inside him burned, yearning for the exploration of

space while flickering somewhat less brightly for the cause of warmonger-
ing machinery. With much passion he wished to assist in the defense of his
country; there's no doubt about that. At the same time, he very much loved
space. But love is always a two-way street, and Korolev needed the bombs
as much as he might have despised them. No bombs would have meant no
missile program. No missiles? No spaceflight. All were links in a chain.

So the exertions continued, his facility slowly becoming a victim of its
own success. At Central Design Bureau No. 1 the fundamental reward
for good work remained *more* work. The collective wanted short-range
missiles and intermediate-range missiles. They wanted waterproof missiles
on submarines. As a result, Korolev's in-box dangerously teetered on the
edge of collapse. And unlike his American counterparts, the chief designer
shouldered great responsibility for his employees' personal needs and living accommodations.

Harried development continued toward orbiting a man. Space suit. Life
support. Ejection seat. Water. Nobody could run so hard at this pace for
much longer.

Newspaper and magazine stories afforded NASA incalculable scads of free advertising. Much of it revolved around the agency's just-announced Mercury
program, aimed at somehow lofting a man into orbit—ideally before the
Reds managed it. But long before seven perfect physical specimens volunteered to become America's first astronauts, an unrelated program was already jelling deep in the endless corridors of Space Technology Labs. Ships
had been built to study Venus.

In fleeting concept it almost sounded preposterous. Here was an otherwise intelligent nation just getting comfortable with Earth orbit and barely
able to claw up there like some mewling kitten's freshman effort to scale a
tree. Every launch was another roll of the dice. And of course, a goodly percentage of the time, their boosters didn't even work. *Snake eyes!* Ten million bucks down the crapper.

But in terms of the next evolutionary step, Venus actually made good
sense. Every eighteen months the two planets are closest, and flights between
them only take about 150 days. In cosmic terms, that's a Sunday drive. On
average this launch window remains open for a handful of weeks—although
the weeks pass quickly when you're trying to plan a trip to Venus.

Cognizant of this approaching opportunity and quite keenly interested in positioning themselves as a leader, Space Technology Labs designed an experiment-laden, solar-powered fiberglass sphere to visit the twinkling goddess of the sky. Like proud parents, STL showed it to their best customer, the air force, who was orchestrating the whole Venus scheme and had ponied up the dough in the first place.

But shortly thereafter, NASA took over space duty and let the wind out of the air force's sails. This whole tangly Venusian ball of twine then fell square into a thoroughly different breed of lap. It wasn't something the new management had been really focused on inheriting, or even looking forward to in the first place. It didn't come emblazoned with a suitably catchy name; in fact, it didn't have any kind of a name at all. It was too new; the ink on the program scope was still drying. And the probes themselves weren't exactly done. So how to proceed? Button each one up and launch them first thing when the Venus window opens? Or table it for a year and a half and get the affairs in order first? Inside loaner offices at 1520 H Street in Washington DC's Dolley Madison House, NASA headquarters had temporarily found a place to shack up. They were still trying to get the mail delivered to the correct places.

Then another kooky situation cropped up: *Luna 1*. Its remarkable flyby churned even more havoc into a toddling NASA—consequently leading to tousled rearrangements of missions and schedules as the United States assumed, in the short-term, a decidedly *reactionary* stance to their Soviet counterparts. Luna's hard volley had the Americans tied up in Prusik knots. What form should the response take? Venus? Moon? It really needed to be a doozy of a shot.

While they pondered, *Luna 2* up and smacked into the Sea of Showers. What a double whammy. "We thought they were way ahead of us," is how former NASA man Robert Gilruth recalled his agency's prevailing opinion of the nervy Soviet achievements. "That was a very good effort for them in those days."

"Russia's Moon shot again demonstrates its lead in space race," blasted the September 20 *New York Times*. Very bad public relations indeed.

Yet STL president Si Ramo definitely saw prolonged opportunities for his company. "The race with the Soviet Union is good!" he quipped of the situation, years later. "Then you have a better chance of being funded!"

If cooler American heads were to prevail in this situation, they didn't have much time to do so. While America kvetched about the Russians launching *this* and probably being able do *that* and who-knows-what-all-might-be-coming-next, Venus insensitively moved out of alignment. Basement mimeographs clunked on, reproducing in that annoyingly fuzzed purplish blue ink documents for American rockets pointed at a suddenly unreachable target. Out at STL sat the ships themselves: half-finished vessels scattered over lab tables, exotic baggage with no place to go.

"STL tried to do too much too soon," criticized the trade journal *Missiles and Rockets.*

Now what? A paltry mishmash of vanilla parts remained. Could they do something with this stuff? Anything? Hurry—American prestige was draining like sand through the hourglass. After much second-guessing, late-night conversations, and reshuffling and repackaging, the absurdly anticlimactic response to Luna materialized on Thanksgiving Day 1959. Wasting no time, STL had quickly obtained approval for a plunderous recycling of their equipment to visit the Moon instead. If the Russkis were going there, well then America was damn well going there too. This resultant offering to the heavens constituted only the first gaffe in a series of panic-driven and half-baked lunar probes that would come to be known more for their odd appearances and watered-down intentions than any outright accomplishment or scientific discoveries.

First out of the chute, *Atlas-Able 4* was also branded *Pioneer P-3*, and either name tended to induce the prickly feeling that it had been chosen in some haste. The mission's objectives went way beyond that of rational optimism: fly into actual lunar orbit, toss in a photo session, measure the solar wind. Achieving even one of these would be a grand slam.

To try to pull it off, STL's revised bundle of experiments came straight from a scientist's dream workshop: multiple radiation scopes, Geiger tubes, magnetometers. A dismally crude still-image television camera resembled what had flown—and then unfortunately come back—on *Pioneer 2*. They added two maneuvering thrusters for in-flight attitude changes, designed, unlike Boris Rauschenbach's Seagull appliance, to be operable for *months* and *months* after launch. Spiking through each pole, the miniature hydrazine-driven nozzles represented a new level of evolution. Steer the thing while it's going!

From the University of Chicago, John Simpson contributed the solar wind experiment. Simpson was no spring chicken—his landmark body of research went all the way back to the Manhattan Project. Since then, he'd established himself as a contemporary of Van Allen's in space physics, and he would come to detest NASA operations in a markedly special way that only someone in his position could.

Trouble had started back around *Pioneer 0*, when Simpson clashed with STL after being informed that he would not be allowed to build his own hardware. Simpson told them to bug off; he'd do it anyway. STL responded with a directive that no experiment would fly until it passed the shake tests. In communicating this, Louis Dunn knew he had an ace up his sleeve: STL owned a shake table and the University of Chicago didn't. Simpson conducted his own shake test by dropping his instrument out a nearby third-floor window into a toy sandbox. It took a licking and kept on ticking, and that's when STL caved.

As *Pioneer P-3* construction wrapped up, the remainder of the glistening ball's surface took on the air of a hippie decorator gone mad. What didn't have a solar panel bolted to it got covered with what looked to be a brutal infestation of monochromatic daisies—or at least some medley of visually stimulating infant toy.

They were actually the latest thing in passive thermal control. White petals edging circles of black. Little flowers. And depending on the local weather up there, a temperature-sensitive coil, which was embedded below each daisy, could spring-rotate the petals to adjust color ratios. It would add more black if Pioneer needed warmth on the inside, white if there was too much. They were lightweight and required no power—space-age mood rings. To this day very few other spacecraft have displayed a similar arrangement.

If these things have souls—and some feel they do—then the first glimmering sphere on deck probably sensed how its aura was not so blessed. This realization may well have arisen when its Atlas booster committed hara-kiri during an engine test. Luckily, *Pioneer P-3* hadn't yet been adhered to the stack. The pad suffered major carnage. Over two hours passed before anybody considered it safe enough for workers to cautiously venture from the blockhouse and start mopping up.

Turning around a launch site does not happen instantaneously; two slogging months elapsed before a new booster was assembled and in place, the probe mounted way up on top like a flagpole sitter. At launch time the giant stack amazingly left the ground, rumbling slowly into a Thanksgiving Day morn. Forty-five seconds later its entire nose shroud peeled off, exfoliated, shredding itself like mouse bedding. Like a cowboy, *Pioneer P-3* rode the wild bull underneath for almost sixty seconds, before unceremoniously being stripped away. All telemetry bottomed out just a minute and forty seconds past launch. Acrid smoke still clung over the tower. The entire opera had cost some $10 million and undoubtedly made for stimulating conversation at the big Turkey Day meal later on.

"U.S. out of space race for at least 2 years," chided the *Washington Star* on November 29.

Boris Chertok ran for his life. Fleeing, he stumbled and fell and wrenched up his knee really badly but got back up *quickly* and hobbled for the trench, and if he could only make it there in time because behind him was war: explosions and dirt thrown up and people screaming. He'd just gotten back to his feet and there was now a firestorm right behind him. Chertok kept laying down tracks: just go-go-go.

Back in the day, he'd been a fair hundred-meter runner.

Reaching the trench, Chertok flipped himself in, on top of dozens of other people who'd already hunkered down inside it. "I said get out!" one woman screamed, officers of many ranks piling on top of her and the others in sheer panic, the very air above them exploding in one thunderclap after another. As Chertok shielded himself, it maybe occurred to him that none of them wanted to do this one in the first place.

That damnable Mstislav Keldysh—he was to blame!

Just a few months back, they'd been putting 1959 to bed with a party. At Korolev's direction, a ninety-minute New Year's Eve meeting was underway; this first one would spark a tradition. Chertok was there along with a handful of others.

Korolev wanted to spin through the upcoming projects. The new R-9 missile needed some attention. And Khrushchev wished to launch a couple of R-7s into the Pacific for no other reason than to tame Eisenhower before an upcoming meeting between the two leaders. More dogs were going to fly. And hmm . . . Keldysh wanted them to repeat the *Luna 3* flight around the Moon. The pictures could have been better; everybody knew that. But

Korolev and the rest doubtlessly preferred moving on to other pastures—like shooting men into space and exploring the planets. Keldysh and Korolev had gone round and round on the Moon photos, though, and Keldysh appeared to be winning.

"Keldysh believes that science will not forgive us," Korolev explained to the group, "if we pass up the opportunity to take better pictures, with the Sun illuminating the Moon at an angle."

A disquieting grumble arose from the men.

Korolev pressed, "It's difficult to argue with Keldysh. He's vice president of the academy." And as the relative merits of all this settled over them, the chief designer alluded to something else.

"As far as I know," he was saying, "we've never gotten manufacturing and testing of the IM into gear."

"It's been in gear for a long time now, Sergei Pavlovich." Roman Turkov jumped in, trying to blunt the issue. He ran the design bureau's production factory. "But we haven't had a chance to move on it yet." The abbreviation IM referred to their Mars probe, a laughingly fanciful whim. They'd been called on to design, build, and fly at least two of them that year.

On a cosmic scale, Mars rubs up against us. But getting to it is still unbelievably tricky, owing in part to celestial mechanics. See, the distances between Earth and Mars are constantly changing, and they vary greatly by roughly a factor of six. At its farthest, the commute is something like 250 million miles, point-to-point. That's when you want to be welding and wiring and testing. About every twenty-six months, then, Mars swings into a twenty-day period of close proximity to Earth. This really is when the trip should happen, because you'll be able to shoot the most amount of spacecraft there and use the least amount of fuel to do it. And guess what: a comparatively short hop of 56 million miles was coming that September. That's the major reason this issue was dogging them—an immovable deadline. And to send a probe to Mars, Turkov reminded his boss that, not only would a new fourth rocket stage have to be invented, but the whole danged interplanetary station itself was still consigned to the preproduction shops. He didn't have half the drawings they needed, and no drawings meant no parts! Korolev acknowledged the situation but informed the group that it had to be done.

"If we aren't ready for the Mars launch in October, we'll have to wait a year for the next launch window!" Korolev smiled big smiles. "Buck up!" Then he wished everyone good health and shook hands all around. To Chertok, he intoned, "Don't forget to pass on my New Year's greetings to Katya today!" And leaving the room, Chertok wondered how best to inform his wife that he'd be working himself to death over the next ten months.

Barely two days later Premier Khrushchev summoned his space program's largest players for a face-to-face. Assembling on that frigid January in 1960, Korolev, Glushko, Keldysh, and others milled about the premier's grandiose office—with its rich, shoulder-height wood paneling—listening as Khrushchev talked of *space* achievements now being as critically important as *any* by the military. A beyond-life-size portrait of Lenin gazed down over all of them.

"Your affairs are not well," Khrushchev stagnantly judged. "Your work is going rather badly. Soon we will have to punish you for falling behind in space." He frowned.

The other attendees stole glances at one another. How could they be behind? Since back in September, they'd figuratively been clinging to an amusement-park ride of emotive delight. It had been the trip of their lives: first satellite, first dog, first lunar impact. And *Luna 3* had already begun tracing a second loop around the Moon that January. Although dead cold and gone, it remained very much alive in the minds of her creators. And the world.

So, what exactly was the problem?

"There's broad and all-out levels of work in the USA, and they'll be able to outstrip us," the premier concluded, pursing his lips.

Was *that* the problem? The competition?

Whatever it was, it sure had little to do with Eisenhower, who held an enthusiastically stale view of space exploration. He failed to rate as its greatest fan to ever walk our planet. Never thrilled at far-fetched prospects of landing men on the Moon, Eisenhower responded to its $35 billion budget estimate with a terse "Well, let's come back to reality," counterproposing a demure "scholarly exploration of space." Not more than a year later, during his final message to Congress, Ike drably orated on how manned flights might rightly end at the close of the Mercury program. A man in space? He'd prefer hemorrhoids.

Befitting answers to the Soviet urgency, then, might well have been found between the dangly ears of Lyndon B. Johnson, who at the time was wearing such varied hats as senate majority leader, admitted space buff, and general critic of the way President Ike went about things. Johnson embraced forward-thinking vision and, in general, operated differently. He'd had a lot to say about the way NASA came together and worked to increase that agency's budget by something on the order of $168 million. Two short and interesting years had passed since Johnson brazenly implored the Democratic Caucus, "Control of space means control of the world. . . . Our national goal and the goal of all free men must be to win and hold that position." Pretty nearly overnight, a Russian translation of the entire address materialized in Soviet hands. And uncapping a full head of steam, Johnson appeared well on the road to following through. Khrushchev seemed to grasp LBJ's words and intentions, and they gave him the willies.

Inside the premier's office, the concussive discussions began. Keldysh wanted at least one more flight of lunar photography, carrying better equipment to produce much higher-resolution pictures. Chertok got to his feet and objected—what about Mars? That alone could consume all their time. They'd already snapped pictures of the Moon; everybody talked about how great a thing it was.

"Don't forget that we also have the Vostok," Korolev mentioned, almost as an aside.

The meeting broke without much of anything having been decided, which might explain why Keldysh called another one five days later to grease the skids of his little pet project. Internally, he referred to it as Ye-2F, although "Luna" stands out as a bit more palatable and definitely facilitates a better understanding of the program's intent. Two *Luna 3* follow-ons, Keldysh intoned, for April. Improved photography remained the only objective. Begrudgingly, the shots went on the schedule. Because Keldysh wanted it—that's why.

After the meeting, Korolev escorted Chertok and another man down to the chief designer's car, where he comfortably seated everyone and then proceeded to ream both of them out for failing to advance the Mars effort. It was stalled; it wasn't going anywhere. They hadn't been working hard enough on it. They hadn't put it on the calendar. They hadn't even decided who was doing what!

A to-be-developed, four-stage Soviet rocket would fly to Mars that September, Korolev explained. It would carry a to-be-developed automatic interplanetary station. It would work.

Because *Korolev* wanted it—that's why.

Giovanni Schiaparelli liked it when the Sun went down. After everyone else in northwestern Italy had retired for the day, after the sky deepened to richly pigmented blacks, the nineteenth-century director of the Milan Observatory habitually set up his telescope in order to contemplate the jewels of the night sky. Though entranced by its entirety, Schiaparelli mainly had eyes for one particular spot. Mars. Training his paired lenses up, over time he compiled scores of observations about that heavenly body.

They went into much detail. Schiaparelli beheld a plethora of splotchy dark areas jumbled about on Mars's surface; he designated the huge features "oceans" and the smaller ones "lakes." Lighter-colored regions could only be landmasses and were promptly termed "continents." By Schiaparelli's assessment, the place must be a lot like our own, just without a nice risotto.

This Italian gent wasn't alone in his studies. The French (and somewhat eccentric) astronomer Camille Flammarion had, for years, been of the overall conclusion that humble Mars was brimming with life. "The actual habitation of Mars by a race superior to our own is in our opinion very probable," he wrote in 1892. Flammarion was convinced of this life, in part, because of these specific patterns of lines that Schiaparelli had noted meandering across the Martian surface like loose threads. The Italian had studied them and mapped them and contemplated them endlessly during the daylight hours, and his chosen word for the lines he obsessed over was "canals." He actually used the Italian *canali*, which can be interpreted as either "canals" or "channels," depending on who is doing the translating. The man from Milan preferred the latter, as it tended not to imply features of artificial origin.

Not long afterward, a Harvard-educated American solidified the concept with his own independently formulated theories. In the early 1900s Percival Lowell hypothesized that desperately thirsty Martian residents must have built these canals as *aqueducts*, channeling water to dry areas from the wet. Irrigation, plain and simple. Martians had probably been doing it long before the ancient Romans. Keep in mind these allegations resulted from educated scientific minds doing the talking—respectably learned folk of the extensively thought-out opinion that this sort of thing really was going on up there.

Initially rejecting the idea, Schiaparelli corresponded with Lowell to the point where he basically acceded and embraced this notion of life on another planet. In popular Italian science magazines, he speculated on the workings of Martian irrigative engineering and even the politics behind such an infrastructure. "Mars must be the paradise of socialists as well [as hydraulic engineers]" Schiaparelli wrote. He disseminated his convictions, and over time an undercurrent of support grew. The more other astronomers scrutinized the red planet, the more they believed these men. Sky searchers worldwide resolved the dark Martian patches through *their own* telescopes with *their own* eyes and distilled *their own* conclusions, which happened to be identical: lakes and oceans, by God. Everybody agreed. They corroborated the peculiar wandering "canals" traversing just about everything up there, and the lines always, *always*, dead-ended at lakes. They just had to be canals—for pete's sake, what else could those things *be*?

"They seem to have been laid down by rule and compass," a 1907 *Scientific American* reported Schiaparelli as having said.

What Lowell and Schiaparelli and Flammarion really wished for was some kind of hocus-pocus method for visiting Mars, for some crazy way to actually go there and check it out. Obviously, no such ability existed then—in Schiaparelli's time, Karl Benz was just about to break nine miles an hour with a two-stroke automobile.

He didn't live to see it, but Percival Lowell might have been shocked to know that many years down the road in 1960, as sugary layers of snow began to melt around the Moscow design bureaus, those inside were beginning to plan out that magic journey.

The prep wasn't going smoothly: "Don't put the launch dates past September," Korolev snarled at Boris Chertok in mid-January, right before heading to a governmental vacation hotel. And Chertok stood there quietly watching him leave, thinking about how the hell they were going to design a new spaceship and put it on a new fourth stage and launch it to Mars inside of the next eight months. Today, by way of comparison, the task is considered feasible within about six years. In response to the chief designer's instructions, Chertok met with the heads of the various institutions and assembled what everyone considered to be a slightly more levelheaded time-table. Then they were supposed to go meet with the boss.

A town of three thousand, Sosny lies about ten miles outside of Minsk in what is now Belarus. Up there in Sosny at the governmental hotel, Korolev received his visitors and eagerly began reviewing their schedule. The men figured he wasn't going to like it very much.

Korolev stared over the pages and then wrinkled his face up into an angrily crumpled ball of rage. Grabbing a heavy writing utensil, he practically tackled the schedule, mauling it, shifting the dates backward in places by as much as three months. He wanted the men to agree on his revisions in writing, with signatures. Chertok's mouth drooped open. The delivered proposal now absurdly included a try for Venus and three ships instead of two! Did Korolev have any clue how unrealistic all this was? But the chief designer brushed the loud concerns aside, singing, "We'll fly to Venus, that goddess of love, in the nude!"

That got everyone's attention. Korolev expanded on his outburst, explaining that he wanted to remove the insulation. "There isn't time to optimize thermal shielding. If there is a failure in the last stage, it'll burn up in the earth's atmosphere anyway. But we'll be able to prove that we are launching spacecraft, not combat missiles!"

So much for being realistic. Chertok and the others felt no option except to sign. Leaving Sosny, he reviewed the documents that now bore his name on the bottom. Caged within these impending Soviet plans, one Mars question begged for answers more than most. Specifically, was there life on it?

Between that edgy meeting and today, the effort to answer this question has already consumed hundreds of millions of dollars. Mars has been orbited, landed on, roved over, drilled into, and photographed more than Marilyn Monroe. Ambulatory, remote-control science labs have done just about everything to it that *can* be done, and so far, no life. There appears to be no irrigation problems, no smartly built canals, no chameleonesque vegetation, no lakes. And no Martians. Waaay beyond a shadow of a doubt.

At least, no *intelligent* life.

Speculation about any kind of living things up there didn't end even after American landers visited Mars through the Viking program of the mid-seventies, which was designed largely to settle the matter once and for all. Right there on the surface of Mars, they ladled scoops of dirt into miniature environmental labs able to slake out any possible evidence of Martians, primitive or otherwise. The inconclusive results are still being argued about more than twenty-five years later.

But in the days before such capable spacecraft, between Lowell and Korolev, the best thing scientists had going was a simple tool known as a spectrometer. When the device is aimed at virtually anything—some body of water, belching Hawaiian volcanoes, or perhaps even Mars—light reflected from this object of interest is processed by the spectrometer, which essentially spells out the kinds of chemicals your target is made of based on the spectra of those reflections.

Then in 1956, around the same time an overworked George Ludwig sat sweating out his festering tape recorder, William Sinton came along and made a discovery of general interest. An enthusiastic American astronomer, Sinton's repetitively exhausting threesomes with Mars and spectrometers churned out quite a possibility: the big red ball could very well be awash in carbon and hydrogen. Jeepers!! Sinton tried to narrow down the problem, focusing in on those dark splotches Schiaparelli always fantasized about. Could they really be vegetation? Wielding a tool the Italian could never have even dreamed about, Sinton got practically the same results while scanning the Martian lakes as he did a cornfield. The thought of what exactly might be up there drove him absolutely batty.

But trying to carry out all this spectrometer work from Earth gets a guy only so far. Being nearer to the target means better results; therefore, Sinton's experiment remained hamstrung with a little proximity issue— even when he held it over his head on tiptoe. Measurements from a nice, intimate spacecraft flyby just might tell the world whether times were right on Mars for planting season.

America's first probe of 1960 incredulously worked as advertised—that is, a functioning booster gave it a *chance* to work. Those picking up the *New York Times* on March 12 found a celebratory headline greeting them: "U.S. Rocket Put into Sun Orbit—Will Be First to Gather Data Deep in Interplanetary Space."

What had everyone in such titters was *Pioneer 5*—thoroughly similar to *P-3*, only retaining a postlaunch intactness so as to execute its intended role. Built once again by STL as a veritable duplicate of the current and very much unfulfilled Pioneer crop, it shucked the daisy motif in favor of a simpler, white-on-black design.

Like her brethren, she was originally built to soar past Venus, but the planet moved out of range before the ship was ready to fly. All that effort can't be just chucked in the dumpster, so the forlorn probe involuntarily underwent brain surgery. Planetary surveyor morphed into student of the Sun.

Regardless of what *Pioneer 5* investigated, the *Times* found extra meaning in the launch: "All men are brothers as they contemplate the mysteries, dangers, and challenges of the vast realm into which *Pioneer 5* is now penetrating," though this could be considered a bit of a stretch.

Just a few days after launch, on the fourteenth, England's Jodrell Bank Observatory announced that space history had been made by tracking *Pioneer 5* to nearly half a million newly traversed miles. The distance bested *Luna 3's* reach of 290,000 miles during her trip around the Moon. *Pioneer 5* subsequently reported on Earth's outer radiation belt, which had been hinted at by earlier Pioneer craft but was now confirmed. Scientists like John Simpson whooped for joy at *Pioneer 5's* record: the solar wind and particle-measuring instruments aboard were all his.

Jodrell Bank itself dated from 1945, when a blasphemously intelligent, cricket-playing Englishman named Bernard Lovell spent time hunting for a quiet spot of ground on which to set up some leftover army radar gear and observe—what else?—cosmic rays. Outfitted with a wicked comb-over, he eventually found a spot that looked like a decent pick. It was nothing more than the side yard butting up against Manchester University's botanical gardens, but nobody else was doing anything there and the enterprising Lovell was able to get his mitts on it. By the summer of 1957 Lovell had scratched together enough spare change to erect a 250-foot radio dish on the grounds. It was fully steerable, resting on gigantic rotating sprockets that only a year ago had been gun turret racks from two different English battleships. The dish anticipatorily went live on October 1 and was promptly christened Mark 1." Three days later Jodrell's new gizmo tracked the *Sputnik 1* booster, enabling both Soviet Russia and the United States to properly realize just how convenient the big honker *was*. Korolev and Maksimov probably felt a tad stung when news of *Luna 1* was generally disbelieved by the world's press, as they'd even disseminated the radio transmission frequencies way ahead of time. It led to no less than furnishing Jodrell Bank with *specific instructions on how to locate and track Luna 2*. Quite a disclosure for the nominally paranoid Soviets.

The facility's liaison with *Pioneer 5* didn't begin with that record-setting distance track. Its dish had been employed shortly after launch when a command went up to effect the separation between *Pioneer 5* and its third stage. Mission control could have done it back in the United States, but what this really did was conduct an important preliminary test. If *Pioneer 5*'s flirty jaunt went according to plan, Jodrell would quickly become the only place with a strong enough voice to yodel that far.

Only one week later *Pioneer 5* clipped along over a million miles from Earth, humming nicely. All systems go. She reported on command during a radio session at two in the morning; the onboard micrometeoroid counter had already logged eighty-seven vulgar dings. Diverting irreplaceable prep time from her wedding, the visiting Princess Margaret ceremoniously transmitted an undoubtedly trivial command to the ship and even heard confirmation almost thirty seconds later. The princess squeaked. Who said royalty bore no skills?

Soon afterward, however, the ruefully undersized towers at Goldstone and Earthquake Valley stations began to lose *Pioneer 5* as its twittering radio incrementally gave way to static. She'd gone too far for them. No choice remained but to have Jodrell assume the reins. The distant orb receded farther, shrinking from reality, blazing herself into who knew what. Telemetry whispered in from seven different experiments, buzzing on a prehistoric and rapidly declining rate of sixty-four bits a second. It continued falling as the mileage dilated between lonely probe and the home it would never see again.

There was more news on March 31: dutifully *Pioneer 5* reported weathering an enormous solar tempest billowing directly at Earth. The very first cosmic weather forecasting. Six hours later meteorologists knew she hadn't been kidding as the broad storm washed over, an invisible cloud of angst disabling communications worldwide. *Pioneer 5* kept going. No looking back, no observance of *Sputnik 3* finally reentering the atmosphere. By April 17 *Pioneer 5* had made it 5 million miles out. Top that, Luna!

April 23, 1960, went down as a fine one for Robert E. Gottfried. When Earthward telemetry divulged a bad diode aboard *Pioneer 5*, the mission's endgame about-faced into limbo. So Gottfried, an engineer at Maryland's

Goddard Space Flight Center, plunked himself down and fashioned a series of instructions to cleanly negate and bypass the fault. He then handed them off to a coworker who promptly radioed the sequence to Manchester for uplink. To everyone's delight *Pioneer 5* received the commands properly at a staggering distance of 5.5 million miles and then executed the whole set precisely on cue. In-flight reprogramming? The world had been in space for just over two years.

Then came the autumn days. By the end of the month, precipitously fading signals obligated a changeover from *Pioneer 5*'s 5-watt transmitter to the separate and relatively colossal 150-watt transmitter. It was her final option now, given the distance. After the new gear sprang to life, bad news wilted back to Jodrell: all twenty-eight batteries had experienced some kind of disabling malfunction and were now uncontrollably venting gases into space. Robert Gottfried couldn't write software to fix that one no matter how much he wanted to. But May 8 came; the barely audible spacecraft had just rolled over 8 million miles, persevering in operation and remaining fairly cooperative. Jodrell commanded on the transmitter a shade past five in the morning for telemetry reception, and *Pioneer 5* obligingly reported back—now down to one bit per second.

The swan song came on June 25. That fine day, Jodrell received a last, hoarsely whispered report of priceless in situ telemetry from 22.5 million miles. And sailing away at twenty-one thousand miles an hour, *Pioneer 5* raced toward the Sun, instruments vainly still operating in spite of the battery failure, solar panels wide in a welcoming embrace. Then she was too far away for anything on Earth to listen.

The film kept coming out wrong. Boris Chertok had arrived in Tyura-Tam on April 7 to find two of the lead camera engineers essentially living in the assembly hangar and about at the end of their rope. Bedraggled, pale, unshaven, they droopily labored over the photographic system. No matter what they seemed to do, the test exposures came out blotchy and unappealing. The whole reason for the mission was to take good pictures, and it couldn't seem to do that even right here on the ground. Korolev had some fresh developing chemicals screaming toward them right now on a Tupolev Tu-104; they were due any time.

Other parts kept failing. Overnight the entire radio came out, went under the knife, and went back in. A couple of nonworking sensors in the rocket were simply excised altogether. The chemicals arrived after two long-distance hops, and the quality of the photographs instantly leaped up. Then it was time to mate the ship to the booster. Men dangled in the rafters like acrobats as it all came together, and then everyone who'd worked straight through the night (again) was sent away at nine in the morning to have a nap and refresh themselves for the late-afternoon launch.

But in these barnstorming days of space exploration, the most heinous enemy of any unmanned probe was the very booster it rode. On that April 15, the stage one boosters on Soviet Russia's next Luna spacecraft worked like they were supposed to, but the machinery above felt less than cooperative. Stage three's engines shut down early, and everything fizzled back through the atmosphere. Another three lousy seconds of pushing and they would've had it. "There won't be any movies," one of the men suggested of their camera—now in the process of becoming garbage.

Along with the others, Chertok crowded around the telemetry in a brain-busting search for the trouble. By the next day, the predicament seemed clear: stage three had run out of propellant. Its tanks hadn't been completely filled! Chertok could only imagine the verbal dissection that the slipshod fueling crew would soon endure.

Luckily, Korolev's entire operation now exhibited a refreshingly advanced level of preparedness for such unwelcome contingencies. Three sleepless days later a replacement booster *and* spacecraft clattered down the railway tracks, groaning, rasping, heaving, and ascending into position. *Clink*, there it stood embraced in the petals of the tyulpan.

At ignition the quintuplet of strap-on boosters failed to build thrust and instead cleaved from their central core. Less than half a second elapsed before each starkly redefined the term "holy terror." One collapsed back onto itself—BOOM. Firing in random directions, the remaining three became lethal snakes ripping acrid lines through the sky, tracing back over one another in dodging rampant confusion while evidently still trying to find orbit. One corkscrewed directly over the barracks. Pad safety officers frantically raced through the detonation sequence as another snake writhingly contorted back at the assembled crowd of spectators, berserkly zipping a hundred feet over their heads at the officially "safe" distance of one mile from the pad, walloping a patch of ground nearby—BOOM. The flashing explosion instantly shattered windows at the very assembly and test building it had recently emerged from. This is when Boris Chertok ran for his life, silently cursing Keldysh along the way as the short-sheeted core booster, upper stages, and diminutive Luna spacecraft finally embedded themselves into a tiny salt lake only half a mile from the launch site. Not a single person died. The pad itself was demolished. Just twenty-four hours later more bleak news arrived: *Pioneer 5* reception—and control!—from 5 million miles out.

Overall, that was not a great day for the Soviets.

As he cowered in the piled-up trench with debris hellishly raining down, Chertok might have taken a moment to reflect on how he got in this position. Near as he could tell, a lot of it seemingly came down to luck. Or fate. Maybe some of both.

Chertok's mother came from a wealthy Jewish family, who unanimously expected her knockout looks to snare a good husband and ultimately a comfortable life. But to the family's horror, she instead received an expulsion from her last year of prep school due to her membership in an illegal revolutionary group. The apparently misguided femme offered no remorse for the situation and became what amounted to a professional revolutionary. Some of her more concerned relatives unearthed a poor yet good-hearted teacher named Yevsey Chertok to marry, and the two settled in Poland. Boris was their only son. When the Germans advanced, the family felt it best to pull up stakes and relocate to the east, in Moscow, where young Boris experienced everything from the tranquility of a nearby apple orchard to the chaos of the revolution. Instead of playing cowboys and Indians, Boris and his friends played Reds and Whites. Nobody wanted to be a White.

Here Boris first sighted the antennae fields of the largest radio station in Russia. Each tower measured sixty feet high with lines strung in between, and barbed wire encircled the whole arrangement. Tours wended through these fields and the station offices, and Boris found the whole scene absolutely mesmerizing. At eleven he flew in a plane, a passenger Junkers visiting a Moscow craft expo. When he learned that airplanes carried radios . . . well . . . that just put it over the top. And Boris Chertok knew the direction he would go.

He rode the bus to school with six kopecks in his pocket for a loaf of french bread to enjoy sometime during the day. It served as his lunch. His parents managed new shoes and clothing for him, but nothing remained for radio parts. His dad did scrounge up enough to buy his son a few magazines, so Boris perused the likes of *Radio for Everyone* and *Radio Enthusiast.*

Soon, Boris's father noticed a few books missing from his son's collection. What happened to his three-volume *Life of Animals*? They'd been a present. Reluctantly Boris confessed that he had sold the books as a way to get radio parts. A classmate had built a receiver, Boris explained, and his own need to assemble a better one outweighed *Life of Animals*. Not long after, he wanted to build another radio but didn't have any money left. Instead, Boris prepared its schematic drawings and, on a whim, sent them in to *Radio for Everyone*, which to his complete amazement featured the design in an article and even put one of Boris's drawings on the cover.

The technical institutes wouldn't let him in; as the son of a white-collar worker, he would have needed a little more grime under the fingernails. "Work about three years and come back," they said. "We'll accept you as a worker, but not as the son of a white-collar worker." So Chertok bided his time as a brick-factory electrician until his application was finally accepted.

Persevering through school, he became a top-notch aircraft electrician and spent a chunk of World War II on the team developing a rocket-powered airplane. And when Germany threw in the towel (again), Chertok headed down to Peenemünde to make available his electronics expertise for the V-2 reconstruction efforts. From there, it was off to the races on Korolev's anaerobic shirttails, from one space project to another, on up to the disastrous scene that Chertok confronted as he stumbled from the trench.

Some time later, Boris Chertok drew up alongside Mikhail Tikhonravov as both arrived for a meeting with Korolev. Lightheartedly they discussed the relative wisdom of entering the chief's office yodeling "Onward to Mars!" It was a line Friedrich Tsander had routinely used in the age of GIRD. They finally decided Korolev wouldn't laugh—he had just come off a jarring, fog-enshrouded plane flight so edgy that the pilot had wanted to divert through Leningrad. S.P. had had to seize the radio and get the air force to override.

"Well, what is one supposed to answer?" Korolev snapped, telling the story of his most recent meeting with the premier. The only subject was Mars, and Khrushchev had kept asking, "Tell me, is it theoretically possible to do this?" Apparently, the premier wasn't really interested in any underlying technological woes. He just wanted results.

Korolev's increasingly gelatinous form puttered about the room. "Of *course* it's all theoretically possible!" he exclaimed, and he had said as much to the head of the Soviet Union. In response, Khrushchev indicated they should then just go do it and wanted to be done talking.

Misha grimaced. Korolev squeezed his hands and tried to calm down a tick. "That's the whole story," he said. "Then they don't give us what we need. But we've still got the assignment that we need to carry out within an insane time frame." Then he castigated the both of them for failing to shift into higher gears.

After leaving the office, Tikhonravov and Chertok were silent for a bit. The last thing Korolev mentioned before dismissing them was that any day now they'd all be facing Khrushchev himself to explain the problems. Chertok smiled to his friend Misha, "When Khrushchev visits us, you will get the opportunity to greet him with Tsander's slogan 'Onward to Mars!'"

After checking their work an incalculable number of times, the navigation group declared September 26 to be the best day for a Mars launch. Two months from right then. Any slipped days beyond the twenty-sixth would necessitate a reduction in payload capacities because the planet would be headed away from Earth.

In Shop No. 44, testing still had not yet begun on the single engineering model. It looked like a big housefly. Even the Communist Party leadership thought so; Central Committee secretary Brezhnev said it during a tour of the facilities. "It would be good if you launched one of these 'bugs' to cause a bit more of a stir," he suggested, implying that nothing of value happened between launches.

A silver cylinder stood six feet tall, the centerpiece of twin swept-back solar panels hanging off each side. Within the shiny walls of the body rode a primitive sequencing computer brain, radio equipment, and insulation. Safely behind a single, thickly glassine eye, a redesigned television system perched with its lens up to a porthole. A carefully selected mix of experiments clung to the outside like parasites. A spectrometer was supposed to go along, taking close-up readings on the amount of organic material in the Martian atmosphere. At last, an answer to the question of life on Mars?

Bugs indeed. Antennae. Eyes. Body. Wings.

The radios hadn't shown up yet. Chertok marched on Korolev's office to let him know, finding the chief designer engaged in a heated phone call over the Kremlin hotline. Recently two dogs, Belka and Strelka, had spent an entire day in space and created something of a worldwide sensation following their safe return. Summarizing his instructions to the mouthpiece, Korolev wanted them segregated from the other mutts in the kennels. He finished shouting and hung up the phone, turning his attention to Chertok, who explained the situation with the radios. His report prompted Korolev to snatch up the hotline again and ring Valery Kalmykov, minister of the radio-technical industry. Korolev harbored few reservations about picking up the phone and yelling at people; here, he bellowed to the minister that the entire Mars effort was going to unhinge in the paws of the radio complex and that they were all going to have to answer to Khrushchev. Then Korolev slammed the phone down and told Chertok they should immediately drive over to the radio facilities.

"We'll have a look and discuss everything there on-site," he indicated, bustling for the car.

Quickly they arrived at the radio plant, which was in the process of being kicked out by the extremely large automobile factory next door. It needed the space. Heading inside, the men found individual Mars radio assemblies scattered about, along with the two individuals in charge. Both looked like the walking dead; they offered no excuses or other deflections and explained to Korolev and Chertok that although the work was inarguably behind schedule, it would all come out okay.

"You all listen," Korolev spoke. "You will perform the integration and testing at *our* facility. Chertok and Ostashev will be in charge."

Boris Chertok couldn't believe it: Korolev had just vacated *all* responsibility for a finished product from these gap-toothed slackers!

One of the radio plant engineers tugged on Chertok's sleeve. "It'll be at least a week before we debug a single unit," he whispered. "We can't send you semifinished products right after soldering."

Back in the car Korolev lit into Chertok. "Boris, you're incorrigible. You think I don't understand that their operation is an utter failure?" he sparked. "But now let them try to tell us that they can't even send us the first unit part by part!" Korolev stomped on the gas and roared off.

On the last day of August, an Antonov-12 cargo plane landed at the Tyura-Tam airfield, slowing down long enough to let two nondescript boxes fall out of the back. Chertok and Arkady Ostashev felt weak as they opened them and tried to sort through all the seemingly random components. Was there even a radio in here? The stuff was more of a secondhand parts jumble.

One month till launch.

Commiserating with the lead booster technician, Chertok's spirits tumbled even further. The guy was telling him about the trouble-plagued new four-stage R-7. They called it the "Seventy-eight": "Forget about that radio unit and all the Mars problems," he blurted. "The first time we won't fly any farther than Siberia!"

But Chertok had radio in the blood. He didn't want to give up. The unit was made to work briefly and then began smoking. They yanked the power and waited for one of the head engineers to show up—one of the very men confronted by Korolev and Chertok at the radio plant a month earlier. He finally materialized, going on no more than two hours of sleep, and crawled inside the Mars bug trailing a wispy soldering iron. On the whole of Planet Earth this man was the only person who truly understood the radio. Day and night he soldered as the launch drew closer and all the State Commission hotshots began arriving. To Korolev, one of them remarked that every time he came through the facility, "I see the same butt sticking out of the spacecraft! Is it going to fly to Mars too?" At that, the guy extricated himself from the ship ready for a tussle. Korolev parted the men as the engineer explained that what he really needed was another four hours.

"I'm already used to your needing twenty-eight hours in every day!" came the retort.

Unfortunately, the rest of the ship wasn't in much better shape. Countless troubles plagued the other internal electronics. They'd get worked on and installed then fail the tests and have to come back out again. A cable testing error put Chertok over the edge, driving him to locate the responsible party in order to find out what happened. He found one seriously fatigued female.

"I confess that I made a mistake," she explained, "after soldering for seventeen hours without taking a break for dinner and breakfast. We gave up having lunch long ago."

The IM was not appreciably ready for testing until September 27. The flight-control system was not working properly. Neither was the television. The State Commission held a meeting there at the launch site; one of the members waved off the effort. "It serves them right," he exclaimed. "It was no use taking on such a project with these deadlines." By the time they left the ground, Mars would be too far away and the ship consequently too heavy to get there. To stand even half a chance, some of the instruments were going to have to swing from the yardarm.

Right away the television camera came out. Someone else suggested removing the spectrometer. But it might work, Korolev insisted, and wasn't this one of the primary reasons for visiting Mars? Finding life? He wanted to test it on the nearby vegetation. In short order, the experiment came out of the spacecraft for transport to a suitably close testing area.

Training the device out over Russian foliage, workers noted the results. "As it turned out," Chertok later decried, "from the recordings we made, there was no response at all!" To the great horror of physicist Aleksandr Lebedinskii, his experiment had bungled the tests. "It was obvious there was no sense in sending that equipment," booed Chertok. "Sending it to Mars, and no life on Earth!" When he contacted Lebedinskii to inform him of the results, the buttoned-down physicist couldn't believe what he was hearing.

"You were lucky!" Chertok salved. "The chances of making it to Mars are virtually nil. So you'll have time to get your instruments into shape."

Lebedinskii didn't seem to be feeling much better.

"At the very least," Chertok continued, "in a year you must prove with your instrument that we do have life here on the steppe."

Two stripped-down Mars probes finally went off within days of each other, on October 10 and 14.

The delicate relationship between probe and booster is analogous to the mating ritual of the golden orb-web spider. Of the two parties, one is exponentially larger. They operate in altogether different ways. Yet both are hopelessly codependent and unite to produce amazing results—followed sometimes by catastrophe.

On that October 10 the mating dance went badly. For golden orb-web spiders, this is usually indicated by the female eating the male. But in Soviet Russia, everyone's first clue was probably when the rocket started coming back down. Frustratingly Chertok recalled that, by his count, *Sputnik* didn't fly until the R-7's sixth time out. But for IM, this was the very first go. No test flights of the four-stage Seventy-eight had ever been made. Not even one.

Their months of toil went for nothing. Both missions failed, one after another. Both probes were lost during launch when their third stages failed to ignite. Both saw Earth from seventy miles up—two helpless insects briefly on top of the world—then pitched over, slumping, sinking, succumbing to unwelcome incineration. Uncontrollable, unstoppable, irreversible, the first attempts at Mars had failed.

A seriously pissed-off John Simpson had NASA's assistant director of space sciences on the phone. Name of Homer Newell. The Chicago professor wanted to give him a piece of his mind. The day before, Simpson had found out what it's like when tumultuous years of hard work are bolted to a totally unproven launch vehicle. He felt it sixty-eight seconds into the flight of *Pioneer P-31* on December 16. Having reached nine miles high, the Able second stage prematurely ignited while the Atlas was running and very much still attached. It flamed; it burned. The stack blew to pieces as *Pioneer P-31* ingloriously tore from its booster. Tiny bits of debris came off in between, looking for all the world like mulch. Within minutes the little nipper of a spacecraft cannonballed into the Atlantic about twelve miles from the cape, settling into a hundred murky feet of water.

Gaa! When was this going to change? Newell had to hold the phone away from his ear.

After the bawling-out, Simpson forewarned Homer Newell that after this latest cock-up he expected to get on more NASA flights. And six months later Newell received another earful by way of a lambasting letter from Simpson, cholerically pointing out NASA's *two* successes over *eight* launchings. Who was making these stupid rockets anyway?

Boris Chertok stood before the State Commission, which was lugubriously trying to determine the design bureau's state of readiness. It was January 5, 1961, and the Soviets' first offering to Venus was being transported from Moscow to Tyura-Tam. Many of the commission members were too skeptical already and didn't ask any questions. One finally did.

"During the Mars launches, we never even determined if the spacecraft itself was reliable," he began. "We never got that far."

Chertok shuffled his feet a little.

"What's the probability," the official continued, "that out of three launches we'll send even one to Venus?"

Assuringly Chertok responded, "One out of three will definitely make it to Venus." He tried to sound upbeat.

The third-brightest object in the sky, Venus occasionally casts shadows on Earth. Only the Sun and Moon rank higher. When people see it, many assume the brilliant pinprick to be nothing beyond a remarkably dazzling star. But it is our closest and most highly reflective planet. And Venus was long thought to be a sort of analogue to Earth. They formed adjacent to one another, measure up almost identically, and contain atmospheres. Russian Mikhail Lomonosov noticed this last aspect as early as 1761 from the St. Petersburg observatory. It was the way Venus passed in front of the Sun—its edge went fuzzy. That *had* to be an atmosphere of one kind or another.

What all might be found on it then? One of the more attractive interpretations depicted Venus as a primitive Earth: warm swamps and vegetation, frogs and flies. Another prevailing attitude saw it as a home to perpetual dust storms—kind of an endless desert hell just one planet over. Many bright minds in planetary science vigorously embraced the possibility of Venus harboring liquid oceans. Most held out hope for the discovery of life. "We were expecting at the beginning to find a twin sister planet of our Earth," remarked Soviet physicist Roald Sagdeev.

Much of this uncertainty related to imaging—though closer than Mars, no telescope could penetrate Venus's unbelievably thick layers of cloud. This befuddling, murky haze was the reason Venus kicks back most every photon of sunlight it receives, explaining that bright glow. Where did the clouds originate from? What made them?

With this new craft, the Soviets intended to answer many questions. Chertok and Maksimov and company called it IVA, and it closely resembled the failed Mars insects. Plenty of design inspiration worked its way into the finished morphology: solar panel wings to pump the batteries full of juice as well as articulating mechanical shutters on the back that could open and close to regulate heat. Once again a host of scientific experiments arrived from outside universities.

This new radio-communications system, however, bordered on art. Long-distance phone calls between probe and Earth would all run through a collapsible six-foot-diameter parabolic antenna scheduled to unfurl midflight. The reflector dish, instead of being solid, was actually comprised of an exquisitely thin copper mesh. Machinists wove it so fine that from six feet away the dish looked like just a bunch of naked umbrella ribs. Harmonized by the stout, fifty-one-year-old Mikhail Riazanskii, the finished utensil interplayed with additional, smaller directional antennae operating at a whopping *one data bit* per second—adequate to report back on the science results as well as its own health.

Riazanskii was the genius who'd wired up the guidance electronics on the first R-7, and his masterful expertise led to IVA's advances in onboard logic. When a command arrived from Earth, the probe's brain instantly parroted back the instructions—and then cooled its heels until specific confirmation returned. *Yes indeed, that's what we want you to do.* Firings of the maneuvering thrusters would then be disabled unless the probe conclusively believed it had a positive lock on its true position relative to the stars. These AIS were becoming smarter.

Originally, IVA's intended destiny was a soft landing on Venus, as evidenced by the addition of a braking system and a layer of instrumentation capable of reporting back about the planet's surface. In case his amazing machine came down in a lake, Gleb Maksimov devised the lander to float. But that was before the grievous Mars probe difficulties got everyone rethinking the actual task before them. Maksimov hunched over his drafting bench trying to streamline. Peeling away the landing ability from his generic bug, he simplified to a more primeval impact probe that would take measurements on the way down. That meant he could yank out the camera. Maksimov did, however, permit another shamelessly anthropocentric commemorative soccer ball to be left in, delivering the socialist manifesto to any Venusians who might be lurking around the impact zone.

Overall, it seemed nice to him!

"What were you guys thinking when you designed it like this?" denounced the launch supervisor. "For this, you designers should have to drop your trousers and get a flogging right here in front of everybody!"

He was upset with Gleb Maksimov, who, design innovations aside, had either overlooked or shrugged off a tantrum-inducing aspect of the new ship. Maksimov would have liked to bolt a computer inside it, but little computers didn't exist. Instead, there rode a sequencer. Think of it as an overblown washing machine dial, where a simple electromechanical gear ratchets along from one position to the next. Each stop triggers a different command. Practically everything depended on this gadget, including the onboard star tracker. It had to be set in a position relative to the launch date—currently February 4—and then the probe could be sealed up inside the nose cone.

Where that all got tricky was if the departure date changed by more than twenty-four hours. Then the mechanicals would have to be repositioned— meaning a disassembly of the *entire* top of the rocket so a technician could pry open the bug and clack the sequencing gear backward. When the launch supervisor found out about this "feature," he was less than thrilled.

"My schedule doesn't include time for a beating demonstration," the supervisor thundered and roared. "I'm not going to complain to Korolev. But if we don't hit Venus, I'll tell him the reason why!"

They made the date. On February 4, *IVA* ascended through dark skies, dropped the strap-ons, continued ascending on stage two, dropped it, went to stage three, and then *click*, things stopped working. Their ship made it to the requisite Earth parking orbit, but that brand-new stage four did not fire. This debilitating failure consequently meant *IVA* wasn't going anywhere near Venus, but it *would* innocuously circle Earth for a couple of weeks until the orbit decayed. Fourteen thousand pounds of *something* looping around up there—including the fully fueled stage four—led many Western intelligence-gatherers to seriously consider whether or not this was a genuine attempt at manned flight.

IVA lasted almost through the end of February, coming down on the twenty-sixth into remote Siberia. And there, the charred debris scattered itself about. Ultimately the official announcement of *IVA* explained it away as "a test of an Earth-orbiting platform from which an interplanetary probe could be launched." A distinctly crude representation of the truth.

Then came this: "The scientific and technical tasks set in launching the sputnik have been accomplished," which was not the case at all. Perhaps the ultimate evidence to contradict this report was the expeditious sequel just eight days later, on February 12. Of identical design and flight characteristics as the initial *IVA*, this bug's blissful fourth stage operation properly sent it scooting along on its way. Was something different between the boosters?

Yes. On the previous flight, an electrical component had been installed in such a way that forced it to operate in the vacuum of space. After the stage failure was uncovered, one quick-thinking technician converted a leftover battery container into a new home for the part, and *whoosh*—it worked. *Venera 1* on the way. And that moniker did not come into favor until later on; at the time, Chertok and Korolev and everyone else repeated their use of "*IVA*" or "the automatic interplanetary station." A good name lends essential street cred to your wayfaring space machine.

In the wake of another sleepless night before launch, the men had gathered for drinks, celebrating this new chance to "rob Venus of her virginity." In certifiably high spirits Korolev hoisted his glass to suggest, "Let's have another dram so that Zeus will forgive us!"

When Jim Burke heard about *Venera*, it ruined his day. Stuffed into a drab office at JPL, Burke had already been blurrily laboring for over a year and a half toward the kinds of grandiose missions that Korolev and Maksimov were already starting to pull off. America held the technical chops. And he was only trying to get to the Moon. Yet Burke remained hobbled and hamstrung by a bureaucratic machine so flatulently obese that nobody had even been able to agree on how much his new spacecraft could weigh. Of course, he had no idea what all the Soviets had gone through to launch, but *Venera* fundamentally jostled him all the same. Buggered, Burke dropped into the overused chair at his desk and tacked up a picture of the insectine *Venera 1*, sent to him by a buddy with connections at the Telegraph Agency of the Soviet Union. Underneath, Burke taped a quote meant to help him focus and get to the meat of his own problems.

It said, "The better is the enemy of the good."

In five short days of travel, *Venera 1* skated over a million miles, temperature normal, 8,750 miles an hour. She looked great for a while, but it didn't last. Only seven days in, the sun sensor overheated and shut down. With stabilized flight gone, *Venera 1* automatically rolled over into a spin-stabilization mode. Then, the temperature-control shutters gave up the ghost. After a telemetry dump on February 17, it fell silent.

A telegram dated May 30 landed on the desks of Alla Masevich and Jouli Khodareo, two Soviet scientists involved in the *Venera* mission. Racing aboard a plane, they beelined to Jodrell and shacked up there for eight solid days. On-site, Masevich axed the secrecy routine, disclosing to mesmerized

British operators how the spacecraft's operation depended on regular house-keeping transmissions—every five days no matter what. Explicitly he detailed the pattern: seventeen minutes of basic locating signals, followed by a long stream of coded information delineating the craft's overall health and condition. Absorbing this, Jodrell's operators fooled with their equipment and strained to hear. Zilch. And by all rights, they *should* have been able to hear the probe, even though its signal coming back from the 112 million-odd miles plunged to something on the order of 10^{-22} of a single watt. Jodrell's equipment was that good.

But nothing. Absolutely nothing.

The British observatory tried again in late June. A few unidentifiable receptions trickled in, but they provided nothing conclusive. Jodrell saw no point in continuing.

"Radio contact with the interplanetary station is therefore lost," commented a Radio Moscow announcer soon after the mission's terminus. "An intensive investigation is now underway to determine the reason for this failure. Sabotage during assembly is not excluded."

Venera 1 only got to within sixty thousand miles of its target. But without question Maksimov and company had laid down a significantly long path of integral stepping stones. They'd embarked from Earth orbit toward another planet. They'd built an automatically orienting probe, capable of aligning itself to the stars and changing stabilization methods if it couldn't. Truly, they created the first spacecraft to another planet. And *Venera 1* still zips about out there to this day, gamely flying with pennants and mesh antenna and all, in a most gentle orbit about the Sun.

10. Job Number MA-11

He was a likeable guy, but might appear
a little nerdy to ordinary people.

Elliot "Joe" Cutting

JPL's in-house newsletter is christened *Universe*; in the issue from August
22, 1997, one distinctive announcement graced its pages:

"Voyager Week" Commemorates Spacecraft's 20th Anniversary
In commemoration of one of JPL's flagship missions, the Lab will celebrate "Voyager
Week" Sept. 2–5.

The JPL celebration will include a special event each day during Labor Day
week. Also, T-shirts and mugs designed specially for the commemoration will
be sold.

On Tuesday at noon in von Karman Auditorium, the Public Affairs Office
will host the "Voyager Bowl," a quiz-show type of event patterned after JPL's
annual Science Bowl. Three teams that include veterans of the Voyager proj-
ect will compete.

A reception and program celebrating the Voyagers' 20 years in space will be
held Thursday at 6 pm at Griffith Observatory. Tickets are $20; for informa-
tion, call the Public Services Office.

Making the rounds, Public Services rep Kay Ferrari slowly pieced together trivia questions for the impending quiz bowl. In the hallways, she
ran across George Textor, who politely asked that her questions cover
"hard" topics that would also be lighthearted and ideally fun. This was
supposed to be an enjoyable thing, after all—nothing to grind teeth over.
Ferrari thus polled the frontline trenchers and soon gathered some excel-
lently obscure factoids. Over two decades, the mission saw a plethora of

head managers come through; as of late, Voyager's reins lay in the hands of Textor.

On Tuesday the fifth, high-level managers and scientists teamed up in JPL's von Karman Auditorium, taking seats on the low front stage as spectators cascaded back in row after row. That day, moderator Craig Leff sat before them reading off the questions and waiting for someone to punch their makeshift buzzer. If the team buzzing in answered that question correctly, they got a bonus question that was related to the first.

As the questions came and went, everyone laughed and joked and consumed large quantities of beverages until just about halfway through. Leff asked the proper name of Voyager's twisty, planet-bouncing course.

One of his teams buzzed in: "The Detour Trajectory," they offered. Their answer was right—now for the bonus.

Leff cleared his throat. "And who first *articulated* the Detour Trajectory?"

Piece of cake. They answered, "Gary Flandro."

Uh-oh, wrong answer. The correct one, according to Leff's card, was "Mike Minovitch." He read the name aloud. "Mike Minovitch."

The mood changed. "No," interrupted someone in the audience. "It *was* Gary Flandro." Donna Shirley was the one piping up, a staff engineer. Murmurs filled the room. Packets of audience members sort of grumbled and shook their heads. Others laughed quietly, a decided awkwardness billowing over the group. Some got it; some were clueless. Down in the front row, Kay Ferrari smiled. She'd been around long enough to understand how this one might be a trick question. So Kay dropped it into the mix on the off chance that it might pop up, and she saw her rewards.

What, pray tell, brought on all the hullabaloo? *Don't* ask Mike Minovitch. To this very day, pungent bitterness saturates many a thought of his offered on the subject. Good memories remain, though they are now outweighed by decades of hardship and seeming betrayal. "JPL decided to deny me credit," is how one of his diatribes began. "And, to make sure it remains covered up, to invent a fraudulent explanation."

He says JPL stole his intellectual property and destroyed his personal records. He claims his bosses lied to him, censured him, sent him threatening memos, took credit for themselves, even tried to buy him off. He insists he effectively got JPL to the end of the solar system and the resultant worldwide fame, only to have the Lab squash him like a bothersome insect. His name is Michael Andrew Minovitch; he is a California resident, devout Catholic, airplane buff. And, oh yeah, a mathematical genius. A lean and soft-spoken yet unnaturally intense man, Mr. Minovitch is one of those people who can look down at a complicated arithmetic problem—one, say, running to several pages of formulas—and perceive something in there that most of us never would. He sees relationships. He sees order, symmetry, and meaning—perhaps even a solution. But something has stopped him from doing more of that. Deep inside, Minovitch carries a vicious, forty-year-old quandary eating his brain like cancer nearly every waking hour. Unmarried and childless—some may assert he's too busy to address either one of those—Mike Minovitch may not be an explorer, but his story is worth exploring. It forms the basis of an ongoing controversy over just who opened up the door to the solar system.

Minovitch lived his first years in New York. He thinks highly of his parents. "My mother was an excellent mother and homemaker," he said through a small mouth, eyes well-protected behind titanic glasses. His dark hair was receding and parted on the side. Functional hair. "Christianity was the foundation of our family."

Sometimes on weekend days, his father would load up Mike and his sister and go visit construction sites in New York City. Minovitch's father was an engineer by trade, introducing his kids to fundamental math concepts at an age probably considered too young by most. It's also the kind of thing good parents do. "He was a loving father," offered Minovitch, "devoted to his family." Some find happiness in giving. The elder Minovitch often showered his family with gifts for the simple joy of experiencing their glee. Dad bought a Lionel electric train, which Mike enjoyed disassembling and re-configuring more than actually operating. One day, Dad brought home a model of the Boeing 314 flying boat. It planted an unseen seed deep within his six-year-old son's brain, a seed for the skies. In 1947 the family unpacked from their move across three time zones to Inglewood, California. Young Minovitch decorated his bedroom with at least fifty models of different air-planes—some built from scrap lumber.

To say Minovitch's airborne interests further bloomed might well constitute a vulgar understatement. Any visitor to preteen Mike's room would invariably have noted the aeronautical charts uninterruptedly circling his walls listing radio beacons, compass points, runway elevations—wallpaper almost. Strolling past the desk would confront the observer with any number of surplus World War II radios lying about, capable of tuning in the local aircraft chatter. On his own, Minovitch figured out how to do all the radio maintenance, and he would indeed become master of solo pursuits.

Crossing to the far wall, this visitor thus gained opportunity to look out the window and perhaps contemplate what on earth might be happening up there with the roof, which tended to leak every time it rained. One clue? Leading away from Mike's window was a trail worn right through the shingles. This path led directly up to an enormous rotating beam antenna of Minovitch's own design. His cyclic ascents, for tweaking or the occasional upgrade, seriously degraded the roof and supporting plywood. "All my activities had a negative impact on the appearance and physical condition of the family home," Minovitch admitted. However, "in all these years in which my behavior caused major damage to the house, my parents never scolded me for anything I did." Years later, he learned that his parents often discussed issues like these in private but consciously decided not to say anything lest they discourage their son's efforts.

Dad Minovitch often chaperoned Mike and his sister to nearby air shows or on trips just out to the airport even when no show was in town. Mike knew the lingo like a pro. His detail-laden conversations among the mechanics and engineers at Los Angeles Airport differentiated him from the other underage rubberneckers, earning Minovitch unparalleled access to the facilities. They let him into the mechanic shops and the cockpits. He even got into the tower itself for observation sessions next to the air traffic controllers working *real live planes up there*—activity that would *never* happen today. This would not be the last time Minovitch's isolate dedication opened passageways nobody knew existed.

High school? A breeze. Minovitch excelled in his academics at George Washington High School, finding enough leftover time to have a go at running cross-country. He made varsity and proudly ran for the Generals.

As graduation approached, Minovitch selected UCLA for his next level of education. Not only nearby, it was insanely cheap at forty-eight bucks a semester. Minovitch inhaled the math and physics like spring air. He immediately moved on to graduate school in 1958 and, a few years in, completed the application for a half-time summer posting at JPL to help cover some bills. The place was regarded as an erotic den of top secret missile and space research. That sounded like a thrill—so Minovitch got in touch with a guy named Tom Hamilton at JPL and eagerly arranged for his interview.

Minovitch had to have felt confident. Despite a basic introversion, his résumé stood about as rock solid as they came. See, the Lab typically doesn't employ many slackers. "People are hired into JPL for their intellectual capabilities," explained Roger Bourke, who headed one of JPL's engineering groups in the late sixties. "Personality quirks and idiosyncrasies, you might say, are not as important as they would be in *some* industries."

Hamilton must have liked what he saw, because he offered the third-year grad student his choice of internship in one of two demanding arenas: theoretical research or trajectory studies. Minovitch knew something between zero and zilch about trajectories. That field involved brain-crunching questions—real ballbusters like juggling celestial mechanics and bodies moving in fluids and mission-hinging story problems.

Minovitch felt exponentially more comfortable studying theory. It was almost watercooler talk for him. Conjecture. "Pure research" is how a scientist would describe theoretical studies. That is, investigation for the sake of learning more, for adding to the collective knowledge. It is unencumbered by any sorts of design requirements or project goals.

All Minovitch saw walking through the Lab gates that first morning in June was the WAC Corporal missile standing there like it was, in front of the admin building. Then Tom Hamilton met up with him for a little bait and switch. The Trajectory Group needed help; their computer program didn't work right. Since Minovitch worked math like Shearing worked the piano, was he interested? Reminding Hamilton of his inexperience in the field, Minovitch nevertheless complied. "I was disappointed but agreed to do this work," he remembered. Practically speaking, the wiry grad student with the big ears didn't have much of a choice.

So the mustached head of the Trajectory Group, Victor Clarke, materialized as Minovitch's boss and set him up with a leftover desk in the same room as his own. Clarke got right down to the uncrackable nuts. No matter how they sifted the code, JPL's balky trajectory program remained loaded with bugs. The same answers never came back twice. The Lab had already been given the go-ahead by NASA to begin work on this program called

Mariner to fly past Venus and Mars. Blueprinting proceeded on schedule, but that didn't do anybody much good unless Clarke and his bandolier of geniuses could figure out where to aim in the first place.

Trajectory is all about celestial navigation. Planets spin and revolve about the Sun. Moons of planets have their *own* spins; some even spin backward. So tell us: when, exactly, *does* that ship need to leave Earth's orbit to perfectly rendezvous with the full Moon in sixty hours? That's where people like Victor Clarke entered the picture, laboring behind the scenes to ensure that people like Bill Pickering would be able to stand up and beam at press conferences, explaining how the spacecraft was perfectly on course.

Clarke maternally cradled his program, comprising dozens of eighty-column punch cards—each about the size of a 4 x 6 photograph. Together the cards added up to a whole, and somewhere within his thousands of instruction lines the program reported values differing greatly from those in the real world. He wasn't sure how to attack. Mike, any thoughts?

A pound of cards, the natural conversation piece. *What do you run this thing on?* Minovitch wanted to know. Realizing visuals might go over better than some windy explanation, Clarke escorted his charge on a short trip over to Building 125 in order to visit nothing less than the world's fastest digital computer—the IBM 7090.

At least, it was the fastest in 1961. Entering the 7090 in a footrace with twenty-first-century technology is a bit like pitting Cugnot's 1769 steam-powered tricycle against a Lotus v-8. Both get you from A to B. But lumping them in the same category?

Thankfully, computers have shrunk over time. While simultaneously being the fastest commercially available computing machine, an IBM 7090 also rated as one of the world's largest. In a custom room, up to *eighty* refrigerator-sized pneumatic tape drives ringed a central operator's console, line printers, and punch card readers. This wasn't a computer; this was furniture. Everything sat on a raised grid floor, set up for the miles of interconnect cable running underneath. Within the equipment itself, magnetic-core storage operated hand in hand with fifty thousand transistors, writing 3 million bits a second through eight data channels. In a 1960 press release, IBM claimed the 7090 could perform 229,000 addition problems in one brief second or less.

Visitors generally saw the machines from behind glass—deafening noise levels reposed them inside specially constructed accommodations that were further reinforced to contain the antarctic temperatures necessary to offset a roomful of machinery drawing upward of a hundred kilowatts. American Airlines owned two of the things, which they used to drive a centralized reservation system called SABRE. One sat in the Time-Life Building where it was rented out to various customers in need of crunching large-scale accounting and inventory-control equations. Another resided in Minovitch's alma mater, UCLA, who shared out time on the box with other universities in the West. Down in Huntsville, Wernher von Braun hammered on a pair of them to validate his rocket designs. JPL had only the one, while the U.S. Air Force laid claim to four. Anybody with $2,898,000 burning a hole in their pocket could order one up from the IBM plant in Poughkeepsie, New York. If the need only lasted a matter of weeks, they rented for $63,500 a month or $1,000 an hour, not including transportation and gratuities.

Minovitch fell in love. "They were capable of solving mathematical problems that were previously impossible to solve," he expounded of the 7090. "I believed that only the world's best mathematicians and scientists were capable of understanding and using them."

The inexperienced grad student did not know it then, but his finest hour approached. In looking back on this moment, Minovitch would consider it his period of transformation into one of the world's great mathematicians. He also did not know of the monster that Victor Clarke would seemingly pupate into, outwardly stealing Minovitch's work and publishing it as his own. Rather, doe-eyed young Mike Minovitch came to work every day that summer, sharing Clarke's office and working diligently.

Over the next many days and weeks, Minovitch's assignment resolved. In addition to fixing the program, Clarke asked if he could please do just one more thing: chart the best course for a probe having a specific amount of time to get from one waypoint in space to another. Although terribly new to guiding spacecraft, Minovitch soon cranked out a slide rule kind of an answer and fobbed off the draft solution to Clarke. Thrilled, Minovitch's boss asked only that the apprentice triple-check his equations before formalizing the work as an external JPL Technical Report. And when Minovitch plunged into the revisions, he noticed something.

To Minovitch's admittedly neonatal eye, the JPL workflow came off as too complicated, too messy. It required a precise arrangement of six extremely complicated variables that stymied him. These factors are comparable to six determined elementary-schoolers running loose in a giant toy store: while keeping tabs on one or two, the others splay out in directions beyond reach. Go for a third and lose the first. Nobody has six hands. Minovitch could think like the JPL program wanted him to, but why? Wasn't there some other, simpler way to approach things? "When I saw how messy the six-element formulation was," Minovitch explained, "I decided to see if I could clean it up." Nobody asked him to do this.

JPL used something called scalar math to calculate the half-dozen plot points representing a ship's position in space. But Minovitch felt it might be easier to address the matter if he did away with all that and instead consolidated the information into three vectors. They worked beautifully for describing multiple quantities in a single figure—things like position, velocity, and acceleration. Near as Minovitch could tell, vector math was just edging onto radar screens in 1961 as an applied way to solve problems, but the eager chap wanted to have a go all the same. Vectors accounted for true three-dimensional space, which definitely made the most sense to him.

Working virtually around the clock, without pesky distractions like bar-hopping, women, sports, or even friends, Mike Minovitch rewrote JPL's six-headed demon in the language of vectors and immediately began walking the shoreline of possibility, feet slapping in the shallows as he hunted a suitable conundrum on which to test and prove it. Headfirst he ultimately dove, into an abyss of tangled, ungraspable problems, wearing only his chewed-up slide rule for a life jacket. The deeper Minovitch swam, the harder the problems got—like . . . stroking . . . through . . . molasses—and the closer he unwittingly drew to an obscure physical concept that occupies his mind to this very day. It was there in the deep black, just beyond his fingertips. But Minovitch didn't grab on, because something else floated past his eyes first. Closing both hands around it, Minovitch surfaced with his catch— really more of a choice than a passing opportunity. For the purposes of evaluating his new vector approach, Minovitch chose a fundamental math problem reputedly so downright unsolvable that in 1887 King Oscar of Sweden set up a prize for anyone who could present him the solution.

It goes by several names, but one of them is the Three Body Problem. It goes like this: in close proximity to one another, two planets spin about. Each possesses a different mass and therefore a different gravitational field as compared to its neighbor. So two variables, which for the average math whiz aren't so hard to reconcile. Isaac Newton figured out the problem of two, but he couldn't get a grip on three, as there is no closed form solution.

Three? Now fling another handy celestial object near the two—like, for example, a dinky little spaceship—and calculate how each of the three will act. Now answer all the other questions that start piling up: Exactly how DO gravitational attractions from each planet affect the other? How do they affect the spaceship? How close can that ship get to a planet before it's affected by gravity? How much of an effect? How much influence does the *second* planet have on the spacecraft depending on how close it is to the *first*? What will be affected there? And when the spacecraft reaches the influence of the *second* planet, how does the influence of the *first* change *over time*? It stumped the best of them since its first description, which did not, incidentally, earmark a spacecraft as the third object. Comets and wayward asteroids originally filled that role.

The problem even entered popular culture. Back in 1951, while Minovitch trundled his way through freshman year of George Washington High School, sci-fi moviegoers enjoyed a B-grade noir flick called *The Day the Earth Stood Still*. A space alien named Klaatu visits Earth to warn humanity of dire consequences if their warring habits persist. This alien, very much resembling British actor Michael Rennie, visits an American scientist who unfortunately isn't home at the time (probably working with cosmic rays). Klaatu deposits his calling card in the form of a blackboard solution to the Three Body Problem.

When the two finally meet up later, the professor asks, "Have you *tested* this theory?"

Klaatu straightforwardly explains, "I find it works well enough to get me from one planet to another."

This tackling of the problem . . . why not? Why not aim for the top? Mike Minovitch opened his notebook to a clean sheet of paper and licked his pencil. Then he unsheathed his slide rule, the only weapon currently available. The voice inside him said, *trust your vectors*. Somebody had to try, didn't they? *Why not*, indeed?

He was up for it.

Isaac Newton didn't have Mike Minovitch, but JPL did.

Come early August, Minovitch let Victor Clarke in on a miniature secret: on his own time, at home, Minovitch's self-directed labors over the slide rule seemed to have produced a working solution to the Three Body Problem—at least, a theoretical solution. Exhaustive checks would prove him right or wrong, and by far the most sensible method of doing so utilized that fancy 7090.

But there was more. Perhaps of greater importance, Minovitch patiently explained to his boss, was the fairly astonishing concept to emerge from the residual data. This was the concept just past his fingertips in that watery abyss.

He still needed to run a confirmation test. BUT if his roomy, tapeworm figures were dependable enough, flying spaceships close to a planet enabled a nearly magical situation to occur. If performed in just the right fashion, tremendous amounts of gravitational energy would be imparted from that planet to the spacecraft itself! That meant additional thrust—*lots* of it. Every planet was a gas station with free fill-ups. The bigger the planet, the bigger the oomph. And then came the pièce de résistance: aimed properly, the spacecraft might well be able to exploit this gravity boost to zip on to the next planet. Another boost there and then on again.

It was all in Minovitch's pages. JPL did their work with little space probes, right? The ship gets launched toward its target, makes observations from close range, then flies off into nothing. Two planets are being studied? That meant sending a probe to each—until now. *Two* planets with *one* probe? Impossible or just really difficult? Two planets? Heck, what about *all* the planets? *Unlimited* propulsion, applied through an excruciatingly careful application of proper trajectories, just might be able to bank a spacecraft all the way out to Pluto—and then back again.

"These things just leaped out at me!" he proclaimed.

Minovitch isolated himself, spending the next many days composing a JPL Technical Memorandum at his own initiative. "I wrote the manuscript in longhand," he explained, "and gave it to one of the JPL technical typists." Buried in the dozens of pages and hundreds of run-on sentences reclined a proposed *seven-planet* mission winding from Earth to Venus, out past Saturn, *way* out by Pluto, and then improbably returning via Jupiter. Intertwined throughout the text lay Minovitch's apparent solution to the Three Body Problem, which directly led to his discovery. This August 23, 1961, document titled "A Method for Determining Interplanetary Free-Fall Reconnaissance Trajectories" is Minovitch's holy grail, his first definitive explanation of "gravity assist."

His finished memorandum worked its way through JPL's Systems Analysis Division, and the attention it received was nil. A couple of the senior engineers, however, took time from their busy days to contemplate the eccentric graduate student who'd been on campus scarcely two months performing work he'd never even heard of before, let alone done. *This kind of thing has already been explored*, they all asserted. *Old story*. Going back to the 1920s at least, physicists and mathematicians of all shapes and sizes and languages and walks of life had already worked the idea practically to death. So why not give it a rest? As Minovitch recalled, "There was nobody at JPL who visited me after the paper was distributed, to discuss it."

Somehow Minovitch found strength in the irritating lack of attention, doggedly posturing to explain his unique approach. Plain and simple, in his eyes it truly was different. All the others, he maintained, saw gravity as a *problem*. It spitefully acted to the mission's detriment, screwing up the intended flight path and angling the ship in unsolicited directions. Until now, Minovitch sermonized, gravity was the *enemy*—an unwelcome force to be overcome by the feeble spacecraft's desperate use of brute-force rocket thrust. And he thought that paradigm might change really soon, provided his numbers all checked out. All he had to do was find someone to vet the work—but he didn't see a lot of hands in the air.

"Clarke refused to assign anyone to program it to conduct any numerical investigation," Minovitch insisted.

Victor Clarke, for his part, superficially appreciated Minovitch's whole approach. Yet Clarke reckoned the guy best get his nose back on the j-o-b —namely, solving those problems JPL actually hired him to do—and not waste the firing of his synapses on little side jaunts.

Even a shallow review of Michael Minovitch's employment record would have unearthed consistent behavior. Two years before, Minovitch had spent the summer at Research Chemicals in Burbank. He found time enough to conduct two discrete and very different research projects at the company while still adequately juggling the requirements of his job description. He got all the assigned work done, so what was the big deal? And why should JPL be any different? "My motivation," Minovitch explained, "was not to receive praise or recognition but to conduct and report original research for the pure love of doing it." No matter that Tom Hamilton had nudged Minovitch away from theoretical research—he was going to do it anyway.

Minovitch spent the remainder of his 1961 summer employment laboring over Clarke's honey-do list—things like how to correct flight paths if the engines burned too long. Clarke even encouraged him to have a whack at using the vector approach if Minovitch preferred. And then summer was over; he'd run out of time for doing anything else on the gravity thing. Glumly Minovitch retreated to UCLA's campus feeling like his *real* work sat half finished. He'd spent the waning days appealing to Clarke. Let him test the idea, Minovitch pleaded, to run programs on the 7090 and conclude whether the method works or not. He always got a thumbs-down. Minovitch felt Clarke was blowing him off. However, accomplishing the task meant creating a wholly original FORTRAN program specifically for the 7090; even then, they still had to get it on the schedule. The giant computer was in use nearly around the clock. Do all that to satiate an intern's whim?

Perhaps in consolation, Clarke did invite young Minovitch to stop back in over Christmas holiday to extract what he could from a smaller computer. The two reunited over the fourteen-day break, whereupon Minovitch resurrected his little scheme of "gravity thrust," as he was now calling it. All the basic principles were there to read off a slide rule. *Wasn't it remarkable?* Clarke felt, well, not exactly the same. Refusing to believe his subordinate, Clarke opined how Minovitch's little brainstorm seemed to violate practically every law of physics Clarke could think of. It wasn't just improbable; it was impossible. *Nice try, kid.*

Minovitch returned to UCLA a touch dejected, but he wasn't completely out of options just yet. The university offered an accelerated course in FORTRAN, which got him to thinking about the idea of playing computer programmer and doing this himself. Zap the middleman. And UCLA possessed a 7090 right there on campus, which sweetened the deal. Competition was something to reckon with, as more than sixty universities jockeyed for hours on a computer able to run only one single program at a time. Minovitch gathered his bottomless notes and notebooks and went over to scenic Royce Hall, visiting a teacher in the Department of Mathematics. Minovitch knew this guy would be familiar with the Three Body Problem and perhaps a bit sympathetic even, but that's all the ammo he had going in.

It was enough. Professor Peter Henrici thought enough of Minovitch's noodle to dial up extension 9236 fifteen minutes later and get the head of 7090 operations on the line. Fred Hollander listened as Henrici hit him up to give this Minovitch guy a little run time once his program came together. Hollander abided by the math prof's recommendation, assigning it job number MA-II and budgeting Minovitch a total of fourteen hours. They'd be parsed out in teeny chunks.

One of the basic questions his program *couldn't* answer related to the known locations of every planet. Already, Minovitch owned *Her Majesty's Nautical Almanac* from England—a reference volume detailing planetary coordinates between 1960 and 1980. "Anybody who's working in astronomy has them on their shelf," explained Roger Bourke.

The almanac began life in 1675 when a decree of King Charles II meant John Flamsteed was in for a bit of extra work. British fleets were getting themselves lost at sea. Flamsteed, the first Astronomer Royal, was tasked with establishing "with the most exact care and diligence," the decree promulgated, "the so much desired longitude of places for perfecting the art of navigation." In short, he was to precisely chart the heavens. This task required only a hundred years and four more astronomer royals before the first edition finally made it to press. Laying out celestial events over time, the almanac contained everything a wayward sailor might require: distances from the Moon's center to the Sun, along with the brightest stars noted *every three hours*. By the early 1900s the work had expanded into several task-specific volumes for everyone from mariners and astronomers to religious groups and film crews. On a shimmering ribbon the almanac's cover logo proclaims "Man Is Not Lost."

Over the course of two weeks, Minovitch manually keypunched nearly the entire almanac onto an immense stack of four thousand data cards, practically big enough to have a gravitational influence of its own. UCLA's 7090 could read the thick stack and bump it over to magnetic tape. Working in concert with the program itself, this database offered answers when the program asked what was where.

Fourteen hours on a 7090 scooted by faster than Wilma Rudolph. Mike could only request one run a day, and it was always limited to ten minutes. But the 7090 took half that time interpreting Minovitch's portly FORTRAN code as machine language. Results never even came until the next day. At this pace, debugging might take forever, and Minovitch had already burdened himself with a full graduate class load.

Then he heard about a secret one-hour time slot every evening at midnight. You could do more than one run then, depending on who wanted in, but the time for each maxed out at five minutes instead of ten. Every little bit would help. He began showing his face, but Minovitch soon hit a wall. The program kept getting hung up in an infinite loop. After ten weeks of exhaustive work his remaining time dwindled to zero. Mike padded back to Hollander and, spreading out the progress thus far, tried to sell him on the idea of overdrawing his account. Maybe by just a teensy-weensy bit?

Mike Minovitch got more than he bargained for, as Hollander proposed that this nonfaculty graduate student actually receive, in writing, *unlimited* 7090 time for pursuit of his home brewed research. The forms came through, and Minovitch couldn't believe what sat before him. Sure, the paperwork cautioned "LOW PRIORITY" and "Run as time available," but there it was in writing all the same. If nobody was signed up, Minovitch could jump on. That meant lots of little sessions; he'd have to nibble through his equations bite by bite. To make things easier, Minovitch laboriously revised his program so as to make it interruptible: if some higher priority job came barging through the door and kicked him off, whatever state the computations were in could now dump out to one of the dozen-odd pneumatic tape drives encircling the main console. Minovitch could then pick back up again later where he'd left off, instead of repeatedly beginning at square one.

By April, around the same time JPL's Jim Burke geared up for another expensive failure, Mike Minovitch swung back around to Vic Clarke—getting him on the phone to breathlessly report the status of his little whiz-bang theory. *Guess what? It's in heavy testing over at UCLA.* A solution to the Three Body Problem seemingly existed—from a program he'd coughed up all by himself. On the off chance Clarke might be interested, JPL could try running it on their own 7090 to verify if Minovitch was indeed doing good math. Oh, and UCLA wouldn't give Minovitch any free paper; they wanted ten bucks a box. Could he get any from JPL? In stark contrast to the previous winter break, Clarke's ears pricked up. Were the kid's numbers really on target?

Yeah, Clarke mentioned, he would indeed like to run it himself. *Take all the paper you want.* And JPL now possessed more hardware. They'd taken on a second 7090 and crammed it into the ground floor of Building 202. It was still being installed.

Two computers! Clarke and Minovitch rendezvoused that Saturday in a conference room on the second floor of Building 202. The wonder boy dished out a few more specifics on his work to date; then out of the blue Clarke announced his interest in buying Minovitch's program. What was the price?

"This question took me somewhat by surprise," Minovitch later commented. He had *never* been thinking along such lines but figured Clarke must certainly now feel more positive about his work than he did a year before. Minovitch looked at him; the guy practically had his checkbook out. "I explained to him that I was not interested in selling the program," Minovitch recalled. "But, like any other scientist, I was interested in preserving my claim on the propulsion concept."

Nevertheless, Minovitch felt secure enough in his relationship with Clarke to hand over his singular copy of the program, bundled into skyscraping layers of manhandled punch cards. The average program used maybe a hundred. Clarke sized up the towers before his eyes and figured he was looking at maybe three thousand. And that was just the program—Minovitch's encoded almanac counted separately. The mathematician left that afternoon with three boxes of paper under his arm and a promise by Victor Clarke to double-check the program. On that day, Michael Minovitch unwittingly left behind what amounted to his career.

Clarke dutifully ran that program to search for errors, but he also made copies, laying out his request to adapt it for general JPL use in a seven-page memo distributed that June. Minovitch's name was on the CC list. When the bloated software kept passing verification runs, Clarke indicated to Minovitch the inherent value of informing Lockheed and MIT about their success. At the time, both entities were heavily entrenched in researching the same issues. One spindly grad student seemed to have upstaged both a giant defense contractor and one of the more prestigious universities at a significantly reduced cost to pretty much every taxpayer.

Evidently ignorant of the duplicative activity, Minovitch raised no known objections to Clarke's memo. Instead, he rejoiced in the face of positive news from JPL's Trajectory Group. Every analysis had confirmed Minovitch's findings. "The tests were a tremendous success," he gleefully reported. This LA math whiz had found a way to explore the solar system by catapulting a spacecraft from one planet to the next using nothing but gravity.

This is a big deal. Yet Michael Minovitch failed to compose any reports or attend even one nearby scientific conference to present a formal paper. For someone like George Ludwig or Konstantin Gringauz, reputation virtually hinges on this critical process. Proper, structured reporting of research findings is a cornerstone of the business. Some do this in person at conferences by making a speech. The other acceptable method is print, within the pages of an accredited research journal. You're only as good as your last couple of papers, and Minovitch was apparently too busy to pause for the formality. Instead he retreated to the labyrinthine graduate studies facing him still and to the endless combinations of planetary missions now possible.

When the spring semester ended, Minovitch took whatever he could get on any of the three local 7090s. Clarke encouragingly made the systems available to Minovitch on the same kind of low-priority, time-available arrangement in place at UCLA. JPL kept track of their 7090 schedule on enormous blackboards, one mounted in each of the computing buildings. Minovitch watched over them like stock tickers, hovering with punch cards and almanac tape in hand. He got in a lot of time at night, flitting between rooms, pounding endless computations on every conceivable multiplanet mission between 1965 and 1980. Minovitch wanted every single possibility.

If time was available, he'd stay in the computer building all night. While the program ran, there was nothing to do except read the 7090's user manual. Minovitch did so, learning every switch and status light and ultimately how to directly operate the machine, while simultaneously culturing himself on IBM's almost ethereal attitude toward their creation. "Intelligent human intervention and supervision can often bring a problem to completion more quickly than computation alone," the user's guide monkishly preaches. "This interpretive approach provides close communication between man and computer when human intervention is desired or necessary."

In short order, Minovitch demonstrated his competence to JPL's surprised head of computing and abruptly won permission to send the regular technician home when Mike's turn came up. Minovitch felt happy because he got to sit behind the wheel and drive. This was done from the central operator's console, arranged relative to the other equipment like a conductor's podium to the orchestra. Sized like an upright piano, its surface controls resembled more pinball game than expensively serious tool. Sixty switches and two hundred lights governed every operational aspect of the machine, telling Minovitch if he was on track to results or blindly stuck in another endless loop.

What about all those results? Mike Minovitch slowly built himself one of the world's finest private collections of accordion-fold computer printouts. Already, by the end of May, his modest office on the UCLA campus bulged with overstuffed boxes. They dominated the room floor to ceiling, stacked on the institutionally metallic gray furniture and anything else available. Nobody could fit in there to visit him. Minovitch himself was on the verge of no longer being able to shoehorn himself into his own room—not to mention a blatant violation of the fire codes. So Mike called up JPL asking for help. Deliverance came in the form of a courier van, removing innumerable heavy boxes to the computer room of one rather unfortunate Ms. Helen Ling, up on the second floor of Building 202. Everything printed out on continuous tractor-fed three-ply paper, leaving someone—usually Minovitch—to separate the pages and to bind and label them all. He often came over and picked at the job during long computer runs.

Summer ended too soon. Making his way down Charles E. Young Drive, the diminutive young man with gravity in his eyes moved back into Dykstra Hall at UCLA for the fall semester. Once again, he signed up for a full course load of math and physics classes. And when the phone rang, it brought him running—flat-out pumping, knees high across the road in the middle of the night to the UCLA computing center, shades of the George Washington Generals.

11. The Science and the Cyclist

We were kind of making it up as we went along.

Bud Schurmeier, detailing JPL's overall Ranger strategy

It was Pearl Harbor Day 1962 when Jim Burke came down the canyon and got his ass fired. Arriving at JPL that morning, he didn't quite know what the flavor of the day might prove to be . . . until his desk phone rang with Bill Pickering on the other end.

The JPL director began by saying, "Sorry, Jim . . ." and immediately Burke shook his head in the knowledge of this bad sign. Pickering didn't call his desk much, and Burke well understood the delicate situation before him. It might change in the next thirty seconds, but as of right then and there, he still held the title of project manager for something called Ranger. And the repulsive, twitchy prickle creeping up his back suggested that this phone call probably had something to do with how all five expensive and very public attempts to crash a Ranger probe into the Moon had gone down the toilet. Every damn time. Burke wasn't in the mood for a showdown and didn't appreciate the coincidence of the date.

It wasn't because he hadn't been *trying*—in Burke's eyes, a distinct lack of cooperation mangled his efforts from practically every angle. On two occasions, the booster rockets had failed. That wasn't his fault. Another time, a bolt deep inside one of the spacecraft unexpectedly worked loose sometime between assembly and launch, allowing two vital power connections to separate and bring the house down. No one maliciously planned *that*. Itty-bitty flakes of loose gold foil ruined another ship plus all the slow months of work that went into it.

And if the mechanical problems hadn't been enough, there were the oozy politics. Only a quick peek was necessary to appreciate the layers of rancid bureaucracy created as NASA, the army, JPL, and the air force all struggled to understand just how in the world to run a lunar program. Burke had signed up for high-tech engineering, not some dysfunctional three-ring circus. What about the other government entities that had shown up at practically the last minute wanting to screw with his payloads? What about those jokers making the booster, who figured its capability wrong? Why couldn't they make stuff right? Could that possibly be *his* fault? Through it all, Mr. James D. Burke, a 1945 Caltech graduate with a degree in mechanical engineering, was pleasant, accommodating, helpful. Now standing there listening with the receiver to his hot, hot ear, it was all about to be taken away from him. At least, that's how it seemed. Burke held the phone with both hands and knew that when it was his turn to speak, he'd have to choose his next words ver-r-ry carefully.

Ranger's lineage traced directly back to NASA's Office of Space Flight Development and the untenable desk of its director, Abe Silverstein. His energetic, vigorous ways made him a perfect match for the job. "I don't think anybody did more ingenious and competent work than Abe Silverstein," NASA administrator James Webb once said of him. And the anaerobic Silverstein was not shy about proclaiming what he wanted and then trying like mad to get it. He was a purebred engineer baptized in the unforgiving fires of the Lewis Aeronautical Labs in Cleveland, a man who loved to make decisions and then make them stick. He wore his dark hair parted too far to the right and horn-rimmed glasses. Unyielding in his decisions—get out of bed and get stuff done—this was Abe Silverstein.

Lately his attention had been most ensnared in the manned Mercury program, which he liked to think was now more or less underway. But Silverstein had a couple of lieutenants on board who were helping him bootstrap the momentum in other areas. One of them was a guy named Oran Nicks, chosen to be head of lunar flight systems. That title was a fancy way of saying his principal responsibility was the wrangling of JPL. As such, he began overseeing the Lab's efforts and in due time was to become Jim Burke's Satan incarnate right here walking the earth. Another Silverstein deputy, Homer Newell, accepted the title assistant director of space sciences. This forty-four-year-old son of an electrical engineer was not a rocket man per se—more of a theoretician who, through the virtue of rocket technology, was in the process of being provided wonderful new tools for exploration.

After signing on at NASA, Newell wasted no time in firing up a couple departments of his own. He opened shop on a Sciences Division with six tributaries blanketing wide subjects like meteorology, astronomy, solar physics, and planetary atmospheres. Newell also initiated a Theoretical Division, complementarily addressing the abstract side of these same disciplines.

In what would become a magnificently serendipitous stroke of luck for planetary scientists, Newell picked a fellow named Robert Jastrow to head up his Theoretical Division. Jastrow was a classic fields-and-particles guy like Van Allen—he understood little of how to investigate good old terra firma. Being in these unfamiliar waters of planetary science, Jastrow packed up a lunch and went to visit Harold Urey, the Nobel laureate chemist, who wished to give him some advice.

Urey was in love—with the *Moon* of all things—and filled his visitor's ears with how much good would come from an understanding of our little gray pal in the sky. Hovering out there and basically untouched like it was, the Moon's careful study would reveal many a hint about the solar system's origins. And to boot, it was more or less right next door, only a quarter-million breezy miles away. Galactic pocket change.

In fourteen years of particle physics, Robert Jastrow had never heard any of this. So, around the same time Carl McIlwain was saying his goodbyes to *Pioneer 3*, Jastrow rolled into a NASA conference lugging Urey in tow. Homer Newell listened as the two men laid out their observation: NASA frankly had no real program of lunar exploration beyond the mishmash collection of ARPA leftovers. It needed a majestic plan.

"The national space program will be open to strong criticism," Jastrow warned in a bristly follow-up memo, "if a very early and vigorous effort is not made in the program of lunar exploration." He had no way of knowing that in five months *Luna 2* would impact the new object of his desire and really goose things into high gear.

Roughly a month later on May 25, Silverstein and Newell put their heads together trying to settle into a plan. Scribbles on paper. Coffee. Morning turns to noon.

A lunar-centric rationale seemed pretty straightforward to them. The Moon lay conveniently in range once every lunar month. JPL saw this as a detriment—somehow as an indicator of the target's second-class stature. Instead, they had designs on flying to Venus and Mars and were lobbying Newell for a go-ahead.

That was a lot to bite off. Not only did lunar opportunities come often, but the flight time from here to there measured in the days, not months. And the nearby target promised fewer communication problems relative to planetary distances. Erase, erase; scribble, scribble. JPL's plan was out. What kinds of missions would fly to the Moon? Connect boxes with lines; doodle flowcharts. They'd need orbiters first—after that, they could try rough-landing an armored capsule of basic experiments. Then they'd bring it all home with a couple of soft-landers, holding more delicate instrumentation? There. They had it. NASA was headed for the Moon. And the word went forth from Silverstein's desk, along with a straightforward directive to cancel whatever blue-sky planetary flight JPL had in mind.

Now, nobody figured going to the Moon would be *easy*. Pioneer's ongoing soap opera handily accentuated the difficulty of Silverstein's professed objectives. And even if a probe was ready, the Vega was giving everyone conniption fits. Stacked on top of an Atlas was supposed to be a new upper stage from General Electric. This was Vega, and money drained into the thing at incomprehensible speed.

But Congress had forgotten what its other hand was already doing. Before the first Vega test flight even rolled around, somebody at NASA realized just how similar it was to a contraption known as Agena that the air force and the CIA were just hammering the kinks out of. They'd been keeping the thing under wraps. It was another upper-stage design for Atlas, but it was supposed to be more capable. A month and a half later came *Luna 2*, and NASA threw Vega in the dumpster. A hurried shunting of money valves now increased the flow of dollars to Agena. Nobody at JPL knew the real story, and others who did talked only under penalty of incarceration. The real support for Agena lay in an unmentionable, top-secret program of eavesdropping.

Just a few days before Christmas 1959, an evidently festive Abe Silverstein signed off on seven new flights to be run by JPL. Any remnants in the tumbledown ARPA queue would be allowed to run their course, but then things would change in a hurry. Silverstein's explanatory letter to Pickering also requested that JPL ascertain the plausibility of carrying along small, token scientific experiments to investigate the translunar environment. Additionally, they were to contemplate safe-landing a modest experiment package able to "survive impact and then transmit significant data." And oh, by the way, could it all get done inside of three years?

Like any new baby, Silverstein's needed a name, something inspiring— adventurous and exploratory. At JPL Cliff Cummings had been the Vega program director until its beheading. Pickering selected him to be the Lab's first director of lunar programs, offering Cummings the unique privilege of influencing spacecraft names. He weighed in with "Ranger," which everyone found to be generally agreeable. Except one man. Crabby Abe Silverstein hated the idea and told everyone so. He held curiously deep-seated reasons that did not involve much of anything: Silverstein had once owned a dog when he was young—a difficult, obstinate, unpredictably stubborn dog named Ranger. Apparently the dredged-up memories would just be too painful every time he got a memo referencing the project. Silverstein didn't like the association, but since the name was already in general use, he would have to get over it.

Nobody at JPL knew what this Ranger thing would look like, how it would get power, or what all would ride aboard. "Environmental investigation and a survival capsule"—that's all anybody knew. It made for a rather stubby leaping-off point. As 1959 was turning into 1960, Pickering laid his hands on the head of one James D. Burke to advise him of his selection as Ranger project manager. Burke would hold responsibility for the project with Cummings functioning as his boss, ally, confidant, and teammate.

In the office that ominous day almost three years later, Pickering's pickling of James Burke was only getting warmed up in a room that had suddenly gone cold.

"I'm afraid you and Cliff Cummings are going to have to do something different," the director explained.

Burke nodded quietly. Career diplomats will explain how in certain situations the most diplomatic thing that can be said is absolutely nothing. It was all unraveling before him.

Many years later Pickering still felt uneasy about how the whole situation went down. There were so many things he wanted to tell his crucified subordinate. "I always had the feeling that we were too hard on Jim

Burke," Pickering said with a touch of remorse. "He was a good man and was made a scapegoat."

Cummings and Burke—the men emerged from similar molds. Both were Caltech boys; Cummings received his diploma in 1944, virtually steps in front of Burke. Mere credits were all that separated them. While Burke was mechanical engineering, Cummings did physics. After graduation the new physicist came directly to work in service of Pickering and the Corporal missile. From there his attention shifted to the Vega booster in '59, not long before it ended in the wastebasket. He was an honest, deeply religious man of the highest moral and ethical standards and therefore had no business whatsoever running government programs. As would be the case with any good and noble man, Cummings embraced structure, personal responsibility, and a sensible chain of command. JPL perfectly suited his kind.

In Burke's mind there also was no greater job, although his arrival traced a more circuitous path. Rather than bum-rush JPL right after college graduation, Burke first joined the military and flew airplanes. Strange for a guy who'd been offered a scholarship to Harvard and turned it down. "The one thing I wanted to do was fly in the U.S. Navy," he explained, "and Caltech seemed to offer better preparation for that." He then went back to Caltech afterward for a master of sciences in aeronautics. Finally, he reached the Lab gates in '49 to begin applying his skills. However, as Burke lightheartedly indicated, before starting work there he had to get hired.

For many people job interviews are milestones of life, but in Burke's case it served as more of an anticlimax. "On the previous, very lively night," he explained, "I had proposed to the lady whom I have now loved for more than fifty-six years." He'd met a certain Caroline while skiing at Mammoth Mountain in the Sierras. "A dancer and a lover of mountains," she would always be his "Lin."

"The next morning she gave me some oatmeal and sent me off," Burke continued, "but I was still a bit, shall we say, *disheveled* for my interview with the dreaded Louis G. Dunn." At that primordial moment in time, Pickering was still lecturing students at Caltech while Dunn ran the whole Pasadena show.

For a while the director pored over Burke's transcripts. Then, in his heavy Afrikaner accent he summarily demanded, "Vot are all dese D's?"

Burke told the truth. They were from the particular school term when historic blankets of snow had covered the mountains near Caltech. "Everybody went skiing," he droopily offered.

Dunn's eyes lit up. "Ve don't vant no playboys JPL!" he screeched.

Years later Burke remembered, "But he hired me anyway!"

For James D. Burke, preparation had met opportunity. As a natural on-his-feet thinker, Burke already coheld a patent on a guidance-and-control system for ballistic missiles. His solid engineering foundation, coupled with a deep comprehension of intricate technological devices, led to his tagging of Ranger at age thirty-five. It's an age when many men are just learning how to get trash to the curb on time. For Burke, the assignment was hog heaven. He surely didn't look like the Ranger project manager in his old U.S. Navy jacket, gliding into work on a bright green motorcycle.

"The Austrian Puch was a silky whisperer," he uttered, almost reverentially, of what was only one in a long string of bikes. Right after getting his naval wings, Burke had finagled a beautiful little Czech 125 that proved to be one of the better cycles he ever owned. "You could pick it up with one hand," he said. "It was the vehicle on which I courted Lin. We rode it a lot together, and after we were married in 1950, I rode it to and from JPL for quite a while."

Burke dived into the Ranger meetings and was always glad to see his pal Bud Schurmeier around the halls. Their friendship went back a ways. "We started as undergraduates at Caltech in the same section of the freshman class," Schurmeier explained of his first encounter with Burke. "He was a really nice guy, and we both had an interest in sailing, and so we used to go down and crew for different people sailing off LA harbor."

Bud Schurmeier was not a Ranger principal. He'd been selected by Pickering to formulate the new Systems Division, and understanding its place means understanding how JPL operates. They termed it a matrix organization. Instead of wholly separate project offices (i.e., one for Explorer, containing a big bunch of specialists; another one for Ranger and all *its* specialists; and so on), the place was split up by discipline: structures, propulsion, communications, guidance. Each one serviced whatever project came along. It's like this today. And in the beginning this kind of arrangement worked okay, but it was not entirely without problems—integration problems. One department made a key, to analogize, while another made the lock to put it in, and yet another fit the lock into the door. But that didn't mean the door always *locked*. Schurmeier's new division aimed to get everyone in sync.

It wouldn't be long before project manager Burke ran headfirst into a sticky net of government interference, which could almost be regarded as a tangible object or being. And the thing could shape-shift at will depending on the circumstances. Sometimes it wasn't a sticky net at all. If the hardware tests kept failing, government was like a playground bully—always in your face. And in the case of the Agena Coordination Board, it was a total black hole.

They'd convened on February 19, 1960, by order of NASA headquarters, because the Agena program—though further along in development than Vega ever got—continued to experience a smorgasbord of hardship. Obviously this was nothing new for high technology, but it was agonizingly vexing all the same. On paper, the board was supposed to organize, coordinate, and resolve any technical dilemma involving the Agena booster. In reality, it turned into a dumping ground for every trivial issue nobody knew how to fix. With alarming swiftness they all piled up on the board's chairman, who held no authority to solve these additional problems.

Compounding this was a managerial division of labor that defied rational thinking: Huntsville's Marshall Spaceflight Center supervised the procurement of Agenas while the air force conducted the actual ordering process. And the force wasn't even building the rockets; Lockheed and General Dynamics were doing it in the private sector under contract. That meant the Huntsville and air force offices were ultimately coresponsible for the delivery of equipment they weren't actually producing.

Major John E. Albert took point as the air force's Agena man in Inglewood, pledging support to Burke. Albert said he believed in Ranger's goals and promised to use his knowledge of air force ways to get things done. And with that, he started swimming upstream. Many over his head felt projects like Ranger only stood in the way of "real" air force work.

"And they were right!" Jim Burke exclaimed—many years later, after learning of the real and true big picture. Unfortunately, Burke only knew what he saw right then, which amounted to scads of trouble and confusion. In order, this is what seemed to occur:

Someone from JPL needs to reconcile a pressing Ranger design issue. The JPL person calls Major Albert with some questions about Agena. "Unofficial" or "tentative" responses are offered up by air force personnel.

"Official requests for and answers to NASA inquiries" are to be routed through the general NASA–air force traffic.

The JPL person calls Lockheed and tries to get an answer from *them*. Lockheed responds that the answer requires "need to know" clearance, which JPL does not possess.

The JPL person is referred by Lockheed back to the air force. Repeat until people stop asking.

Lather, rinse, repeat. For the love of mercy, what in the world was going on? The eventual truth must have almost been a relief. "At the time, the other Agena program was black," explained Jim Burke, after spending years unaware there even *was* another Agena program. Back then, under severe penalties of law, nobody would *ever* have mentioned the word "Corona." Shh! The word *itself* was confidential, leading to the use of "Discoverer" as a code name for a code name. Riding Agena's nose, a snoopy camera platform orbited at 160 miles. Fat rolls of 70 mm Eastman Kodak film wound through twin stereoscopic lenses capable of shooting exceptionally high-resolution pictures of pretty much anything its controllers desired. Ground resolution approached six feet. Corona could and eventually did image practically every Soviet military installation that was known, plus a few that weren't. The covert program received first-class, no-compromise, red-carpet treatment and was considered *far* more important than any dopey moon probe.

Of course, Jim Burke knew nothing of Corona during Ranger, but he did know the Agena situation had a bad smell about it. Burke was trying to design a spacecraft. It starts with a weight limit and overall shape. Both conditions are largely dependent on what booster is doing the lifting, and Ranger's original specs were all derived from Vega, which purported to lift half a ton. And Vega had been molded around a hexagonal shape, so Ranger's preliminary sketches all reflected a six-sided superstructure.

Now here came Agena to throw the doors wide open. It promised only seven to eight hundred pounds of payload capacity. Beyond that, nobody was 100 percent certain of the configuration anymore. Should Ranger be a cylinder, like an Explorer? A thick Pioneer pancake? Burke played with his toddler's shape-sorter game, turning the pieces over in his hands. *Sphere or square, ovoid, cone?* Any one of them could be made to fit a rocket; only one would fit *best*—provided you knew what it had to fit in. He couldn't even pick the shape. But JPL designers had already moved on to higher-level issues and choices, many of which took their cue from the hexagon. They knew it was perfect for almost anything. The decision finally came down to saving time and effort; hexagonal it would be.

Perfect for almost anything? Right from the beginning, designers realized the inherent value of planning for any one of three targets: Moon, Mars, or Venus. Sure, the experiments might be different, but what about the thing they're bolted to? That's when it hit them: go vanilla. So they planned on a common power supply, radio, and control thrusters—each in modular compartments to bolt on the chassis, with room left over for science. Ergo, they would have created a solid, generic probe you could take in virtually any direction. They called it the "Ranger bus," with the idea that everything riding on it comprised the passengers.

Spacecraft destined for the Moon and planets would each need a radio antenna dish. The signal from it didn't have to be as strong as a terrestrial radio station's, but the dish *did* always have to point toward home base and retain a minimum signal strength. During launch it could fold up underneath the bus. For the Moon trips, extra bandwidth in the radio signal could be used for television feeds. Building from there, nearly any flight beyond Earth orbit would also require course corrections along the way—little twists of the steering wheel. Both of these issues—communication and guidance—meant the craft would have to align itself in space across three axes. Accomplishing that promised to be supercomplicated but really the only way to go.

With all that decided, what about power? Voyages to Mars and Venus require hundreds of days. The best batteries last a matter of weeks. The only way to do it, then, would be with solar power. The scope of the experiments would provide baseline electrical requirements, which in turn would partly dictate the size of Ranger's central command-and-control module. And once *that* was known, the dimensions of the solar panels could be figured. They'd probably have to be on wings and folded up for launch.

The bus could only be so far across, as restricted by Agena. So to pro-

20. Jim Burke headlining a press conference, 1960. That's a Ranger model on the table. Burke's relaxed demeanor proves this was taken *before* any of his spacecraft flew. (Courtesy of NASA/JPL-Caltech)

vide breathing room for the more sensitive instruments, Ranger's framework sprouted a tower extending up from its middle—stretching high, filling out the remainder of space in the nose fairing. Hexagon with a cone on top; Ranger had a shape.

Next, there had to be something for it to carry—that was Homer Newell's department. He thus set out on a one-man publicity tour to hopefully scare up some action. At the academic conferences, Newell made presentations on the opportunities available. He wrote letters to universities filled with the promise of study in literally far-flung locations. He sipped coffee at a million meetings to talk about what NASA was up to. And when the time came to choose, Newell found himself staring down a mountain of proposals. If every one of them were green-lit, they'd be flying Rangers until he died.

Oran Nicks flew in to see how the prototype frame was shaping up, and something caught his attention. Rising up from the thick main hexagon, four tubes slowly drew together, ending underneath what was to be the low-gain antenna mount. Nicks slowly circled the mock-up, thoughts wandering back to those confounding days in engineering class. He'd endured dizzying rounds of structural design, materials, statics—and this thing in front of him didn't look right. Almost any kind of load on these tubes, angled like they were, would bend them like twist ties.

"Where are the diagonal members to react against torsion?" he wanted to know. One of the young technicians cocked his head and squinted. Nicks could picture gears turning in the guy's brain.

"Oh," the tech finally said, "there won't be any torsion loads."

Nicks nodded but didn't say anything right away. He sensed a little hostility; it was in the air. Plenty of JPL workers felt stung by their perception of NASA's "takeover" of the place, so getting off on the wrong foot was no good. Nicks diplomatically offered up some words of concern to the tech and moved on, pacing himself to keep with the day's bloated schedule.

Later on? "I think he blew his stack," was how Charles Sonett described Nicks's true reaction once safely off-Lab. "That was terrible, the way things were done."

New to NASA, Sonett had come on board with headquarters as the chief scientist of lunar and planetary programs. Before working at NASA, he'd designed experiments at Space Technology Labs for their Pioneer spacecraft. And Sonett's first impression of JPL was not overly positive. "JPL didn't know how to manage in the first place," he indicated, "and they had some outlandish ideas on how to do things." This faux pas with the Ranger prototype structure was only one example. "Abe Silverstein just about went off his bonker. I mean, that was a sign of just complete incompetence."

Around the country, Ranger management positions ballooned in unfettered expansion. Every involved entity named their own managers: Burke at JPL, Nicks for NASA, Albert in the air force. Any time confusion arose as to responsibility, or oversight, the solution was always a heavy-hammer approach—put more people on it. Lockheed added someone. NASA launch operations put a man on. Huntsville now had at least four, including their own rep at Lockheed. Burke didn't know who half these people were. The entire web of departments, people, and responsibilities was nigh on impossible to visualize; so on paper Cliff Cummings diagrammed out what he understood the organizational flow to be. When finished, he looked down on a virtually incomprehensible arrangement of boxes and dotted lines connecting offices, departments, and individuals scattered across at least three time zones.

An Agena mock-up showed up at the beginning of September with a hexagonal spacecraft adapter tacked inside. As soon as the wrappers were off, JPL inspectors rolled their eyes and sent it back to Lockheed because the workmanship was half-assed. Lockheed failed to mention Corona as their source of distraction. Numerous other design problems rose up to bite. Soon Burke was logging sixty-hour weeks, living off coffee and the Lab's vending machines after the Sun went down. He flew to Lockheed some weeks later to review the exact status of the Agena-Ranger separation mechanism. Nobody knew what the status was. Back at JPL, prototype Rangers were failing their vibration tests. Wonder of wonders, severe torsion was buckling the cone-shaped frame. How could *that* be happening? Diagonal braces would have to be created, countering the invisible force but also adding weight.

Oran Nicks learned of the torsion problem and shook his head. As Nicks later wrote, "There was so much high technology associated with the conduct of a space mission that JPL project officials didn't spend time worrying about freshman-level design problems."

"Those people were living in a dream world," Sonnet suggested of JPL. "That, in fact, was part of the problem."

Jim Burke sat on the phone with Pickering that day in December 1962, wondering when he'd hear the actual decree and what form it would take. He knew of plenty of folks in Washington DC who didn't like the dustups that happened when Burke sat in on meetings, who didn't like when he said things like "that's outside of the project scope."

"That was why he was looked at by those guys as the problem," recalled Bud Schurmeier. "Because he wouldn't do what they wanted."

But did any one individual in particular call for Burke's head? Who might that person have been?

"If someone called for my replacement as Ranger project manager, the name is unknown to me," Burke explained forty years on.

Pickering's voice came back into the handset, finally telling him that the NASA boys were done looking things over. The review report, of which everyone at JPL waited in crashing fear, had reached Pickering's desk. And it stooped to some pretty low descriptions of the Ranger program. Things like "shoot and hope."

Was that what they actually said?

The JPL director kept talking: "This doesn't necessarily mean firing people. We'll just put you on something else in the Lab."

And that's when Jim Burke knew NASA always got what they wanted. And today, they wanted him out of the captain's seat.

The first couple missions had already been laid out. Both would test the waters. No lunar impact was planned, nor even a close flyby—just straight up into Earth parking orbit, then restart Agena to shake down the three-axis stabilization. Each craft would end up in eternal solar orbit. Accordingly, Newell chose to include experiments suited for broad-spectrum measurements between Earth and the Sun. They called this first set of missions Block 1. They would act as a proof of the overall concept.

While both Block 1s were presumably flying, an upgraded Block 2 would be in the metal-cutting stage back at JPL. On these, the midcourse-control engine and lunar experiments would be added in. Block 2 was to look very different as compared to its predecessor, owing to a humongous silvery-metal basketball perched on the very end of the tower. Despite appearances, this orb was the main experiment: a lunar rough-landing capsule intended to hit hard but alive, firing up a battery-powered seismometer to record moonquakes.

This ambitious undertaking was foolishly welcomed by an appendage of the Ford Motor Company known as Aeronutronic. Supposedly the division had been conceived as a way for the car manufacturer to evolve from V-8 Thunderbirds into all this up-and-coming space hoopla. In February of 1960 Aeronutronic received $3.6 million and nineteen months to produce a trio of landing balls, each of which would need to be kept within a three hundred–pound weight limit. To head this up, Aeronutronic selected Frank Denison. The guy knew plenty of Ranger folks over at JPL because he'd worked there for a while. Coincidentally, he'd also been Jim Burke's first boss, years before in the Lab's cro-magnon days of experimental missilery.

Aeronutronic faced a daunting measure of untilled ground. Ranger's bus would have to contain a special altimeter for triggering capsule separation, along with purpose-built electronics to link them together. The '61 Galaxie Starliner had options but nothing like these. The capsule itself was supposed

21. Pretty, it ain't: JPL's first Ranger begins taking shape. The lackadaisical, garage-like surroundings would quickly become an issue. (Courtesy of NASA/JPL-Caltech)

to eject with Ranger still a few miles off the surface and then rely on a bantam retro-rocket to slow it down. The best place for such a cork-popper was definitely up on top of the tower. But that would be tricky placement, because the more sensitive experiments and low-gain antenna had already commandeered this prime spot. Yet no place else made any sense.

The basic capsule plan was to encase one sphere inside another, filling the in-between space with water. When Denison's orb hit pay dirt, the four-inch outer shell would probably fracture, giving its life to the survival of the inner parts. Aeronutronic got their hands on an aircraft and flew out over the Mojave Desert, where engineers threw mock-ups out to see what kind of shape they landed in. The first few rounds didn't go so well—every one of them shattered like china plates. Nestled at the core sat batteries, antenna, and seismometer proper, but after impact, the rig tended to not switch on. It'd been turned into hamburger.

Denison reviewed the mission parameters: near the very end, their ball would be screaming toward lunar destiny at 136 miles an hour as it slammed home with three thousand g-forces. If they could just get it to take that single wallop, even one time, then all the experiment had to do was send back usable data for perhaps two weeks. The team tried endless combinations of materials for the thick outer ball, and not a one of them could take the pain. Somebody suggested balsa wood.

Forever popular with model builders and fishermen, the lightweight stuff—actually categorized as a deciduous hardwood—ended up being as good of a match for the job as anyone could find. But it wasn't perfect. Final qualification tests were done in the LA area with Aeronutronic engineers strapping their balsa creations onto a rocket sled. Once up to speed, the sled would instantly decelerate and impart three thousand g-forces on the equipment. Things still weren't coming out so great.

Chuck Sonett was in his lilliputian office when the results came in. "Oran showed me the report," he began, "and it was ludicrous!" He burst into laughter and, calming himself, explained, "I get a laugh when I think of it! They said that they fired the first one and it didn't work, and they fired the second one and *it* didn't work. And I don't remember what the problems were, but the conclusion was, 'Well, now we're done with those tests; let's get on with building the thing.'" Sarcasm crescendoed in Sonett's voice.

"I'm not kidding, and the attitude in our office was, 'The hell with you guys. You're doing this all over again and gonna do it right!' No . . . they were just going to go ahead and keep building! Oh, it was impossible!" Sonett was aghast at the incompetence. "Anybody running a program that would put out a test report like that, and then the conclusion should be that 'we keep going,' is some kind of an idiot!"

Sonett wondered aloud to Nicks about the wisdom of continuing the project as currently defined. "Here's this plan to put the seismometer on the Moon," he clucked, "and they couldn't do it!"

Cummings and Burke now figured the time had come to remind all involved parties of the main objectives as JPL saw them. Cummings did up a memo prioritizing four basic concepts and disseminated it out to the combat lines. It was meant as policy. *First*, get the spacecraft working properly. *Second*, keep the launchings on schedule. *Third*, firm up the procedures for joint NASA-JPL interplanetary flights. Fourth and last on the list was science. No matter what Homer Newell or Bob Jastrow thought, Cummings and Burke stuck to their guns. It didn't make any sense to expect jack from all these mashed-in experiments until they had the spacecraft working right. Even Abe Silverstein backed them up.

There was, however, a surprising amount of interference that came from left field. On April 13, a bunch of nondescript individuals waltzed into Pasadena from Los Alamos Lab and the Sandia Corporation. Collectively they represented no less than the interests of the Atomic Energy Commission (AEC) and were dutifully intrigued by the tour Jim Burke gave them of the spacecraft assembly areas. They saw *Ranger 1* coming together. They heard about the science experiments. A Los Alamos representative then explained the group's cultivation of their own little experiment—specifically, a device to recognize aboveground nuclear explosions. "I was interested," Burke remembered, although probably unaware of what was coming.

"It's called 'Vela Hotel,'" one of the men quietly explained. Top secret. And their most passionate of desires was to place one aboard a Ranger craft and fly it in space.

Burke dug a toe into the floor, diplomatically explaining how any and all science packages had already been approved by their customer—NASA—and the design was about to freeze based on what they had. Sorry.

The way Burke saw it, Ranger was maxed out. Already, it lacked space in the trunk for redundancy. The guys were able to squeeze in a backup radio transmitter with its own batteries, but that was about all that would fit. So to even think of adding more scientific instruments? Every experiment affected Ranger engineering in some fashion or other. They weighed something, took up space, consumed power, got hot, and lined up to use the radio. They changed Ranger's center of gravity and therefore its onboard control systems. So JPL should tamper with an already fragile ship, teetering on the bleeding edge of technology?

The visitors repeated their request. Burke squared his feet and politely told them, "No additional experiments could be accommodated in the program as now planned."

The response in the room was, of course, frustration. *Didn't Burke care about the Russkies and their nukes?* "I cared a lot," Burke asserted, "but about their *Lunas.*" Yet the comment struck directly at a patriotic motivation that had helped push him through all the turmoil. Burke rolled it around in his head for a minute and in consolation finally mentioned that all of them were "welcome to examine my results with our instruments," but the nonanswer failed to please. A little while later the Vela Hotel group left and went right over Burke's head.

Like mosquitoes, telegrams and letters and memos flew about at the upper echelons of NASA, ARPA, and any brigadier general who might have something to say about it. The situation backed up to Homer Newell and Abe Silverstein, who interestingly enough didn't say no. They didn't say yes either. JPL should ask to see a proposal, they explained, similar to one that every other scientific investigator in the program had submitted. JPL would give them due process. That was kind of like building a low wall— just enough of an obstruction to throw Vela Hotel off balance. The paper-work might kill it, because nobody would want to go through all the has-sles; then Silverstein wouldn't be in any trouble for saying no. That night, Burke motored home on the Puch, nudging the throttle, ground blurring into abstract greens and browns. He and Lin lived on the edge of the can-yon near JPL, and Burke usually rode a trail in.

"I don't recall thinking about work while riding," he claimed, "or think-ing about riding while at work. Maybe that's why I never got a case of as-phalt rash. And all of us riders always wore helmets. This modern protest against them is silly."

A horsewoman appeared and looked down at him. Respectfully, Burke halted the cycle and killed it.

"Do you do this regularly?" she asked.

Burke said, "Yeah. For many years."

She told him, "Well this is your last trip."

George Hobby called up one April day and told Jim Burke how his delicate lunar machine was going to get baked in an oven for twenty-four hours at 257 degrees Fahrenheit. Burke gritted his teeth but already knew this was on the way. Cummings knew it too and had drafted up a list of Ranger parts he wanted exempt because they literally couldn't take the heat.

Hobby was a biologist over in the JPL Space Sciences Division, tasked with supervising this unorthodox procedure. Sterilizing the spacecraft? Burke didn't think so highly of it—his list of priorities ranked sterility way down in fourteenth position. Number one? *Reliability.*

The underlying issue, however, remained: what happens if a spacecraft lands on a world that has its own biological species and we contaminate it with Earthly microbes? The topic had been lingering in the shadows all the way back to Vanguard. Back then, a symposium had been held, and the resulting discussion and debate spilled out into the rounds of the National Academy of Sciences. The proceedings coalesced into a report strongly recommending that all Earth nations adhere to methods of sterilizing *any* spacecraft being sent to contact another celestial body in any way. And at the time, nobody knew the Moon was just a big gray ball of dust. So by September 1959, after Ranger's inception but prior to its commencement, a letter flew across the desk of a NASA administrator named Keith Glennan that advised him to literally clean up his probes. Not being landers, the first two Ranger machines were exempt. But this lone requirement would almost single-handedly ruin the program.

The Vela Hotel proposal came back two weeks later and was surprisingly deferential. Ranger's apparent level of design maturity rated highly enough, the pitch began, to preclude the idea of completely reworking it. So Los Alamos proposed adding its own computing brain and power supply for the sensors. This way, Ranger programmers and engineers could avoid the frazzling exercise of trying to shoehorn one more slice of pie into the bus. The experiment itself wasn't much—some tiny radiation detectors and lead shielding, no more than twelve pounds at most. Newell and Silverstein looked it over and had trouble finding any objections. Then it went out to Cliff Cummings for a once-over. He spent his time with the proposal and concluded that a seat for Vela Hotel could in fact be made available on *Rangers 1* and *2*. It was a stunning admission that said much about his character. If the numbers told Cummings that Vela Hotel was smartly designed and could fly, then the man's conscience would never permit his reporting otherwise.

Without protest he forwarded his conclusions on to Burke, who, at the beginning of June, delicately advised the troops by memo that "it is probable we will be requested to incorporate the experiment." *Let's plan for it now,* he basically explained. Burke went on to suggest that strategizing "will ease the pain of incorporating it considerably." Vela Hotel was in.

Then Lockheed managed to ruin Jim Burke's perfectly fine summer day on July 11, 1960, by announcing that Agena . . . um . . . wasn't able to carry as much as they originally thought. The lifting estimates were off by about seventy-five pounds—not a whole lot really, about the heft of a nine-year-old. But it would force Ranger onto a crash diet.

A problem like that soaked deeply into the cracks. It affected the parking orbit, which in turn affected the trajectory. These mind-bending formulas had already been worked out. Now JPL would be starting a whole new round of computer runs on the 7090. The launch dates were already on the calendar for July and October of next year. Actually, Burke thought to himself, the change might not affect the two shakedown flights because they weren't tipping the scales yet anyway. But flight three, oh boy. That baby needed all eight hundred pounds they'd been promised. Now seventy-five of them were getting taken away? It would be the worst seventy-five pounds anyone at JPL would have to lose for a long time.

Burke picked up his phone and told Bud Schurmeier of the news. Although struggling to crank the Systems Division into gear, Schurmeier vowed any possible assistance he could offer, which partly involved rehashing the day's issues after work at a local hangout. "It was called the Blue Fox," Schurmeier explained of the place accessible by JPL's back gate, on the Altadena side. "Where you'd go for a beer or two after leaving work if you didn't have something else you had to do. And so there were always a lot of JPL guys there. And girls." Farther down Lake Street stood the topless bar others enjoyed visiting sometimes, but Schurmeier and Burke preferred the ambiance and proximity of the Blue Fox in which to air their woes. How to get the fat off Ranger? It was already a skeleton.

Out went the aluminum hexagon and fittings, replaced by weight-thrift magnesium. The gauge of the electrical wiring thinned. Cliff Cummings directed machinists to drill holes in the electronics boxes wherever possible. The scale groaned anyway. Then Frank Denison called: their landing capsule was headed over its own three hundred–pound limit. The AEC called back because now the Vela Hotel group was all hot and bothered to fly on *Rangers 3, 4,* and *5,* too.

Burke tried to smile. He was losing.

"He was fighting the battle of weight and performance," Schurmeier remembered, watching from the sidelines as his friend sprinted from one emergency to the next. "That was a continual battle Jim fought the whole time."

Ranger project manager James D. Burke finally put up his hands and called a time-out. The Los Alamos guys were wearing out their welcome at an incomprehensible speed. They wanted *more* missions? And the air force with all their damned security and privacy. This was a team effort here, right? Did it have to be a fight every time? And this problem of Ranger being too heavy—just how reliable were these weight figures, anyway? Usable answers weren't available for six taut months, torpedoing the *Ranger 1* launch period.

Burke didn't like it one bit—especially when launch schedules slipped. The big picture went far beyond Ranger, to Mars and Venus. He clearly viewed Ranger as a precursor to planetary missions, whose launch dates were irrevocably set by celestial mechanics. "We *had* to learn how to carry out a complex development leading to an immovable launch window," he insisted.

Cummings generated a letter to NASA about Vela Hotel: couldn't those guys just be told there was no room at the inn? Down in Los Alamos the AEC got the hint and quietly pulled out. Vela Hotel would not ride on Block 2.

The year 1960 had been spent in meetings. Now winter was almost half over. Competition struck: JPL employees arrived at work on February 12 and gloomily took note of how an unnamed group of individuals in the Soviet Union had launched their *Venera* toward Venus. Four days later an uptight Burke cranked out a memo for Bud Schurmeier in reference to their ongoing weight-control problem. Should he have his guys down on the shop floor building parts when nobody was sure how much they could even weigh at the end?

"He probably said a lot of swear words at times," Schurmeier put it. "He could never tie down Lockheed for the launch vehicle performance, so he was kind of in the middle of two big uncertainties."

Time for another memo from the project manager. "We must begin removing items from the spacecraft," Burke prefaced the strongly worded document titled RA-3 *Weight Situation*. "RA" stood for "Ranger-Atlas." Burke listed a tough decision: drop the backup radio transmitter and the battery that would run it. Since this would still not be enough, "Review the entire instrumentation schedule," his instructions continued, "and remove a portion of the equipment so as to save weight. At the expense of creating a higher-risk situation." The last sentence was as painful to write as it was to decide on, but the crummy state of affairs had forced him to draw hard lines. Every science instrument stayed in. The design froze.

All through spring of 1961, they continued rebuilding. New magnesium tubes underwent skillful manipulation into a perfect hexagon. From engineering drawings came wiring schematics, which translated into actual wiring bundles that began to fill the specially constructed channels and raceways of the Ranger bus. The electronics boxes were populated with circuit boards. The telemetry system came together. The electronic brain, the experiments, the radio—they'd all been poked, shaken, banged on, quaffed in a vacuum chamber. There was a certain feeling in the air, a positive one—that it was *real* this time, that what they were building was actually going to fly.

A few months downstream, Lockheed's revised performance figures came through and, in keeping with Burke's lurking suspicion, reflected an improved boost capacity. They'd get a whopping 164 pounds of additional payload weight. And Cummings had been making swiss cheese out of the hardware boxes.

Deeply interested in the sanity of himself and his crew, Burke was not about to unfreeze the design again—no matter how much more weight they now had to build with. That would have sent Ranger into a dead man's spiral of changes, revisions, blueprints, tests, scheduling adjustments, launching delays. There would have to be more trajectory runs. Those 164 pounds? "Unexploitable at this late date," Burke explained in another terse memo.

Although the complement of onboard experiments remained excellent, the ship now lacked any redundancy whatsoever. But it was all Burke could do to stay on schedule. They had to launch *something*. As Bud Schurmeier clarified, "You did your design and development, you know on paper," he said, "and in addition you had a flight test program. You carried those along in parallel because some of the answers you didn't get until you flew."

By May, Oran Nicks had finished examining *Ranger 1*'s build history and finally, almost miraculously, proclaimed it ready for transport. A month later, the craft survived its womblike trip to Florida inside a specially constructed, air-conditioned, air-*suspended* van. The route was laid out to bypass hazards like low bridges and sporting events. There at the cape, the ship was mated with its nose fairing, which in turn linked to Agena with Atlas below. This was known as the stack. And at the end of June, the *Ranger 1* stack waited out there on the pad.

During the fourth launch attempt, *Ranger 1*'s countdown finally reached the point of a routine test. The task at hand involved a final calibration of the science experiments, and that meant applying power to the bus. For such activities, a small test port had been designed into the Agena's side. The size of a doggie door, it unlatched to reveal a bunch of electronic couplers. This let ground workers plug a Ranger into an extension cord and not drain its auxiliary battery.

No sooner had the console switch been flipped than blockhouse data screens began lighting up. *Ranger 1* had come alive. Somewhere deep inside, an electrical malfunction had made the spacecraft think it was already in orbit. *Ranger 1* fired explosive bolts to separate from the Agena and commanded all the experiments on. The twin solar panels unlatched and tried to unfold, banging into the sides of the Agena shroud. People leaped out of their chairs, but really there was nothing to race for, since the ship had already finished activating itself and waited for the next exciting thing to happen.

The entire nose would have to be opened up. *Ranger 1* came off the top, enjoying a short ride back into the hangar for a bit of going-over and reworking. They pried the covers off the hexagon and found some parts inside that had fried themselves.

Oran Nicks tried a positive spin. "At least it was not in the ocean," he pointed out.

The guys crawling around inside *Ranger 1* were thinking that one of the science packages was responsible for the odd, high-voltage discharge that set their craft in motion like it did. This diagnosis wasn't 100 percent certain, yet the news made Burke growl. *A science package?* At its very core, *Ranger 1* was supposed to be an *engineering test*. The experiments themselves were along only because of space. He'd been bitten by his own decision. "We could not tell in advance which flight or flights might succeed," Burke explained of their inclusion. "And engineering development of the instruments was as important as that of any other subsystem." The dissonance pulsed through his cerebrum. He knew that including the science packages made good sense—they *should* have been on. However, Burke didn't really *want* them on. "Jim always called 'em," Schurmeier recalled with a bit of levity, "and that probably frosted 'em off, they wanted to add these scientific 'trinkets.'"

The next time they tried was on August 23, and the team watched as *Ranger 1* got stuck in Earth orbit because the Agena failed to restart. Burke turned away from the data terminal and clenched his fists. They weren't even asking for the Moon, just a chance to exercise Ranger's design out to perhaps half a million miles. *Was that just too much to ask?* Oran Nicks later said he spent the whole flight back to DC just staring out the window.

Then Burke heard about a fumble on the play: two tracking ships, one in San Juan Harbor and another out by the Ascension Island station, had quite literally battled for control of *Ranger*'s signal during ascent. In the heat of the melee, they irretrievably lost minutes of crucial telemetry. It was like losing the black box from a crashed airliner.

Jim Burke didn't get canned because of *Ranger 1*'s failure. Like *Explorer 2* and *Pioneer 1* and many others by this point, the probe itself had worked to a tee.

He didn't get canned because of *Ranger 2*'s failure, either. It showed up at the cape while Burke and Nicks and the others were sorting out what could be done with the in-orbit *Ranger 1*. Even though multiple launches were planned, Nicks impressed upon his JPL cohorts the importance of each try. *Pretend it's the only one we have*, he told them. "It got into some shouting matches, you know," remembered Bud Schurmeier, "about things that Oran had stated: 'THIS IS THE WAY IT'S GOING TO BE!'" A real finger-poking-the-chest bawling out.

Schurmeier paused briefly. There was something else he wished to say about Nicks: "But he can be pretty dictatorial." And of the inflamed tension between Nicks and Burke, Schurmeier offered, "They just didn't mesh together very well."

Ranger 2 finally went up on November 18, meeting the same fate as its cousin—Agena restart failure. Burke rubbed his forehead and knew it would end just like the last one. Most of the data coming back from any experiment would be useless. For example, the particle detectors—courtesy Van Allen—quickly got flash fried from too much sunlight. An exhausted Jim

Burke flew back to LA on the nineteenth. The twin failures cost 60 million bucks.

While Burke and Cummings had been sweating out Ranger's weight and configuration in the summer of '61, a series of events transpired in the White House that would alter JPL's fate in ways nobody could have predicted. Notwithstanding fizzlers like *Venera* or the IM attempts, President John F. Kennedy had been contemplating the Soviet Union's space achievements even more than Jim Burke. And he had the willies.

Kennedy's trepidation ballooned when they put the man in the can. Young Soviet fighter pilot Yuri Gagarin had actually *ridden* this monstrosity of Soviet rocketry, orbiting once on April 12 before a safe return. The news hit Capitol Hill and went off like a grenade. What next, Khrushchev's Red Army descending from space onto the Pentagon? Kennedy was likely unaware of the R-7's abysmal success rate or of how Gagarin's manual controls were actually sealed with a combination lock. Nevertheless, on April 20 Kennedy dashed off a short memo to Vice President Lyndon Johnson that would become the literal genesis of America's space exploits for the remainder of the decade.

It asked, in part, "Do we have a chance of beating the Soviets by putting a laboratory in space, or by a trip around the moon, or by a rocket to land on the moon, or by a rocket to go to the moon and back with a man. Is there any other space program which promises dramatic results in which we could win?"

A threadbare document, it would have been less than half a page without such huge left and right margins. Kennedy's stark memo is brazenly shallow in scope and casts blinding light on what the president didn't know about his own space program. The word "science" is completely absent from this document. No thirst exists to advance the world's knowledge base. There was not even some far-fetched, half-baked eloquent desire to "explore." Some of the twentieth century's greatest engineering feats began with his crudely juvenile concept.

Johnson got back to Kennedy eight days later. The laconic reply advanced such concepts as "the head start of the Soviets" and contained other illuminating statements like, "If we do not make the strong effort now, the time will soon be reached when the margin of control over space and over men's

22. Van Allen shows his Russian visitors around the Iowa physics department, 1959. (Courtesy of the University of Iowa Department of Physics and Astronomy, Van Allen Collection)

minds through space accomplishments will have swung so far on the Russian side that we will not be able to catch up, let alone assume leadership."

Johnson added this of the tea-sipping Reds: "They also have the booster capability of making a soft landing on the moon with a payload of instruments, although we do not know how much preparation they have made for such a project. . . . We cannot expect the Russians to transfer the benefits of their experiences or the advantages of their capabilities to us."

The validity of this last statement depends on who is being asked. After graduating Iowa, Larry Cahill went on to a physics professorship at the University of New Hampshire and sponsored a visiting Soviet scientist. "He was surprised that I let him read all of our papers and work unsupervised in my lab," Cahill said. "When he left after a few months, he had a suitcase full of Xerox copies of documents from our lab and our physics library!"

It didn't end there. Later, Van Allen journeyed to "enemy" territory in July 1959 for an international physics conference. "They were very cordial," Van Allen later recalled of the Soviet scientists. "I thought they were almost eager to the point of not being gracious, in the extent to which they hogged the program and described what they were doing."

Such hospitality by foreign competitors was shortly returned. Van Allen hosted Leonid Sedov and Anatoli Blagonravov, who stopped over in Iowa City during their first visit to America.

Sedov's English was fair enough; so at Van Allen's invitation, he gave a talk in MacBride Auditorium, one of the university's gargantuan lecture halls. Lots of theoretical discourse poured out on the Iowa campus that day—talk of rocket technology, flights to the Moon, controlled unmanned landings, and sample returns. Later on Sedov delivered a small-group lecture in the physics department. He dished out virtually everything that had been learned from *Object D* along with important findings from other Soviet missions.

The Iowa visit went on for a couple of days. The group had a grand old time together. While the rest of the nation bedeviled themselves ad nauseam with such abstractions as Communist infiltration, these Soviet guests were treated like royalty in no less a place than the heartland of America—with an elaborate dinner at the Van Allen home, selflessly crafted by Abigail. The Van Allens told their kids, "Hey kids, these men are from the Soviet Union," and none of them worried about it. "They were very kind, very nice people," recalled Van Allen.

Wernher von Braun had also been on Kennedy's CC list for the April 20 memo, and he responded one day after Johnson did: "It becomes readily apparent that the Soviet carrier rocket should be capable of . . . soft-landing a substantial payload on the moon." Von Braun estimated that whatever the Reds might try to land, there was no way it could be much above 1,400 pounds. "But it is entirely adequate for a powerful radio transmitter," von Braun continued, "which would relay lunar data back to earth and which would be *abandoned* on the lunar surface after completion of this mission. A similar mission is planned for our 'Ranger' project." Von Braun figured that the United States held a "sporting chance," as he put it, of alighting some sort of Ranger-delivered unmanned research station first. Considering the unprecedented Luna efforts, his opinion was that the Soviets could well accomplish an unmanned rough-landing "at any time." Von Braun also raised the notion of placing real live people on the lunar surface—but not before dispatching simple probes ahead of time "to determine the environmental conditions man will find there."

All this rhetoric, kneaded together and formed like bread dough, was enough for the president to go on. That May 25, in a speech to a joint session of Congress entitled "Special Message to the Congress on Urgent National Needs," the president discussed a variety of topics including economic progress and national self-defense. But in a major portion of the speech, John Kennedy had something to say about America's response to the Soviet Union. He stuck his neck out in the direction that had been indicated to him: "I believe that this nation should commit itself to achieving the goal, before this decade is out, of landing a man on the moon and returning him safely to the earth. No single space project in this period will be more impressive to mankind or more important for the long-range exploration of space, and none will be so difficult or expensive to accomplish."

Most reporting of this event cuts it off right there. But Kennedy wasn't finished just yet; he still had a little more to read off—words and challenges that would backstroke through JPL and tumble right into Jim Burke's corner pocket: "We propose additional funds for other engine development and for unmanned explorations—explorations which are particularly important for one purpose which this nation will never overlook: the survival of the man who first makes this daring flight."

Conceptually, Kennedy had laid out a very simple problem on the blackboard. But launching people in the absence of hard environmental data ranked less than advisable. He—*America*, really—was going to need an army of scouts: riderless carriages to go first, to poke and measure and scoop and take a picture or two and, in sum, find out whether or not a man-carrying lander could even survive touchdown.

On the very day of Kennedy's speech, most every existing unmanned space program was deposed, falling subservient to a new master—a manned lunar landing.

12. Get Off the Bus

I'm not sure that a lot of the fault wasn't Jim Burke's.

Charles Sonett, on the Ranger debacle

The American Moon program was called Apollo. Today, it remains such an extraordinary mobilization of resources, money, manpower, and engineering brilliance that it is widely regarded as another wonder of the world.

Charles Sonett didn't find much value in Kennedy's plan. "They really weren't interested in science; they were interested in manned flight," he said of the forces driving Apollo. And of the people heading it up Sonnet offered, "They were real dizzy . . . they had no idea what to really do, because they had no sense of science. Not one of 'em knew anything about the Moon. I'm not sure they knew how to spell 'moon.'"

In many ways, this conspicuous imbalance between manned and unmanned flight came down to glamour. Bathtub faucets had more allure than a space probe. People loved astronauts. *Explorer I* never signed a single autograph. *Life* magazine wouldn't be doing a write-up on what *Ranger 3* did with free time on the weekends. Forget about it. Alan Shepard's Mercury capsule was going on display in a museum, but none of the probes were coming back. Humans need closure. How can you be a fan of something that doesn't even come back?

Not coincidentally, NASA's new master plan of launches came off the press the exact same day that Kennedy gave his speech. The programs in support of Apollo—understand this wasn't Apollo itself, just the supporting programs—were already laid out in great detail. This unintentionally shone bright lights on the extensive backstage goings-on *before* the president ever spoke. Ranger and its infant cousin Surveyor made the list, but their original reasons for being had been erased with new wording inserted. Both were now identified as serving in direct support of the manned Apollo flights, and to that end, Ranger would move forward by omitting most of the science packages. What would fly instead? Television. Aeronutronic's balsawood impact spheres would stay in the queue, to then be followed by this new stuff. Abe Silverstein tacked four televised Ranger flights onto the roster, upping the total to nine.

Big changes indeed. As best he could, NASA deputy administrator Hugh Dryden explained them to the Senate on June 8: "We want to know something about the character of the surface on which the landing is to be made," he articulated, "and obtain just as much information as we can before man actually gets there." So in very short order, Apollo required close-up, high-resolution images of the Moon's surface. Funding approval sailed on through in a matter of weeks.

The next day after Dryden's chat with the Senate, formal directives arrived on Pickering's desk from Oran Nicks, detailing the abrupt shift in direction. Jim Burke looked over the changes and figured Vela Hotel's intrusions had been chicken feed in comparison. The quintuplet of additional flights, in Nicks's words, would "afford a better opportunity for Ranger project success, and make a corresponding contribution to national prestige during the early phases of our lunar program."

Slaves to Apollo.

Shoot detailed pictures. Identify surface conditions. Whatever they barbecued up for actually *doing* that had to fit somewhere on the machine before them. Where to slap a television camera? Maybe on the high tower? With Block 2 the tower was being largely repurposed as a throne for the seismometer ball. Put the camera up there for the new Block 3?

What Ranger almost needed was a near-complete rework. Yet Burke reminded everyone that the schedule, oddly enough, hadn't changed. It hovered over them like storm clouds, and in no way could they afford to chuck major design elements out the window. The only changes Burke agreed to let through would be the ones directly involving the cameras.

Over the course of subtractions and additions Ranger gained fifty pounds, mostly for redundant components. Backup hardware improved Burke's quality of sleep, but his entrenched Agena concerns still led him to conclude they'd only have one good flight in Block 2—and probably the same in

Block 3 with the video setup. He was killing himself in pursuit of a 15 per-cent success rate.

This germ-killing process could easily make that worse. Already built, *Ranger 3* was going through its integration tests and failing most of them. It had just surfaced from a battery of inhumane environmental extremes designed to remove as many earthborn contaminants as possible. Ah, the sterility phase.

In the cavernous assembly areas, JPL embraced a new and impossibly high standard of cleanliness: rooms, floors, transport dollies, work surfaces—they all had to be perfectly clean. Even the ceiling hoists were sterilized.

What the facilities got was nothing compared to the poor ship. Beginning down on the level of subassemblies, jumbles of parts were baked until ster-ile. Not just clean, but sterile. Any component surviving this regimen would next be quarantined in a separate, germ-free test area. Here the pieces finally began to connect. Before any two came in contact they received a swabbing with alcohol. Afterward, these individually baked parts got a *second* baking. And their sanitary condition was expected to persevere through component assembly on up, until the whole 725-pound spacecraft was all snapped to-gether and unpolluted as the fresh mountain snow. Ranger's virginal pu-rity was even supposed to hold while road-tripping to the other end of the continent in that special van.

No chance for respite after making it to the Cape; before launch, the craft received a final toxic soaking in ethylene oxide gas. Realistically, it would never be totally pure—a statistical impossibility. Some JPL estimates quantified the sterility of the final ready-to-fly product at roughly 90 per-cent, tops.

The heating guidelines said 257 degrees. At that temperature, wiring turned brittle, solder joints cracked and separated, and capacitors leaked. So let's get this straight . . . a television camera was supposed to keep work-ing after all that? Aeronutronic's landing ball was miraculously now able to dash its brains out on the Mojave rocks and still have the instruments come to life. But the story changed abruptly after it came out of the oven—total bonk on the retest. They'd swap out wires and circuit parts and try it all over again in a never-ending Möbius loop.

Ranger 3's engineering model hadn't encountered many showstopping problems, but failure after failure kept piling up on the flight article. With

23. "Cliff was never hopelessly confused about anything," as Jim Burke put it. Here, Cummings does a stand-up for NBC Television, 1962. Behind him sits a Block 2 Ranger with landing ball on the top. (Courtesy of NASA/JPL-Caltech)

interest Burke noted how the only difference between the two was that pre-scribed oven roasting. And half a length behind, *Ranger 4* was slowly wend-ing its way through the pipeline. After it emerged from heat sterilization, the assemblers looked over the framework and noticed how it was warped. Immediately Burke and Cummings petitioned NASA for a whole slew of new waivers for various Ranger parts. They wanted the batteries, the retro-rocket, all separation explosives, and a multitude of electronics to bypass the heating shtick or all bets were off.

Burke sipped coffee, pushing through the week's tribulations. He ap-proached things in much the same way as Cummings: "One of his meth-ods is a lifetime habit of mine," Burke said. "Namely, the use of three-by-five index cards to sort, assign, and deal with each day's tasks." When all the cards were gone, that must mean the program was over, but lately Burke had been constantly asking one of the secretaries to please go fetch him an-other pack.

After awhile, Cummings himself walked up to Burke and needed to talk. He said, *Look, we don't have any time to put together a bid spec for the TV and do the dance of competitive bidding. That'll take forever. Let's just pick one good company and go with them.* Burke rested his java mug and gave Cliff both ears. Cummings liked RCA, who had even done a bunch of work already on a Ranger television setup back when it was going to be just another piddly experiment on the roster. That year, the company had triumphantly introduced the world's first completely transistorized video camera. So if anybody could make a space camera, it was RCA. Burke liked the plan. Cummings then flogged his single-source idea to NASA man Ed Cortright, who helped keep pace by immediately signing off.

Assertively waltzing in on July 5, the RCA folks methodically laid out their proposition. This would not be some *Luna 3* rip-off that shot film and mechanically developed it on board. RCA said they could blow Luna's fuzzy-sponge pictures clean out of the water. For Ranger a vidicon-tube arrangement seemed to make the most sense; it would be pretty similar to the equipment at above-average television stations. But RCA wanted to take even that to the next level. They recommended having several cameras on board—at least three—innovatively operating in a round-robin machine-gun series. The first one snaps a picture, and while its data zips through Ranger's electronics to the high-gain antenna, the next camera has fired off already, which sends, then followed again by the next. They'd overlap like that and get total coverage of the impact site. RCA finalized the design in September and settled on a battery of six cameras operating sequentially.

The next issue was something of a gray area: how detailed must the pictures be? What do they have to show? A sensible baseline reference seemed to be the resolution possible with ground-based telescopes. That should be the jumping-off point—no sense in starting below it. On the best days, with the best telescopes, lunar details could be resolved down to about a thousand feet. So to justify Ranger cameras, everybody agreed that they had to show features down to at least a one hundred–foot resolution.

Was that still good enough? The first manned Apollo lander could easily wipe out on a ninety-eight-foot boulder. The RCA guys finally said, *Well, if everything else works right, our cameras will detail the surface down to about six feet.*

Tom Kelley warmly welcomed the news. Kelley was at Grumman Aircraft Corporation out in Long Island. They were creating the actual ship for men to land in, and along with a million other details, Grumman needed to settle on a design for the landing gear. By this time, Ranger was already supposed to have provided them with enough data to get a handle on the surface conditions. But Grumman had zilch; so for the time being, they worked off of best estimates. If and when Ranger finally worked and if its cameras really could see down to six feet, Grumman could then design with confidence.

Freshly installed NASA administrator Jim Webb, who could definitely see the forest for the trees, was in the process of experiencing cognitive dissonance. While Apollo support had been personally guaranteed by him to President Kennedy, Webb sharply appreciated the valued returns from a robust science program. Back and forth it went in his head. Was America calling for science? Gee, no. Not even close. America was calling for the Moon. Webb could practically hear them chanting for it outside his window. They didn't give a rip about science. But how could anybody deny it? He weighed the relative importance of Apollo and concluded NASA's framework had to be shaken up.

On November 1, 1961, Webb's massive NASA reorganization went into effect. Abe Silverstein and his whole office were out. Instead, Webb conjured the Office of Manned Space Flight. Having a separate human flight department conveyed its stature, but also in his rework Webb slipped a little science under the door. Homer Newell found himself running the gleamy new Office of Space Sciences. JPL now reported to Newell, with Cortright as his deputy riding shotgun. Below Cortright, Oran Nicks took over as director of lunar and planetary programs. Even after the divorce, Newell's office was still formally regarded as "scientific support of Apollo."

Homer Newell wasn't sure what to make of all this change—every program now being viewed through Apollo-colored lenses, no matter what its true origins or intentions were. As Apollo stampeded toward glory, Newell would spend a fair percentage of his waking hours toiling to influence the popular media's perception and reporting of these other programs. Newell's own superiors blasphemously subordinated them as "pathfinders" or "advance scouts"—mere road builders. He did not want to be in charge of the Apollo road crew.

But Newell repeated the name of his division—Office of Space Sciences—and let it roll around in his mouth like a big, fruity jawbreaker. There was a nice, distinctive ring to a name like that. He'd picked it out himself, and even though his proud new department was the size of fly scat compared to Apollo, it could still do loads of good. Who knew what groundbreaking scientific discoveries could be made while serving Apollo? To the Jim Van Allens and the John Simpsons, Newell pledged every possible effort to make sure they could somehow tag along on every step of the journey to the Moon, providing more than just suggestions for where to get out and walk around up there.

Oran Nicks was certainly determined to fit more science into NASA; he really wanted to set a benchmark with Ranger, although in his eyes, the one and only James D. Burke always seemed eager to strip it back out. What was he going to do with those Pasadenans?

Until the day he died—in a glider built by his own hands—Oran W. Nicks was a man continually spellbound by things that flew. It was this wonderment that led to his quest for an aeronautical engineering degree. By sheer coincidence, the man's post–secondary education paralleled London's rampant attack by v-2 rockets, which got Nicks to thinking that situation over. Contemplation of senseless death aside, he spent more time noodling about the fascinating way in which the Germans were able to shoot a rocket up to the edge of space and navigate it at 3,500 miles an hour. It was fascinating stuff: Nicks even got the opportunity to lay his hands on actual v-2 hardware during a stint at North American Aviation after the war.

In spite of the Apollo business, Ranger was still going to fly the seismometer on Block 2. It was a great experiment, if those Ford Boys could ever get it to stop failing tests left and right. Nicks finished moving into his cramped new office and then asked Burke and Cummings to look more into the situation.

They did and found that really, most every hang-up with the landing sphere had nothing to do with design. Their prime suspect? NASA's cadaverous heat-sterilization procedure. Out in Newport Beach, Aeronutronic guys were working double shifts, chasing their tails in search of the trouble when it was already known. It got so bad that Frank Denison went out on the family boat one evening and took his shoes off. He swallowed a load of pills and then lay down in the bunk and figured he'd never wake up. Hours later Denison's inert form was discovered and rushed to the hospital, where they managed to pump life back into him. Frank Denison's wife, Ruth, knew Jim Burke from way back and called him to ask for help. Immediately Burke said he'd do anything. Ruth wanted him to go out to the boat and get Frank's shoes, and Burke dropped everything to go get them.

In the wake of this calamity, Aeronutronic only persevered, attaining a fresh level of dedication far above the original contract terms. First, they relieved Denison of command. Then they redid the cleanliness procedures and began assembling their flight hardware inside large, clear chambers that looked more suited for working with radioactive thorium than a seismometer ball. Each side of the box had portholes with heavy rubber gloves mounted in them for a technician to walk up, stick his hands in, and get busy. California had seismometers all up and down the San Andreas Fault, and none of them had ever been put together like this. Only three days later Burke sailed a report over to Nicks telling him things were under control.

Ranger 3 launched on the afternoon of January 26, 1962, and almost immediately began to malfunction. Its landing capsule was fine; the seismometer, perfect. The *Atlas booster* began to lose its head when the internal radio guidance system failed almost as soon as it left the pad. By the time Atlas expended itself, the stack had been driven to a much faster speed than planned. The Agena performed beautifully. However, *Ranger 3* would get to its rendezvous point in space with the Moon still twenty thousand miles away. The colossal spread was beyond what the ship's midcourse engine could handle.

Yet James Burke looked at the incoming telemetry and knew he'd done his job. The solar panels unfolded right on cue, the bus powered up, and *Ranger 3* found Earth, dialing itself in for the long trip with a perfect communications lock and perfect spacecraft health. The gamma-ray spectrometer, for example, had great data surging back from it. So Burke got on a plane and flew back to California. He went directly from the airport to JPL in order to start postulating what in the world they might do with a spacecraft that would never hit the end zone.

That same January day in 1962, as *Ranger 3* zipped farther and farther away, nothing seemed to be going very well anymore. The radio connection began to fade. What Jim Burke did *not* do was hop on a plane and fly through snowstorms with ten of his favorite engineers. No, Burke stayed in the JPL control room and began to troubleshoot the high-gain antenna. It seemed to be working correctly, so the ferreting-out process moved backward and upstream through each component, one by one. Everything checked out until they reached the central bus electronics. Near as anyone could tell, they had suffered a catastrophic failure, allowing *Ranger 3* to drift out of position. Most of the resultant finger-pointing was directed at the heat sterilization.

The signal faded in and out for a while before it finally dropped off altogether. The Ranger program was zero for three. They'd try again; they had to. *Ranger 4* was already together and being checked out. It was supposed to leave for Florida by the end of the month.

Oran Nicks took a deep, cleansing breath and tried to look at it from all sides. First came his own. Burke needed to focus on engineering management but was too distracted with whatever the scientists were up to. That was almost the bigger problem right there. Scientists, in Nicks's opinion, required constant herding. Spread nationwide among universities, they gallivanted about, unfettered and insubordinate—whimsically dancing naked in the daisies, free from America's warm and comforting military-industrial structure. This, Nicks simply despised. He wanted nothing more than to round up every scientist involved in the NASA programs and place them all on staff, planted in Washington DC where he could more easily monitor and regulate their behavior. Perhaps then Burke could do his job better and the Rangers might start to work.

"I can understand," Charles Sonett offered of Nicks's attitude toward the scientists. "I mean, Oran came from a very different background of culture."

By mid-March, Homer Newell's infantry had reviewed almost ten new experiments and felt most of them worthy for inclusion in the Ranger program. This was part of Newell's response to the plethora of hate mail clogging his desk from jilted scientists whose instruments had either died on Ranger flights or been yanked off in the wake of Apollo. Burke protested that adding these new experiments was only going to make things harder than they already were. What about the schedule? JPL hadn't budgeted any time to deal with these new packages. Nobody down in assembly and integration would say with a straight face that they could easily plug new hardware into the bus. JPL hadn't budgeted more people for this either, and some of the current staff already practically lived at the Lab. The maintenance crews had even set up a pitch black trailer next to the Operations Room and filled it with beds.

The new recipient of all Jim Burke's emotive complaining was William Cunningham, who officed over with Nicks and Sonett as NASA's Ranger program chief. The space agency had only recently gotten around to creating the position, and with equanimity the new appointee listened to Burke's every word. Above and beyond Cunningham's decidedly scientific upbringing—mainly physics and meteorology—he evidently benefited from 2 million years of cooperative evolution in his family's gene pool, which translated into Cunningham's possession of fundamental people skills that applied well in tight situations.

After Burke finished his outburst, he stood there waiting for a response. In diplomatic terms, then, Cunningham patiently clarified NASA's position. The discussion was over. More science experiments would go onto the Rangers. In his next memo, Burke seethed, "if they wanted to do an exercise like this, they had half of 1961 . . . and now it is pretty late."

Oran Nicks sure liked coming to work with Bill Cunningham. In Nicks's eyes, though, the guy was too soft. "Bill's dedication and loyalty were unmatched," Nicks later wrote. "Although I sometimes felt he was too forgiving and informal in his dealings with JPL, I knew that he provided qualities complementary to my more serious and sometimes tyrannical methods." And Cunningham got results with the Mr. Nice approach. People responded well to it. When Burke's replacement eventually came along, Cunningham quickly forged a healthy relationship with him, too.

Nicks wrote Jim Burke a letter on March 20 regarding the disputed new toys. "Take steps to integrate these experiments into the spacecraft bus," he instructed. Burke was furious. He jetted to a meeting at NASA headquarters on March 28 ready for war. By this time he was flying back and forth so much that at one point a stewardess approached Burke and returned his U.S. Navy greatcoat, saying "Did you know that you and I have been wearing each other's coats for a while?"

On the other side of the table that March day sat Nicks, Sonnett, and Cunningham, along with a few others. This meeting was supposed to address the particulars of adding the new experiments NASA wanted, but it degenerated into hostile verbal skirmishes. "The Ranger effort already was in serious technical trouble," Burke pointed out years later. "That was the key thing. And what skilled manpower we could command was busy fixing the technical troubles that existed. But . . . I just couldn't understand— I couldn't comprehend the thinking which would suddenly load on eight additional trinkets, with all their demands and interactions, right at the moment when we were perhaps approaching our first payoff!" In addition, these new experiments were *not* lunar related. "The short, direct flight to the Moon would not provide much opportunity for magnetospheric investigations anyway," Burke later justified of his actions.

Burke tried to say as much and more that day, but Oran Nicks got in his face. *Ranger was designed for this*, Nicks argued, *and you are the vendor. We are the client, and we are running this program. Things will be done at our direction. You are flying a science mission going through an interesting part of Earth's environment. Let's make some measurements along the way, okay? And don't ever let us catch you calling them "trinkets" again.* Nicks didn't ask Burke—he *ordered* him—to get a plan together for this new round of experiments, and by golly he'd better be ready to discuss it at the next meeting.

Early in April JPL courteously unrolled its revised plan during a visit to their number one customer. It carefully folded the new science experiments into the existing round of agreed-upon flights. Third-party contractors had signed on to beef up the manpower. Burke put a spin on his argument against whimsical spacecraft additions: he reminded everyone at the meeting how Ranger's new mission was first and foremost the *photography*, and all this monkey business goofing around with science fair projects could throw a wrench into Ranger's Apollo support. It worked; he'd shot straight down the party line.

Jim Burke flew back to Pasadena, where the shop floor hummed with activity. The assemblers were sweating it out, trying to hang all this extra hardware on the bus and finding new problems almost as soon as they solved the old ones. A sort of universal bracket had already been tacked on to help bear the burden of the added gear, but Ranger's power situation was now in tatters. The new roster of packages commanded such an insatiably large current draw that the twin solar panels wouldn't be able to keep all of it powered and still have enough to run the craft itself—oh yeah, and the television cameras, too. That was on July 13, and Burke found himself contemplating some rough options. They could run across campus and steal the bigger solar panels off the new Mariner spacecraft, which was still coming together. The change would gain ampere-hours for Ranger but send JPL deeper into engineering hell. Or they could cut down the number of science experiments. Burke swallowed hard and knew the second decision was the right one, but it might also be the hardest to sell. If he presented an engineering rationale for deletion, would Oran Nicks think he was just trying to pull a fast one?

Many years later Ranger project member Gordon Kautz would offer clarifying words in Burke's defense. He claims there was never any untoward subterfuge to remove science experiments; that wasn't the kind of person the Ranger manager was. "He knew the tradeoff that was being made," Kautz suggested about Jim Burke. "I'm not at all sure that headquarters ever really knew of the effort that was made to get them on in good faith."

As their next Ranger launch period approached, the locus of attention shifted from Pasadena to Cape Canaveral. During these final hectic weeks, everyone—the air force, JPL, Lockheed, NASA—convened every day for a status meeting. And both Oran Nicks and Bud Schurmeier noticed how as of late the work atmosphere seemed to be on the upswing.

The meetings *used* to begin with what Schurmeier considered a highly interesting facet: no formal agenda. If somebody had something to say, they just blurted it out. And responsibilities confusingly overlapped. Schurmeier had noticed it the very first time. "That," he later opined, "was a real eye-opener in a sense. The thing that depressed me was that so many different groups and organizations were all involved in various ways. It reminded me of a bunch of ants on a log floating downstream, each ant thinking he was steering."

At some point in these meetings, one of the involved parties typically had a problem to disclose. Everyone then *quickly* dispersed. Nicks and Schurmeier would later deduce that the exodus was a result of others in the meeting not wishing to report on their own tribulations. With that day's fall guy handily singled out, the remaining contractors were essentially gifted another twenty-four precious hours to fix their own little glitches. In due time, the game evolved to include bluffing; utilizing vague language, a group could sometimes hold out long enough for somebody else to fess up before they did.

Thankfully, Schurmeier noted, things seemed to be improving.

Ranger 4 launched shortly before four o'clock in the afternoon on April

23, and as soon as it cleared parking orbit, everyone knew the mission would never go. Everything Agena related had been 100 percent all the way, but after separation, *Ranger 4*'s telemetry suddenly flatlined like somebody had thrown a switch. No one knew for sure what the spacecraft was doing. The transponder signal ramped up and down, suggesting that poor *Ranger 4* was tumbling. That meant the solar panels were probably still folded up tight.

On the way home Burke stopped over in Miami by way of an air force plane, and he quickly signed off on a bunch of commands to be radioed up from the Johannesburg station. They didn't help. It looked like the computer was dead. Over the course of *Ranger 4*'s ground tests, it had been run in excess of seven hundred hours with not one single burp. There was a little more telemetry coming from space, which might be of help, but the signal had to be recorded on tape in Florida and then go by plane to JPL.

Turned out, the kiss of death happened just at the moment of Agena separation. The two had ridden up plugged together, male pins from the bottom of *Ranger 4* sticking into a ring of female sockets on Agena's top end. Rocket staging always involves a sprinkle or two of violence and debris, and in those uppity moments after separation, a tiny fleck of insulating foil drifted onto two Ranger pins. When they touched, the bus completely shorted itself out and killed *Ranger 4*.

On April 26 Burke entered the Goldstone tracking facility in the Mojave Desert, along with Cliff Cummings, Bill Pickering, and Oran Nicks. It was the middle of the night. They straightened up when Jim Webb came into the control room. Webb had gone to LA to give a speech but detoured through the site, because today was the day *Ranger 4* was supposed to impact. Although inert, it had kept flying. The Goldstone workers did what they could for entertainment. Somebody rigged up a television camera with a telephoto lens to feed video of the Moon into a monitor hanging in the control room. The seismometer had its own separate power supply, and the sound of the ball's transmitter was patched into the main speakers so everyone could hear. It wasn't much but did make for a compassionate gesture from the station crew. Then *Ranger 4* disappeared behind the Moon, and its transmitter audio cut out. Webb straightened his trousers, gathered himself, and walked from master control over into a side room where a few reporters had been hanging out.

Eulogizing on the "long strides forward in space" made by this mission, Webb called attention to the comparative superiority of American technol-

24. The press conference at Goldstone. Cliff Cummings stands at far left as Jack Albert turns the microphone over to Bill Pickering. Webb, Burke, and Donal Duncan complete the group. Duncan took over Frank Denison's position at Aeronutronic. (Courtesy of NASA/JPL-Caltech)

ogy over *Luna 2*. Then Bill Pickering sidled up next to Webb to predict how *Ranger 5* would really bring it on home.

How wrong he was.

Nikita Khrushchev read the words from Pickering and Webb and couldn't resist a few digs. He staged his own little press conference to orate, "The Americans have tried several times to hit the Moon with their rockets. They have proclaimed for all the world to hear that they launched rockets to the Moon, but they missed every time!" He probably laughed out loud when *Mariner 1* went up on July 22 and then came back down about five minutes later. JPL went idle to try to figure out what all happened.

Mike Minovitch spun the downtime to his advantage. Less than sixty minutes after *Mariner 1*'s descent, Mike leaped on both of JPL's 7090s. Nobody was on the schedule; they'd be his for the entire weekend. One computing run went thirty-seven hours with no interruptions.

As the story of *Mariner 1* goes these days, it has apparently become something of a fable—a fairy-tale yarn with good guys, bad guys, and lessons learned—though *not* among the planetary scientists who warmly welcomed this first attempt at a Venusian flyby. Rather, the mission's outcome is legendary among the docile fraternity of computer programmers.

Within any group, there are those who pine for superlatives in their work—compiling entertaining statistics on "the longest" or "the toughest," and on and on. People writing software are guilty of similar pleasures, including their assessments of the worst computer bugs of all time. Of course, "computer bug" refers to a kind of typo, or oversight, in the tens of thousands of individual instructions that collectively form a computer program's "code." And the inherent nature of human imperfection means the code isn't always perfect. Forever lost somewhere inside sits an error or two, waiting to spring out and yank your pants down.

Sometimes these glitches merely force a computer application to stop running. Other times, the bug is a proximate cause of appallingly humorous consequences. For example, in 1989 a computer of the French court mistakenly cited forty-one thousand Parisians with offenses ranging from indecency to extortion to drug trafficking.

Sometimes a computer bug makes things blow to kingdom come, things like *Mariner 1.* The mission's programming code once received the decidedly unofficial Worst Computer Bug of All Time award from programmers worldwide. Here's what happened: JPL's first Venus mission was all systems go and high in the clouds four minutes after launch. Then the Atlas booster's antenna failed, and radio guidance data from the ground stopped coming in. A rocket this big simply *must* have a guidance system.

Atlas had a roomful of smart engineers behind its design, and they'd loaded up the onboard computer with backup software that could be tapped in an instant. Smartly, the Atlas clicked over as soon as it realized the ground link was down. The only problem was that quietly nestled within those thousands of programming commands lay an inadvertent omission totaling one hyphen.

Now, one teeny hyphen may not sound like much, but in this unique case, it performed an important—nay critical—function. Sans hyphen, the remaining program code regarded tiny velocity changes as serious errors, leading the missile to overcompensate for its actions. Atlas began flying off course and, therefore, precipitated an unfortunate and irreversible series of actions. The range-safety officer lifted the hinged, protective cover

over his button of last resort and firmly pushed it down while he still had six seconds left to do so. The button initiated a radio command that shot through the console right up the communications antenna to the thundering Atlas and detonated a package of high explosives laid right alongside the rocket's working parts. It blew itself to bits in an instant—which was the idea—and that was the end of *Mariner 1*. Six brief seconds later and the Atlas would have staged.

Oran Nicks watched the *Mariner 1* opera play out in real time while standing about a mile away on the blockhouse roof of Mercury Pad 14, a dangerously nearby location. Nicks accompanied two congressmen who were patently interested in seeing a launch close-up. Their presence was strictly against the rules; when guards showed up to kick them out, Representative James Fulton let his position become known and advised the security men to obtain a second opinion. The guards took off and located a telephone, rousing the air force base commander out of predawn sleep to ask him what they should do. The guy wasn't able to make up his mind in time to stop the launch and clear everyone back past a three-mile radius. *Mariner 1* lifted off.

Nicks wasn't wild about standing on the exposed roof of a building, with the unbridled wind shoving him about, and furthermore possessed an educational background suited to appreciating the risk of being so near. But Fulton had muscled a number of space appropriations through Congress, and when he said he wanted to see the launch close-up, well, Nicks understandably had to produce. The other representative, Joe Karth of Minnesota, was something of a newcomer to Congress; nobody knew much about the guy or where he stood on issues. But today Nicks had drawn babysitting detail and Karth was just along for the ride.

The congressmen and their chaperone stood there on the roof unplugged and bare, as it were—no squawk box from mission control reporting the countdown status. There were no phones or radios or any sort of device to tell them what was going on. All they knew was that the Atlas lit up at 9:20 a.m. And when it blew four minutes later Karth asked, "What happened?" Casually Nicks mumbled that the launch vehicle had exploded, although a *very* slim chance remained that the booster had maybe staged beforehand. If that happened, *Mariner 1* and its Agena could still be heading up and out of view. Soon Nicks found out that was not the case. Things had come apart at the destructive hands of a vicious hyphen.

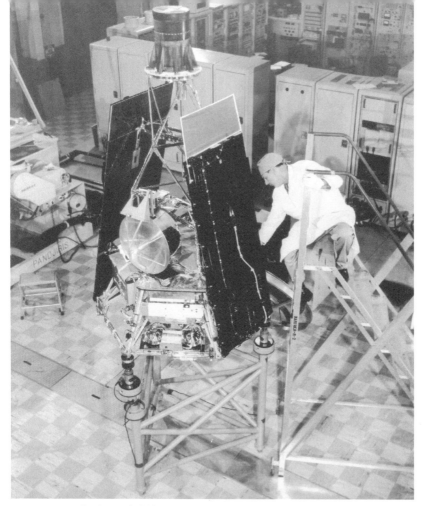

25. *Mariner 2* with solar panels folded. In a crowd it could pass for Ranger easily. Sterile conditions weren't necessary for a flyby mission. (Courtesy of NASA/JPL-Caltech)

The men somberly made their way down the blockhouse steps to ground level and got in a car. Fulton wanted to actually walk out on the steaming launch pad, so they drove over. While strolling around, he began cramming his pockets with remnants of wire or tape or other bits of junk left behind—anything that he happened to notice. Fulton attended nearly every launch and had a particular habit of doing this; he packed it all home for citizens in his district as souvenirs of America's space program.

When Fulton's pockets could hold no more, he turned back to Karth and Nicks, asking them to do the same. The trio then made their way to a cafeteria on the base. A number of disgruntled Mariner engineers had collected there to down endless cups of coffee and pore through their dead bird's telemetry, attempting to draw a bead on what in the world might have happened.

With some disgust, Oran Nicks calculated that, since arriving at headquarters, his tenure had grown to envelop seven complete mission failures. Technicians were already racing to prep *Mariner 2*. Nicks hoped it wouldn't make eight.

13. Swing in Time

We called the Rangers "American Kamikazes."

Oleg Ivanovsky, Soviet space engineer

Aeronutronic had been calling Jim Burke, wondering if they should free the seismometer team after delivering *Ranger 5*'s payload. Burke didn't know what to tell them. Texas congressman Albert Thomas, with his long serious face and bow tie, was wondering aloud why NASA should bother to renew its contract with Caltech and the Jet Propulsion Laboratory. Shouldn't JPL, the Texas Democrat reasoned, be more like other NASA field centers instead of some bizarre end run around the typical government framework? He wanted to strip the Lab off from Caltech and run it as a separate NASA underling. Bipartisan support appeared from Republican James Fulton, patently embarrassed by the atrocious record of nonsuccess. "We should be beyond this stage," Fulton asserted, presumably while unloading his pockets. And teetering on top of it all, civil war threatened to break out at NASA between Homer Newell's scientific objectives and Brainerd Holmes, who was commanding the Office of Manned Space Flight. Both seemed intent on hijacking Ranger for their respective departments. Weary of the dance and now stuck in the middle, Pickering knew his responsibility had never been to decide what Ranger was *for*. Even Burke had to agree.

Bill Pickering finally sought a gathering of all concerned parties to settle the issue once and for all. If Ranger was going to be in support of Apollo, then let's happily go fly an Apollo support mission and shut the heck up. On the other hand, was NASA ultimately after good science returns? Then let's *get* them and drop the manned support. At a face-to-face meeting on October 11, NASA's two divisions agreed to work together and share Ranger equally. Pickering flew home with a minor sense of accomplishment.

Obviously, major issues remained. Rolf Hastrup had been making a lot of noise these days. The JPL man in charge of heat sterilization faced a peculiar situation: the better he did his job, the worse things performed. Yeah, all this operating-room cleanliness was a good thing, but if anybody ever wanted the stuff to work, then they needed to drop this cooked-in-an-oven business.

On July 25 Hastrup floated a memo over to George Hobby, who'd written up the original tortuous procedures. "The application of sterilization procedures such as the dry heat cycle presents a serious risk to reliability," he argued. JPL's assembled *Ranger 5* had already endured the pain and suffering, so Hastrup wasn't feeling real chipper about it coming off any better than the others. Already Jim Burke had NASA knee-deep in waivers for additional parts that could easily be damaged by the heat. The entire central computer made the list along with more cabling and all six television cameras. The waivers got rubber-stamped in no time, which came as a relief, although the cloud of uncertainty would still hang over *Ranger 5*.

Jim and Lin Burke decided to finally stop putting off a European vacation. JPL commanded an extremely meager slice of the Cape Canaveral facilities, and while Mariner was out there hogging it all nothing more could be done on Burke's project. The couple visited Denmark and Germany then moseyed on into England.

Mariner 2 launched from the Cape on August 27, and as soon as its tower cooled the crews rushed in at top speed. They needed to turn the pad around so that *Ranger 5* could be stacked up for launch. Then after *it* got off the ground, both could be tracked simultaneously by the same facilities. And in case the slightest confusion existed anywhere within the decision-making tree, a note about priorities came down from JPL deputy director Brian Sparks the week before launch: if a tracking station's capability became saturated with both missions, the priority would be *Ranger 5*. Period. *Ranger 5* would only be out there for three days anyway, versus *Mariner 2*'s protracted flight.

After the liftoff, Mariner project manager Jack James thoughtfully dispatched a cable to Grosvenor House in London, where Jim and Lin were staying. "Isn't it too bad," Burke later suggested, "that Western Union will never again deliver one of those pieces of paper with little words on it?"

Burke opened it after stepping into the elevator. "That upward ride is a vivid memory for us," he recalled, "especially now that our admired and beloved friend Jack James is gone." He read:

MARINER 2 LAUNCHED SUCCESSFULLY.
MIDCOURSE MANEUVER PENDING.
WE'RE ON OUR WAY TO VENUS.

"wow!" Burke cried with glee, slapping his hands together, hugging Lin. She wrapped her arms around him and held on tightly. The eruption almost gave the lift operator a heart attack, who finally got in on the news after Burke explained.

"Good show!" sang the operator.

Jim and Lin went on to their room. Minutes later knocking was heard on the Burkes' door. Champagne and flowers rolled in on a cart and fine linen, compliments of the hotel. Jack James didn't give details in his cablegram, but the *Mariner 2* launch contained a real white-knuckle moment. The Atlas rolled thirty-six times during its short operating life.

On the whole, Mariner was experiencing considerably fewer growing pains relative to Ranger. Yet both were in production at roughly the same time, using the identical pool of JPL resources and talent. Remember the matrix organization? Every project had access to the same technical departments. But the experience was altogether different. Things *worked* on Mariner. Deadlines were met. People got more sleep and didn't yell so much and never got their asses fired. What was going on?

The difference had everything to do with Jack James himself. When Mariner came his way, the schedule called for nine months of paper development prior to actually building any hardware. Before he did anything else, even call one darn meeting, James sat down and wrote a big book of rules governing how every single little thing was going to happen. He laid out requirements for the schedule and materials handling and testing environments and design reviews and failure reporting and everything else he could possibly think of. The rules were fair and reasonable and, above all, formalized. Anybody in doubt could leaf through a copy and find exactly what they wanted in black and white. It contrasted greatly with Ranger, where the system of failure reporting, for example, was largely undefined and relied heavily on informal, personal contact. Proper follow-up really came down to the individual's mood that day.

Like a good parent with simple, unambiguous tenets, James's structure created an environment in which the spacecraft and the people who built them could flourish. And as the engineering matured and durable components were built and passed all their tests, James politely and successfully deflected repeat requests to glom on more and more science experiments. It could only be so complicated, he sensibly argued.

Jack James was a Texas man who'd drifted for a while after high school, finally tackling the electrical engineering program at Southern Methodist University. This particular discipline was maybe not his first choice, he once explained to Oran Nicks, but his vague understanding of what an electrical engineer did made it somehow attractive. Civil engineers did things like build highway interchanges and wastewater treatment plants. That was not the type of thing for Jack James. Mechanical engineers designed new furnaces or elaborate trade-show signage. Ditto. The only thing left was in the electrical department, so James went for that one and clicked into it nicely.

His *Mariner 2* established a pattern for America's planetary exploration over the next decade and a half: when the infrequent launch period is open, hedge your bet and send two. Besides, never keep a lady waiting, especially one as bejeweled—and exclusive—as perhaps the most beautifully exotic of them all, fair Venus.

Her launch period makes things tough, with nineteen months between opportunities. "Period" refers to JPL's term for the overall launch aperture—those thirty days or so when the source and destination bodies are at their closest. Curtain call for the '62 period happened within a month of opening night, and each individual day therein afforded less than an hour firing window to get the stack blasted off and headed in the right direction. Somewhere in the many journalistic column inches along the way, these distinctions combined into the more popular term "launch window." And the immovable dilemma that it is keeps people on their toes. During those frantic days in 1962, the period drew closer and closer to the end of its run until they finally squeezed *Mariner 2* off the ground.

Oran Nicks and Jack James had visited the Mariner blockhouse during a built-in half-hour hold about sixty minutes before launch. The breather was planned in advance in order to cue a sort of flag-waving appearance by the NASA brass. The idea had come from James. Personally, Nicks also felt a calm visit like this—when nothing was on fire—might help if the probe upchucked somewhere down the road.

Their ship left Earth parking orbit and settled in, locking onto Canopus for the extended cruise. Its protracted itinerary called for, in essence, a fundamentally absolute kind of road map. To keep *Mariner 2* going down the middle of the highway over these next months, its bus had been carefully programmed (including any necessary hyphens) to search out and lock on to three distinct celestial objects. Triangulating its position relative to these markers would tell the self-reliant spacecraft exactly where it was. First came the Sun for its obvious presence and brightness on any day of the week. Next was Earth. Feeding back data from *Mariner 2* required a constant lock on the planet anyway. But Earth also made things tricky, confounding even the most intelligent space platform with the way it kept moving in orbit and withered ever smaller as time went on. At great distances the correct Earth location could easily be confused with the Moon or even Mars. So as a triple check, *Mariner 2* looked out among the heavens, patiently scanning until it found Canopus—the second-brightest star visible from our planet.

Being of similar lineage, Mariner and Ranger shared a common physiology. At the core of the bus sat the central computer and sequencer—a squat, unremarkable box six by six by ten inches around. Today's standards would classify its intelligence on the level of a Happy Meal toy. Even so, it handled every aspect of the probe's operations (from course correction to event timing, experiment operation, and data relay back to Earth), all in a decidedly hands-free way. Major elements of the flight program were already stashed away, waiting for the right count—things like acquiring the positions of the Sun and Earth and locking its radio system to home. Engineers also loaded it up with the program for Venus encounter just in case something went haywire with the radio at that weighty moment. Three little electronic glove compartments rounded out the onboard system. These were configured to store individual ground commands for later use at a specific time.

That was the brain, and it relied on a heart: one simple oscillator. Immediately prior to launch, *Mariner 2* technicians initiated its clock, whose sole yet critical function was to count out a measured beat. Like music, flight plans are scored down to the second, and this spacecraft issued commands when a specific count had been reached.

Primitive electronics relegated the computer to only twelve Earth-based instructions. A dozen commands don't go far. Single-use orders—say, to unlatch the solar panels—were therefore tacked onto others when the consequences didn't matter. In one command, for example, the beating clock triggers a directive to find the Sun, and riding piggyback on top of it is the command to unlatch the solar panels. They've been essentially written as a single instruction. If *Mariner 2* fails to lock on the Sun, the command gets sent up again. Over and over, it can be reissued with no adverse consequences because the explosive bolts already fired and the solar panels already unfolded. Telling the ship to release them again makes no difference. And so equipped, *Mariner 2* flew on, with twelve commands filling his hip pockets, aiming down the road for a date with his lady.

The ship underwent a midcourse correction on September 4, and midcourse corrections always play with fire. The whole problem begins as soon as the spacecraft launches, because it doesn't go in a straight line. It feels the tug of gravity instead, obeying the same laws of physics as any other celestial body. In this case, *Mariner 2* therefore began describing a lazy arc between Earth and Venus. These behaviors have been well thought out and charted in advance. And the trajectory planning is nearly always good enough to put the spacecraft *close* to its intended path. But close doesn't always cut it. The onboard science instruments assume a given range between ship and port of call, beyond which their effectiveness falls off a cliff. If necessary, the probe's trek through space can be reworked, altering course slightly to get the vessel back on track. That's called a midcourse correction, which gets flight controllers really nervous and sweaty.

Here's the problem: Your spacecraft passed all the ground tests and then reached orbit after the booster rocket avoided blowing up. Thousands of lightweight, handcrafted parts didn't burn themselves out due to chintzy electronics, power failure, heat sterilization, or some stupid oversight. Now, after years of planning and late nights and fighting with contractors and bad-weather postponements, it's finally on its way to a strange planet humans have observed for tens of thousands of years but know little about. The metal explorer flies perfectly, but *you* want to see if its course can be "slightly improved." To do this, your precious spacecraft will receive commands to unlock its position, dropping two of its three celestial reference points and—as if that's not edgy enough—briefly powering off the solar panels and high-

26. "He was straight as an arrow. Absolutely straight," said Chuck Sonett. Here's Oran Nicks, early 1960s. (Courtesy of NASA)

gain antenna. The machine starts to drift, and for many moments, your pride and joy dangles on a thin string along with your reputation.

Now in the vacuum of space, you will be tooting the gas jets to nudge the ship just a fraction of a bit, just a smidgen. This grand leap of faith requires confidence in the totality of the design . . . not to mention confidence in the donut-smacking college graduates who told you what to do and when. Afterward the machine supposedly reacquires the two reference points it dropped, twisting the solar panels back in line and on-line, turning its huge antenna back in place, continuing flight like nothing ever happened. This all takes half an hour from start to finish.

If it works, you've still got your mission. If it doesn't, then you've got junk.

From approximately 1.5 million miles out, *Mariner 2*'s midcourse correction came off without a hitch. Oran Nicks tried breathing easy once again and paused to straighten his tie before receiving people from Hughes Aircraft. They were due any minute.

Hughes wasn't in the neighborhood for a weenie roast; they'd shown up to carp about how their newborn "Surveyor Orbiter" was infected with a slew of problems that were heaping up faster than the solutions. Nicks was supposed to do something about it.

He stole a glance out the window. He had a headache building. One of *Mariner 2*'s tracking sites had recently lost power as well as ninety minutes of data. The power company had shrugged and told them stuff happens. And recently Nicks had learned that the otherwise dependable Jack James had quietly snuck a tiny American flag between *Mariner 2*'s layers of insulating blankets. He'd done it just prior to launch. Yikes! Nicks feared certain political backlash if the symbolism was misinterpreted.

He refocused by flipping open the disquietingly plump Orbiter file. Although the project was being managed by JPL, NASA had mandated that the spacecraft itself be furnished by a third party—leading to the selection of Hughes Aircraft as the outside vendor.

"This was not looked on with favor by most JPL officials," Nicks later suggested.

Surveyor Orbiter aimed, at least on paper, to accomplish what heretofore was considered fantasy. The unmanned lunar machine sported two parts. The first was an orbiting mother ship that would reconnoiter the surface and shoot thousands of images, partly for the lunar cartographers but mostly for Apollo. Then, following detailed appraisals, a lander would gently drop onto one of the most promising sites for ground photography and even a bit of soil sampling. But the more lines the engineers drew and the more what-ifs they discussed, the more it seemed too huge. Nobody had been able to smack the Moon while taking a close-up picture. Now they were supposed to orbit, photograph, and soft-land a picnic basket of experiments? Nicks's own excruciatingly researched Surveyor Orbiter appraisal put the odds of success at no better than one in three.

Ranger 5 finally got off the pad after noon on October 18. And it would be the last chance for the impact ball. Glued to his telemetry screens, Jim Burke hardly even blinked. Then it happened—the same exact Atlas guidance problem that ruined *Mariner 1*'s day. Burke held his breath as backup commands screamed up through the sky and found Atlas, scurrying into its computer. Not every ground station could uplink commands, however; so this better work. Green light. Staging. Parking orbit. Second Agena burn read nominal. There was no pin shorting—they'd been redesigned after *Ranger 4*. All according to plan. Whew.

Burke retucked his button-down shirt, mopped his brow, then walked outside through the overcast afternoon. He joined a line with William Cunningham, the Cape's Kurt Debus, and some others while reporters lobbed questions over to them like lazy softballs.

Thousands of miles above the gathering—at about the same time Burke was noncommittally saying, "All of us are keeping our fingers crossed"—a small screw deep within *Ranger 5* was in the process of loosening itself. The screw joined two critical terminals in the power switching and logic unit; once they separated, nothing would work right. Sixty minutes in, *Ranger 5* was numb and cartwheeling.

When he got the news, Bud Schurmeier said, "Man, we're in deep poo-poo here."

Jim Burke flew home, pecked Lin on the cheek, and then sat on the ugly couch in his living room. Burke hadn't been spending much time in the living room lately, so he stayed for a while—upright, hands on his knees, not moving at all. And within minutes every ambient sound except one faded away to nothing. The only thing Burke heard was their clock up on the wall. And the ticking grew louder, thumping, filling his head.

In 1962, with American radar encampments listening from nearby Turkey, every single one of the three Soviet attempts to reach Venus failed. Things weren't going well in Sergei Korolev's personal life either. Daughter Natasha excelled in her studies and was about to emerge from Elementary School #243 with a gold medal cum laude. From there she'd dash off to First Moscow Medical Institute, ultimately persevering in her journey to become a lung surgeon. They lived no more than a few hours apart, she and he, but Natasha never called. Miles didn't separate father and daughter the way hard feelings did. It was the divorce that started it all downhill—back in 1948 when Sergei had called it off with Xenia. The delicate situation further chafed Natasha when Dad's new flame Nina entered the picture. At that point Sergei Korolev's own mother no longer wanted anything to do with her son; Natasha felt the same. She never even invited dear old father to her wedding.

Natasha finally visited in 1962. He was many things: Hero of Soviet Union, Hero of Socialist Labor, Soviet chief designer, space pioneer. Never Dad. That day Natasha brought along three-year-old son Andrei, warmly welcomed by a pair of unfamiliar eyes. "We will be good friends," the chief designer often spoke of his grandson thereafter. But they weren't.

Korolev's tribulations continued through mid-September, at which point the Venus window quietly closed on his efforts. But shortly thereafter opened the one for Mars. Readying near-identical machinery as that for Venus, they built three winged Martian spacecraft: 2M. Had everything worked, the first one might have also been called *Mars 1*. It destroyed itself shortly after reaching orbit. Leaky seals remain a prime culprit. However it happened, the stack broke into a cloud of debris.

Instead of being called *Mars 1*, the confluence of pieces remained large enough to trip the Alaskan sensors of the Ballistic Missile Early Warning System right smack in the middle of the Cuban Missile Crisis. On-site American military personnel had to decide whether or not this was a nuclear attack, and the Soviet government's typical reluctance to disclose launch failures did little to allay American fears. Then the air force computers rethought the data and concluded the anomaly wasn't an ICBM after all.

The launch almost didn't happen. At Tyura-Tam, Boris Chertok watched as the deputy range chief barged in on his preparations. The colonel, one Anatoli Kirillov, impatiently ordered Chertok to immediately get the Mars rocket off the pad so it could be replaced with a live nuclear missile.

"In case of attack by U.S. paratroopers," Chertok recalled being told.

Tyura-Tam's facilities only had two pads at the time. An entirely separate second launch facility had recently come on-line, over at Plesetsk, but it only had one place to launch. "All of the phone links were being used by the military," remembered Chertok, "so I couldn't get through to Korolev, who was at his home in Moscow, ill with a cold." Sergei Pavlovich wasn't having the most healthy of years. Already that summer, he'd been carted to the hospital during a lengthy, intense attack of stomach pain. Deep in his gut, a growing tumor festered that nobody knew about.

So Chertok diplomatically explained to this colonel how tough it would be to displace the mammoth rocket; they were all ready to fly. It's not like yanking the plug from the wall. Kirillov threatened court-martial then instructed his own men to begin readying what they could of the ICBM at its current location. Chertok didn't have any aces up his sleeve—only Korolev could get the bizarre order cancelled. A few of the Mars rocket team finally traipsed over to the house Korolev used when on-site. Chertok went along. "I remember that we ate a watermelon and thought we might be waiting for a U.S. thermonuclear strike," he said. Fortunately, Kirillov soon materialized at the house door telling everyone to stand down, the order cancelled.

Mars 1, insect #2, finally left the pad on November 1. Though identical to the alarm-raiser of October, this one safely escaped its boosters—and Earth—heading toward Martian territory with just under two thousand pounds of spacecraft. Solar wings popped open; the guidance system trained in on the Sun. Information came streaming back to Russia, and that's when everyone knew they had a problem. A little gas thruster adorning the ship seemed to be leaky. Nothing blew up like before, but one leaky thruster will slowly push a ship off course. Endlessly laboring to correct its position, *Mars 1* would slowly consume its fuel and most certainly die. The ship lacked any ability to receive instructions from the ground to close off the leak.

They tried launching again just days later and once again met failure: booster rocket destruction while transitioning out from Earth orbit. Again the U.S. Air Force noted the presence of something in space the USSR didn't feel like identifying. These were not good times for the Soviets, as *Mars 1* gently fell silent in March 1963 at a distance of 65 million miles.

Bad luck endured, catastrophic failures running all over the map: excessive vibration, leaky fuel inlets, valve failure, power failure. They weathered a dismal 40 percent success rate. As 1962 drew closer to the end, Korolev could only look back on a dozen blown missions and wonder when or if his fortunes might change. The next year tallied four launch attempts to the Moon and Venus. All failed for a variety of reasons, most involving the booster.

Despite one of two solar panels conking out on Halloween, *Mariner 2* was flying textbook. Yet Bill Pickering knew there was still going to be a hell of a price to pay in light of five straight Ranger goose eggs that had emptied 80 million bucks from the nation's coffers. On top of it all, the Reds had launched that probe to Mars.

Homer Newell put a review board together, and its findings were not real complementary to JPL's operation. "Marginally designed equipment," the report cynically accused, led to a "shoot and hope" mentality in a program frantically clawing at too many diverse goals. At any point in its mission, Ranger was supposed to conduct science experiments, advance design knowledge for planetary missions, and scout Apollo landing sites—all while staying 100 percent germfree? Impossible. The final report version was approved, sent through the mimeograph, and then delivered to Pickering's hands in record time. *Shoot and hope?* Even skimming the low points, Pickering knew what he had to do.

And so that fateful day came on December 7, 1962, when James D. Burke answered his phone to hear the news he undoubtedly realized was coming.

"I can recall it as a cloudy day," Burke remarked, "and a day of much concern about the missiles in Cuba." And after he heard the word from Pickering, there was a bit of silence, which is to be expected in these kinds of situations.

Burke finally said, "Do you want to hear from me who ought to be the future project manager?"

"Sure," Pickering told him.

Burke hadn't made up a name right then and there. He'd been mulling it over for a while, appreciating how his number was probably up. And Burke had a name in mind. There are those who may be tempted to use such ailing situations in an attempt to somehow engineer vindication by sliding the mess over to a party ill-suited for the task, hoping to heck they crash and burn and end up making the deposed come out looking like a sacrificial holy lamb. James Burke wasn't the type to do that, though; he therefore recommended the name of his good friend and Caltech classmate, who he honestly felt could steer Ranger where it needed to be.

Burke said, "I think it ought to be Schurmeier."

"Well, you scored," Pickering told him, "because that's who it is." James D. Burke would never again head a JPL project.

One of the Ranger secretaries was in tears as she left that afternoon. "Cliff is out! Jim Burke, too." Cummings had also gotten the noose.

Before Pickering had rung up Burke that day, he'd first called Bud Schurmeier.

"Can you come over here now?" Pickering had asked. And Schurmeier bustled over to the director's office from across campus in another building. With much genuine surprise, he received an invitation to command Ranger. Schurmeier wasn't all that enamored with the idea.

"The reason I was reluctant," he clarified, "although I was concerned about replacing my good friend Jim Burke—I think I was more concerned that I had not completed the job of creating and organizing this Systems Division." Schurmeier didn't feel it was far enough along just yet to let go.

Pickering dialed up the flame. They really needed a whip-smart fellow to take on this intractable stepchild and get a picture of the Moon.

"Well gee," Schurmeier waffled, "I sure want to finish this job you gave me. But I'm not done."

"Well, just take on this Ranger thing, and then you can go *back* to do that," Pickering had suggested to him, making Schurmeier sound more like a pinch hitter than a true replacement.

"Okay," Schurmeier agreed. And no surprise, he never did make it back to the Systems Division.

After leaving Pickering's office Schurmeier drove home and waved at his neighbor—Jim Burke. "We actually had built houses that were only about three doors apart," he explained. They weren't anything fancy—a thousand square feet, maybe, with two bedrooms plus a single-car garage. But the abodes looked down over the canyon edge to spectacular views.

Burke came up to Schurmeier and told him, "Look, I'm . . . I wanna do whatever you think I can do to help you make this thing work."

"What job do you want?" Schurmeier asked.

"I'll take *any* job," Burke said. "But the one I'm good at is making sure the interfaces among all the components, like the spacecraft and launch vehicle, fit right." Nodding, Schurmeier indicated they'd definitely be talking more and then went inside.

"I wouldn't say he seemed upset, was my recollection," Schurmeier later remembered. "If he was, he hid it very well I think, because I'm sure he *was* upset."

Over dinner Schurmeier shared the news with his wife BJ, who quickly grew concerned about the switcheroo. "I think she was more worried about what it would do to the family relationship," qualified Schurmeier of his wife's anxiety. They did so much with the Burke family, such as trips to the beach and mountains. Would all that change? "Our kids were kind of growing up together, and we lived close together. So I think she was more worried about how it would affect that aspect."

Within ten days a revolution swept through Ranger as two hastily convened face-to-face NASA meetings approved a thundering round of changes. While Jim Burke sat fidgeting on a chair outside the DC conference room, Bud Schurmeier hammered out the particulars of Ranger's new way, including an end to heat treating and removal of all the nonlunar science trinkets.

27. Looks good in a suit: Bud Schurmeier models before a Block 3 Ranger. Compare the ship's design with *Ranger 1*. (Courtesy of NASA/JPL-Caltech)

The latter detail alone liberated over fifty pounds for redundancy features on the bus. Even the final treatment of ethylene oxide gas was cut. *Ranger 6*, whenever it finally left Earth, would only be taking pictures.

By now, JPL had come to expect NASA to request these meetings be held in Washington almost every time. "One of the reasons they did it that way is that the government guys at NASA didn't have many travel funds," Schurmeier clarified. "So the way they'd solve it is they'd get *us* to travel!"

Bud Schurmeier always figured he'd end up at Boeing. Years before, he'd even driven up to the plant on his honeymoon to speak with the brass about coming aboard. "But it turned out that '49 was a very poor year for engineers," as Schurmeier explained it, "so I didn't get hired there." He spent a while figuring out what in the heck it was that he was going to do for work, with a new wife and all, when the phone rang. It was a guy from JPL who said, "Hey, we have a new supersonic wind tunnel just being delivered, and we've got to put it together and calibrate it." Was Schurmeier interested? Yeah, he was.

"To make a long story short," Schurmeier said, "every time I thought I'd learned about all, they found something else interesting for me to do, so I ended up staying there for my whole career."

Ranger's overhaul moved along, but in short order matchmaker Bill Pickering was distracted by far-flung courtship. Temporarily shelving the putrid mess before him, JPL's boss turned his attention skyward. Oran Nicks sided up next to him, also looking at the heavens—to the diva. *Mariner 2* was getting close, and the encounter promised to be a nail-biter for many reasons. Their craft was running a high fever. In fact, nearly half of *Mariner 2*'s onboard thermometers had pegged. Something deep inside the central computer had gone way off its feed. The area with the problem contained in excess of 160 resistors, 51 transistors, 40 diodes, and 29 capacitors. Any one of them could be at fault.

But their man in the sky hadn't flown all this way just for a letdown. Millions of miles from Earth, the instruments were all tuned up as the crowd quieted down and the music slowly began to rise. *Shhh . . . it was happening.*

The lady in question had long finished her hair. She'd finished adjusting her sequined gown and shoes and long gloves. She'd been waiting eons for this time to finally come, and in 1962 on the magical night of December 14, it finally did. Stepping out for the debutante ball, Venus coyly extended her hand to *Mariner 2* and accepted a dance. To any human the expanse may have been great, but this bejeweled lady considered the twenty-two thousand miles between herself and her gleaming suitor to be the most intimate of distances. Tonight was the night.

He received her gloved offering gracefully; then, retreating a step, *Mariner 2* smiled while taking the lead, escorting his lady across the dance floor and beginning the slow swing in time to each other—their first and only. In the great blackness watched their audience of three—Earth, Sun, and Canopus alike beamed down on the couple they'd worked to unite. Back on Earth, a giant, lone dish strained its ears to pick up the music. And it really was music. Telemetry sang the 940 megahertz range, just below a soprano's high C. *Mariner 2* worked the floor, notes pittering up and down slightly as they were broadcast through the entire JPL control center and uplinked to NASA headquarters for a live press briefing. Oran Nicks, stalwart curmudgeon that he was, let his eyelids drop, soaked it all in, and even swayed in time just a little bit himself.

Two danced as one. Music swelling, he twirled her close, then back out, and again closer still—whispering in her ear, *Your secrets. How I long for your secrets.* The fever and gimpy limb did nothing to limit his steps. Venus, she threw her head back and laughed, *Oh, to be courted by such brazen masculinity! The embarrassment, the audacity of it all! What would Mercury say? And yet it's been so long, so long that I've waited, and now here tonight I feel so comfortable, so easy in your gaze that confiding myself becomes the most natural of emotions.* So Venus giggled as they moved in sync, beginning to whisper what ageless men have always looked skyward to contemplate. All they've ever had to do was come out here and present themselves and ask. Bobbing, working the floor, *Mariner 2* learned the habits of his intended: how she spun backward and ran thick with clouds, such massive clouds going straight up for miles. He tilted his head to one side in disbelief—*Such amazing news! So mind-numbing!* This ungainly suitor, built for performance and never for looks, was unable to keep such things to himself and had no choice but to tell the others. He did so not from disrespect of this lady but because it was in his nature. Tens of millions of miles away, his California friends were now in on the secrets. More came shortly: the clouds resulted in unimaginable temperatures and pressures down by the surface—perhaps eight hundred degrees and fifty atmospheres. Indeed, on the night of his great encounter with the lady, *Mariner 2* seemingly snuck out the back door to gossip with friends.

Seven hours later the party was over. Music and the dance both ceased, *Mariner 2* retreating toward unavoidable solar orbit, transmitting his final utterances just after New Year's Day. Venus took off her gloves and hung up the gown. Oran Nicks would later write of *Mariner 2* as nothing less than "a beloved partner, feverish and slightly confused at times, not entirely obedient, but always endearing."

Even Oran Nicks had watched the dance.

And had been moved.

Bud Schurmeier walked out to the assembly shop and spent a long time taking in what lay before him. Everything looked the same as it did last week: *Rangers 7, 8*, and *9* all splayed out in different stages of assembly. All three had a variety of components aboard that unfortunately had already been heat treated. Somebody was going to have to play a painful game of scavenger hunt to figure out what all needed replacing. And Jim Burke's precious schedule? The schedule was out the window.

Yes, Bud Schurmeier said to himself, *no matter how things look around here, change is in the air.*

At least Burke was still around. He could have gotten on his motorcycle and ridden away forever. Stoically, maturely, he chose to hang out on the program, tackling the interconnect designs between Agena and Ranger's bus—just what he'd asked to do.

"That eased the transition in command considerably," Schurmeier recalled, "and much of what we accomplished in succeeding months I owe to Jim. He was absolutely dedicated to making Ranger work, and he had what it took to stay and help, when another would have turned his back and walked away."

Over in Washington DC, Homer Newell's desk was beginning to fill up (again) with angry letters from the Ranger scientists, many of whom (again) felt like their throats had been cut in the middle of the night. The word had been formally delivered in a meeting on January 21 by William Cunningham. Not much compares to uninviting tasks like informing a tableful of scientists that their carefully prepared experiments were going on the shelf. *Rangers 6–9*, they solemnly learned, would carry only television cameras. "The majority of the experimenters were understandably disappointed," Cunningham later advised Newell in a memo.

RCA had a lot of their problems ticked off the list. They'd properly built to their own design, making it fit the restrictive power, weight, and configuration requirements of Ranger's bus. They'd been mostly on schedule. There was also the possibility they might even make money on the job. But engineers were still having a heck of a time with the imagers.

A television camera of this vintage works on the principle of light being received and processed by an imaging tube called a vidicon. It was invented in 1950. Television tubes are manufactured in enormous batches; while all of them are built to the same specs, differences always exist. Anybody who watched a football game on television at home in the sixties could notice how the blacks and whites were always a little different from one camera to the next. And that was when a technician futzed around with the controls that whole morning before opening kickoff. Similar technical difficulties befell Ranger television, and it wasn't even on the air yet.

RCA could get the individual camera imagers to work as advertised; they looked great. But Ranger used half a dozen, each snapping every sixth frame of the movie; each tube was seeing brightness levels and contrast in a distinctive way. Every little variance just leaped out.

Subcontracts flew out to three additional tube manufacturers, and endless batch runs failed to isolate one single North American television tube manufacturer in business who was able to produce six measly tubes that all looked the same.

The Schurmeier Tornado rolled on. By February 5, ten critical issues had been identified, such as integration of the bigger Mariner solar panels and the backup explosives needed to deploy them in case something went goofy up there. The changes were all green-lit in no time, which meant *Ranger 6* could quite possibly be ready for launch within eleven months.

A distraction now materialized as *Luna 4* lifted from Tyura-Tam on April 2 and prepared to upstage JPL. An unnamed Russian astronomer boasted to the American press how this craft would "send back detailed reports on the most topical issue—what the Moon's surface is like." Radio Moscow scheduled a special broadcast for prime time called *Hitting the Moon*. But when the malfunctioning probe yawed off into space five thousand miles away from its target, Radio Moscow yanked the program and shoved lame elevator music on instead. Bud Schurmeier was almost too busy to notice. By mid-June, *thirteen* simulated lunar missions had been carried out by a proof test model in JPL's vacuum chamber. Each test ran almost seventy hours, and only one part of the entire system had croaked. Ranger could go, and they started building the actual flight article on July 1. With a whole lot of work, it might even shove off in December, just before the holiday. Schurmeier knew what *he* wanted for Christmas.

About the same time, a tidbit of news appeared—a nearly forgotten traveler had rounded the horn and was back near Earth, having completed the first lap of many to follow for eternity. The day was August 10, and *Pioneer 5* was once again close enough for a session of radio contact. The opportunity to call in and gather fascinating information on the long-term performance of a space probe would ultimately not arise. After several attempts at both listening for and speaking to *Pioneer 5*, it was called off. They were speaking to a dead probe.

RCA wrapped its testing and shipped the television assembly to JPL in mid-August. It was much later than originally planned, but Schurmeier reworked the schedule to clear some other minor details that always seemed to pop up. The one thing that hadn't been fixed, however, was the camera-tube matching. While their children had been enjoying the nice weather all summer, RCA engineers pretzeled their brains trying to make half a dozen imagers all look the same. The best results came by way of combing through repeated manufacturing runs to handpick tubes that most closely matched. But all that legwork pushed the launch back to the following year.

With *Ranger 6* spending its holidays in the shop, there would be no Christmas present of Ranger pictures in Bud Schurmeier's stocking or Homer Newell's or anyone else's. To make things worse, another NASA review board scathingly recommended JPL turn Ranger construction over to an outside contractor.

Not even more contractors would be salvation for *Ranger 6*. It would never ever return any pictures.

14. The Meeting and the Mechta

Our many years of work may go for naught.

Sergei Korolev to his engineers, 1963

As Mike Minovitch looked the situation over, things just didn't seem real hot. Everything he tried doing to spread the covenant of gravity thrust always fell on deaf ears. In February 1963 he'd presented a couple of one-hour JPL seminars to the engineering section, with an emphatically lukewarm reception.

The job status bugged him. Within JPL's matrix his project was unquestioningly illegitimate. It had no official JPL job number, or official JPL contact person, or even an unofficial one. Vic Clarke couldn't say he was in charge of Minovitch's project because he wasn't; Clarke only was responsible for tasks he himself actually delegated. The computing department never scheduled Minovitch's time—he just snuck in whenever the 7090s weren't in use. Debugging assistance? Forget about it. That required an official Request for Programming, which Minovitch found to be quite out of reach. First, get a job number . . . only given to valid projects . . . and on it went. So, despite gobbling hundreds of precious computing hours with no distinct terminus, Minovitch's gravity-thrust project operated wholly off the grid.

Then a magazine article caught the guy's attention, front and center. From the March 4, 1963, issue of *Aviation Week and Space Technology*, page 56 exploded in his face like a moldy cream pie. "Manned Venus-Mars Fly-by in 1970 Studied," headlined the article. The news caught Minovitch's breath halfway. It described significant technological mission advances by . . . him? Nope. By MIT and Lockheed. Complex mathematic computations, the article explained, promised rewarding spacecraft flight "beyond one single destination." Yikes . . . that was a little too close and personal. He read on. Minovitch's heart dropped and then stopped. Performed correctly, the article continued, *several planets* were theoretically reachable during a lengthy *single* mission. *How could this be?* Minovitch wondered almost aloud. Skimming over those names again begot a faint little whiff of familiarity. Hadn't Victor Clarke offhand said something to him at some point about sharing his work around? And wasn't there a memo . . . ?

When confronted, Clarke possibly felt the sticky threat of imminent bodily harm. He somewhat diplomatically attempted to unpaint himself from this troubling corner, explaining to a boiling Mike Minovitch—hotter than he'd ever seen the guy—that yes indeed, both outfits named in that article had received intimate details of Minovitch's work. And yes indeedy, Clarke was the one who did it, just like he said he was going to do way back when.

"Clarke did not care that others were claiming the credit," Minovitch asserted. Though Vic Clarke had mentioned plans to *communicate* with the two outside companies, Minovitch supposedly remained unaware the guy also handed out party bags loaded with magnetic tape and punch cards. "I did not give JPL permission to use this program in any way whatsoever," he later insisted. "Clarke just took it. This is how JPL acquired the technical means to explore most of the solar system." He maintains that Clarke gifted the program to Lockheed and MIT outright—for any use they saw fit—while Minovitch never even got a Hallmark card of thanks.

"This is what JPL has been keeping secret from NASA and the American people for over forty years," Minovitch argued, claiming he knew nothing of these "gifts" until some twenty years later. "They have been collecting millions of dollars from NASA to explore most of the solar system that was made possible by my invention—while denying me the credit and honor."

Strangely enough, Minovitch kept working anyway. By August 1963 he decided to expand his already mountainous endeavor of computing multiple-planet flight opportunities. This lofty new goal encompassed *ten thousand* possible mission combinations extending from 1965 to 1980. It would be fabulous.

Then the boy wonder caught wind of JPL's newly minted planetary almanac running clear through to the year 2000. To boot, somebody had already encoded it on three big reels of computer tape, ready to spool into an IBM 7090. Huzzah! So Minovitch hoofed it over to the JPL computer building, chatting up one of the programmers there about getting his hands on this grand new reference. The guy told him, *sure, all anybody needed was a job number. What was his?* Minovitch stumbled, quietly volunteering his absent legitimacy. So, um, where could Minovitch get one of those handy numbers? He was told to try William Melbourne.

After finding the man and inquiring about the updated planetary almanac, Melbourne told Minovitch the book was available like any other reference. *Whom was the project for?* Minovitch had no answer. Melbourne was really asking, "What do I associate this request with? Who's in charge?" But Minovitch's need really wasn't for *anybody*, anywhere on the entire JPL campus. Nor was it even for anybody at Caltech or UCLA for that matter. It was for a Minovitchian whim. But this man wasn't coming out to "play computer." Some rather heavy stuff was in progress here, and it could only assist the Lab's efforts. He needed in. Couldn't they just let him in?

What NASA division oversaw the project? Melbourne wanted to know—he was confused. Minovitch certainly knew the answer: none of them. NASA had no involvement whatsoever. *Well then*, came Melbourne's explanation, *you have to be able to bill it to a tangible project.* And Michael A. Minovitch had thus impacted his first bureaucratic brick wall. It was a real slap in the face. Yet Bill Melbourne harbored no evil intent; his responsibilities enveloped proper accounting. He was doing his job. Not having careened into many walls before, Minovitch tried an end run. "I asked him if I could simply charge the work to one of *his* job numbers, or someone else's." But no deal. Melbourne explained how that kind of piggybacking broke the rules and would only get their butts in trouble.

Ensconced in his Dykstra Hall dorm room, Minovitch set aside his major frustrations to doggedly compile drafts of his latest technical report laying out the work to date. These were long papers, running dozens upon dozens of desiccated, typewritten pages, ornamented with comparably huge numbers of diagrams and references—the geometry problem from hell. But unlike many of his halfhearted attempts at publicity, this one actually went out the door. The UCLA computing department read their copy and asked for a little presentation. He gave it in the chemistry building during a campus-wide academic exposition showcasing all of UCLA's offerings, including classical music and ballet. In front of six hundred spectators, Minovitch outlined his concept and used a humongous blackboard to sketch out his fantasy: a lengthy mission skirting every single planet except Pluto. He took a few questions at the end and then strolled around campus the rest of the afternoon taking in some of the other happenings.

Over at JPL, that windy report caught the learned eyes of Elliot "Joe" Cutting. When Victor Clarke shuffled jobs to merge with the engineering staff of *Mariner 10*, Joe Cutting took over his spot in Trajectory. Over the next many weeks, Cutting exchanged a friendly, encouraging series of memos and brief letters with the young gravity master, taking delight in his profound work. Additionally, the fatherly Cutting offered suggestions for condensing and refining Minovitch's text—and not just for sport. This process might well improve the guy's chances for publication in the professional journals—and by extension, acceptance by the professional community. *Didn't he wish to submit?* Mike Minovitch never took the time to do so.

Instead, he took his PhD exams and bombed. Computing runs, paper management, and endless revisions to the FORTRAN gravity program had brusquely infected the sparse weeks of review time Minovitch set aside for test preparation. It just wasn't enough. And, rudely, his failure coincided with the end of the 1964 spring semester. He was on to another few summer months of JPL paychecks—certainly a welcoming presence, but they served to usurp his formal education all the same.

Back at the Lab, Minovitch came upon IBM technicians ripping out the 7090 hardware, replacing it with up-to-date 7094s. They were three times faster than what Minovitch had been working on. He marveled at their improvements, but the machines introduced another obstacle—they didn't think enough like the old ones. He was going to have to rework the program yet again.

Summer went well enough for Mike Minovitch—better than it did for some. Black Americans called it the freedom summer, an overdue period of nonviolent demonstrations and sit-ins designed to further the cause of racial equality. Massive numbers traveled to the South in participation. And over the course of the freedom summer, thousands of students were arrested during peaceful protests of segregated lunch counters and bathrooms. Plenty got their heads bashed in. Three young men were murdered in Mississippi while organizing for black voting rights, in spite of Mississippi law already having given it to them almost a hundred years before.

Mike Minovitch spent most of the freedom summer hashing out endless

Earth–Venus missions along with a few thousand Earth–Jupiters and then a few thousand more that were designed to wing past Jupiter and escape the solar system completely. He did it with six fully distinct versions of the program, taking long enough breaks to start work on what would dilate into an obscenely long hundred-page technical paper itemizing the results. His overtaxed JPL office sat five floors up in Building 180: the tallest, most prominent, and probably most aesthetically pleasing building on campus. Seeing how it's built into a hillside, the JPL grounds present unique architectural challenges. Often, buildings have ground-floor entrances on two different floors. A long, graceful marble staircase adorns the entryway of Building 180 and connects the first-floor entrance with the one up on floor two.

Building 180 never saw a cubicle; everyone benefited from decently sized offices with walls going all the way up and real doors that opened and closed. Minovitch's own door read Bay 504 on the nameplate, and one August afternoon a visitor stopped by in the form of William Sjogren, longtime JPL engineer. Sjogren expended most of his waking hours on how to best establish the position of a Ranger spacecraft in Earth orbit—nauseatingly precise information that was then used to try to make one of them hit the Moon. His was known as the Orbit Determination Group. "I came in on the evenings and on the weekends and tried to get my stuff done *and* make a good impression on the boss," Sjogren offered. "But Mike was in there working all the time!"

Poster-sized diagrams lay spread about in Minovitch's office like unrolled lengths of wallpaper. After the hellos, Sjogren asked what was going on, and Minovitch eagerly filled him in. The two worked in completely separate Groups—meaning Sjogren knew little about calculating trajectories. Intrigued by such ideas, the visitor examined Minovitch's beloved plot of eleven discrete mission possibilities spanning an upcoming period between 1967 and 1978. All began with Earth-to-Jupiter trajectories and then branched off to diverse combinations of outer planets.

The two glanced over Minovitch's flight profiles. One of the more interesting possibilities went Earth–Jupiter–Saturn–Uranus–Neptune. They could get there; he'd proven it. If Minovitch had a spaceship, he could have reached out and plucked Neptune off the tree.

"I would think a *lot* of people would've seen that stuff," Sjogren later commented. But nobody had. And at that point, the busy Sjogren excused himself to resume work.

A few weeks later, Joe Cutting asked Minovitch whether he'd be coming back for the summer of '65. The Lab could definitely use any available help. Minovitch reported that he felt a bit drained from the whole experience as of late and probably would just focus on getting across the plate with his doctorate. It still hung over his back like dead weight, and he'd come too far to just chuck it all in the garbage. Next summer would have to be devoted to getting the PhD wrapped.

Before departing that fall, Mike Minovitch arranged to store his roomfuls of trajectory printouts in the basement of Building 180. He hadn't been gone long before somebody down there needed to free up a little space and began throwing them away.

Central Design Bureau No. 1 endured three costly *Venera* washouts between January and March of 1964. Then the Venus launch window clanged shut. They took a whack at the Moon, but never managed to reach orbit. Then two *more* lunar landings were attempted in April and June. Both ended up as miscarriages. None were due to the probe itself. Really, the only half-decent luck Korolev enjoyed was with his car-sized, insectile *Zond 1*, which, after a refreshingly successful launch, headed to Venus on April 2—about the time those first busloads of rallying Americans got stopped at the Mississippi border. *Zond 1* was made to study the character of space in between planets along the way. It also carried an unpretentious little urn of science meant to unhook from the main ship and set down on Venus. The lander resembled a bulbous oil filter: a spherical body converging on a flat, enclosed top, allowing the sensors and sniffers to jut out a smidgen. Water landings remained a possibility—so this descent module was built to float and carried temperature and pressure sensors as well as a device to sample the air. It held cosmic ray detectors, a stash of batteries to run it all, and just enough room was found for a little Soviet pennant. On the downside, the module wasn't built nearly as tough as Venus demanded: it would've been crushed by the atmospheric pressure.

During transit, *Zond 1*'s bitsy pressurized operations module developed a pinhole leak. That's the same as having a tiny and rather unwanted rocket thruster pointed in the wrong direction. *Zond 1* progressively drifted off course no matter *what* the corrective maneuvers; it would never reach Venus. Optimism remained that the transit experiments might continue running in the near vacuum. But then the radio switched on, and with the half atmosphere inside—*zap!* It created a miniature lightning storm and fried all the electronics. Instant ghost ship.

Zond 1's descent capsule, however—being a separate module with its own power supply, minicomputer, and radio—hung in there to keep the fight alive. Built-in redundancy allowed the entire aggregate craft to be controlled using only this part of the ship. Its design strategy kept the science experiments running. For a month and a half, it sent back everything it found until falling silent at the end of May. She'd missed Venus by sixty thousand miles, and *Zond 1* was finished. "Contact lost," the official report concluded, which is something of a tight condensation of the facts.

What happened with all those other launches? The situation boiled down to Korolev and Tikhonravov having a few growing pains with their Molniya launch vehicle. The name means "lightning," and it had been in the works for almost five years. The whole thing had started with another ubiquitous Korolev missive to advantageously balance two high-altitude stages on top of the classic R-7 base. They needed a lot of rocket to nail Venus (or Mars), and Molniya could handle this load. Except there was a catch.

What catch? Starting a liquid-fuel rocket engine when it's *weightless*. On Earth, gravity means that fuel stays in the bottom of a car's gas tank. Fuel pumps draw from there, so the gasoline typically comes when called.

But no fuel tank of any kind can ever be completely full. And if a rocket engine is two hundred miles up in Earth orbit when its turbopumps start sucking air, the whole thing is going nowhere fast. Over in the West, the U.S. Air Force and Space Technology Labs collectively worked their own designs on boost from orbit; every time, they ended up staring down the exact same problem: how do you get the fuel to the bottom of the tank?

The crowning Soviet fix—which became everyone's—incorporated tiny solid-fuel rockets ringing the base of the orbiting stage. They'd fire seconds before final stage reignition, providing just enough oomph to settle fuel at the bottom of the tanks and keep the feed lines full. Today they're called ullage motors. "Ullage" is an old brewer's term for empty space at the bottom of the beer barrel.

Korolev's rough streak wasn't over; that June, another Zond went belly-up. Then *Zond 2*, using the last teeny slice of the November 1964 Mars window, somehow failed to unfold one of its solar panels. Right on its heels came *another* triplet of failed missions, all lunar landers, draining Korolev of precious rubles and energy. After that, *Zond 2*'s twin was finally ready, but it was much too late for the Mars window. They went anyway in July 1965 to wring out the overall system design and enjoyed rich yet shallow dividends in the form of vastly improved pictures of the lunar farside.

A day after the October 12 launch of the three-man *Voskhod* mission, Nikita Khrushchev became irate when nobody called him to report the blast-off, as was custom. He finally got the chairman of the Military-Industrial Commission on the line, a Chairman Smirnov, and demanded of him, "What's going on with Korolev's launch? Why haven't you informed me?"

Smirnov mumbled something about not having any time to call, but that wasn't the whole truth. Khrushchev, seemingly, was pretty much the last to know he was no longer in power. His heavy (and public) drinking, cantankerous ways, and bullying nature, all made the Soviet Union look like they had less of a leader and more of a court jester. He'd put up the Berlin Wall. He'd cancelled a world summit meeting when Gary Powers's U-2 got shot down. He introduced nuclear weapons to Cuba, and then, his critics argued, he mishandled the resulting hairy standoff with the United States. He could have done more to help the country. His poorly thought-out agrarian policies, in particular, had been a devastatingly complete failure. The great Soviet Union, with all its power to the workers, had been importing steadily increasing amounts of grain from other countries. This village idiot simply had to go. And Nikita Khrushchev's enemies were capable of vastly more than benign responses like disapproving speeches. They plucked him out of office and confined the guy to house arrest. Leonid Brezhnev stepped forward to take the reins.

With that, a competitor of Korolev's, named Vladimir Chelomey, was about to reach the end of his own personal gravy train. For years Chelomey's entirely parallel rocket development had only served to mooch rubles from Korolev and undermine the progress of both men. This was competition, in a socialist economy. He won the favors of Khrushchev, who had taken quite a liking to this man.

But no longer did Vladimir Chelomey have the head of the country in his pocket. By the end of the month his R-7 competitor, the UR-200, had been wiped off the schedule. Many years later, a high Soviet official stated that Chelomey's diversion of resources from Korolev cost the Soviets five solid years of progress.

Had Vladimir Chelomey been handy with a crystal ball, he might have gladly given the UR-200 to spare himself of what was coming down the pipe. Brezhnev's new leadership brought Chelomey and all his programs at Central Design Bureau No. 52 under a powerfully awkward spotlight. The Chelomey empire now measured twice the size of Korolev's. The guy had his own completely separate assembly, fueling, and test facilities at Tyura-Tam. There, construction progressed on attached living quarters for ten thousand people. Was all that necessary? How well did he keep his bureau's secrets? How did he keep the books? How thick was the carpet in his dacha? Everything came under microscopic review.

Long ranked as one of Chelomey's subsidiaries was a diminutive place called the Lavochkin Design Bureau, located just northwest of Moscow at Khimki, right on the Moscow Canal. They'd started with airplane design back in World War II, progressing to cruise missiles by the fifties. For many years, the dense habitation of aerospace firms in Khimki left the city decidedly off-limits to any foreign visitor. Today it has a basketball stadium and an Ikea.

Vladimir Chelomey used Lavochkin to sublet out heaps of his "busy" work—things like antiship artillery missiles—thus leaving his days more freed up to sight directly on the intercontinental missiles and heavy launchers. Not even twenty-four hours had passed after Khrushchev's overthrow before Lavochkin effectively seceded from Chelomey. The emperor's wings had been clipped.

That gave Sergei Pavlovich an idea. For some time, Korolev had noodled around a few possibilities regarding his own design bureau, and Lavochkin's sudden availability must have attracted him to it. Undoubtedly to clear his plate for the upcoming push to land cosmonauts on the Moon, Korolev began slicing up his organization and gifting out major divisions of it to other entities. *Reconnaissance* satellites went one direction; *communication* satellites, another. Those were the easy bits, far from his heart. But as the Moon race heated up and his back pressed more firmly against the wall, Korolev finally, involuntarily, addressed the projects that meant so much more to him. He was losing control of it all. Hyperstressed engineers caused basic mistakes, and the slipshod work drained victory through his fingers like grains of sand.

To this end, Korolev finally knew what he must do.

Until humans had begun flying in space, the cosmic rockets and the interplanetary stations were dearest to Sergei Pavlovich. Each was a little child of his, like a piece of his own intrinsic being gone off to explore. Now his gaze turned to the northwest, past the Moscow Canal into Khimki. He contemplated Lavochkin and realized it was time for his beloved planetary machines to start hatching in a surrogate nest.

By early March 1965 Lavochkin had officially become their own free-standing entity, run by a confident and respected fifty-year-old radio technician named Georgiy Babakin. And the place was hungry. The organization's namesake had died just a few years before, leaving the melting pot of old and young designers adrift on a logjam of unfinished work. Their last pet project had involved a sort of space aircraft that could take off, enter lunar orbit, and then return to land on a runway. But the nation's change of leadership was putting it all in dry dock.

Sergei Korolev made his first and last trip to Lavochkin that July. They'd readied a good show for him that day, tacking sketches of their brainstorms around the walls of Babakin's office. Designer Vladimir Perminov stood close to the drawing of their proposed Mars lander as Korolev and Babakin entered.

Lavochkin already had a strong work ethic in place. Young, fresh graduates from Moscow State Technical University often came straight from final exams to the bureau, cherry-picked by Lavochkin himself. Single and highly motivated, they never failed to set a blistering pace. Most of the facility ran straight through the small hours without so much as a break. For those who foolishly tried to sneak home for a pot of tea and sack time, a special car—complete with driver—ricocheted through the streets collecting those particular few who might be able to solve the problem of the moment.

When Babakin took over, he continued the trend—lights on late in the office, hunkered over the predicament, looking for a way out of it. And after Perminov joined up, he failed to elude the system. "Quite often I was awakened during the night and was taken to work from a warm bed," he expounded many years later of the car-at-night arrangement. "There were wild nights, when problems appeared in several divisions simultaneously. Then, my wife worked like the secretary and by telephone told where one could find me."

Vladimir Perminov was no spring chicken. In 1954 the cherub-faced youth graduated Kazan University—a dense yet heady aviation-centric institute sitting on the Volga River, east of Moscow by hundreds of miles. Like so many others, he joined up with Lavochkin straight away and plunked down to work on blueprinting the space plane. Now, that gimcrack had kicked the bucket and here Perminov was, waiting to hear about designing a Mars craft. He couldn't believe it.

Korolev glanced around at the posters. Lavochkin had a whole bloodline of planetary orbiter-lander combinations in the works, along with a crop of lunar landers—a bit of everything. Some of their dizzying pent-up energy went into overhauling Korolev's own existing Mars and Venus probes and deftly removing dozens of pounds from his designs. They aimed to soft-land a camera on the Moon by early 1966—a specific goal of Korolev's he hadn't yet been able to squeeze out. In sum, they'd really blitzed the chief designer.

"One could see that Korolev was thinking about something quite important to himself," Perminov recalled of the man's behavior. Nobody knew that in six months S.P. would be dead. He and Babakin approached Perminov, and the chief designer studied one of the nearby posters for an interminably long time. The picture showed Lavochkin's plate-shaped Martian lander slung underneath a parachute. Finally Korolev mumbled, "The landing should be performed by the engines, without parachutes." It was the quiet sort of phrase not intended for others to hear; he had audibly spoken to himself. Perminov swallowed hard and stoutly brought himself to remind the chief designer that Mars had a thin atmosphere, so they figured on using it to their advantage. Korolev shot him a look and then ponderously assumed a place at Babakin's grandiose meeting table, bringing the affairs to order. Quickly Perminov sat down and didn't say another word.

Looking out at approximately ten men before him—Gleb Maksimov, Babakin, Perminov, and some others—Korolev drew a heavy breath and silently contemplated the character of Taras Bulba, from Nikolai Gogol's famed novel. In this old tale of Russian Cossacks, Taras kills his son for betraying the country.

Gesturing to Maksimov, S.P. indicated they'd been working themselves almost to ruin on the interplanetary stations, slamming into problem after problem. There'd been some success for sure, but the weight of the moon race had ultimately proven him unable to manage it all. There was too much, just too much. It was time for someone else.

Reverently he said, "I hand over to you the most valuable possession—my dream."

Nobody spoke.

"I expect you to work hard," Korolev continued. "But if my faith is not rewarded, I'll do as Taras Bulba once said: 'I gave you life and I'll take your life.'"

And then the meeting was over.

15. Think Like Gravity

They just said, "Why don't you go ahead
and work on that." So I did.

Gary Flandro, on reporting to JPL in 1965

Grocery shelves first welcomed SpaghettiOs. Ali knocked out Liston in one round for the heavyweight championship. Sony introduced the home video recorder for a thousand bucks. U.S. president Johnson authorized offensive ground forces in Vietnam. The Rolling Stones recorded "Satisfaction." And that was all before July.

The watershed summer of 1965 may well be remembered for many things, but one of the unsung moments certainly involves Gary Flandro, at the time a Caltech PhD student and part-time JPL employee. At the beginning of that summer, on a warm June morning, Gary sat before two unremarkable pieces of eleven-by-seventeen-inch graph paper and used them in making a discovery that would revolutionize our knowledge of the solar system. Mike Minovitch commandeered a trio of multimillion-dollar computers, but Flandro had his graph paper with the little orange lines.

Back in March was when Flandro's supervisor, Joe Cutting, began doling out a variety of new jobs to his underlings. As a seasonal member of Cutting's Trajectory Group, Flandro's latest chore involved looking ahead to see where possible, worthwhile exploratory missions might lie. And Gary Flandro was in a mood. This whole stupid assignment had left him clammy in the first place. Gary wasn't the type to come out and tell anyone that directly, but all the same, he figured this task was your typical, inane busy-work project. Until that morning, he felt any "real" JPL work involved the Venus and Mars preparations.

Up on the third floor of Building 180, Flandro hunched over his rumpled pages, futzing with his graph paper this way and that. It was nice and sunny outside. In the back corners of his mind sat the idea that one day, when his thesis on rocket combustion instability was all finished, he'd be doing something perhaps a little more . . . *worthwhile.*

His bachelor's degree read, "University of Utah, Mechanical Engineering, 1957." With Flandro's cap and gown barely off, the Sperry Corporation picked him up right there in Salt Lake City. Flandro spent more than a year thrashing out Sergeant missile designs and flight trajectories. But unfinished business always nagged; he wanted to go back for more school. The company harbored endless JPL connections, so Flandro used them after calling a hiatus from Sperry. He joined JPL for part-time work while pursuing a master of science degree at Caltech. And thus Gary began operating in a rhythm: full-time school during the academic year and JPL during the summers. Flandro got his master's degree in hand, but it wasn't enough. He lingered in California to begin work on a doctorate, again managing to pick up extra cash with spare hours at JPL. "I probably have the record for the most times being hired, then going back to school," Flandro explained in a deeply rich, wizened voice that would not be out of place on a radio announcer. By June 1965 his PhD still lay off in the distance.

Gary's first problem that day involved figuring out not where the planets *were* but where they *would be* in years down the road. It's a little like the problem facing a football quarterback, who must throw the ball not to where his receiver is but to where his receiver is going to ideally end up in the next moment or two.

Joe Cutting hurried to finish instructing Flandro and set him loose, returning to his own work. Cutting had other irons in the fire. Taking a cue from Mike Minovitch, his goals lay in reaching Mercury—but first going by Venus. With a close enough flyby and pinpoint timing, a little ship could theoretically be slowed enough at Venus to get throttled along to Mercury at just the right speed. They were finally going to test it out. Maybe Gary would want to apply similar techniques to the outer planets?

What, like Uranus? Neptune? They were so far away that it almost wasn't worth the effort. Neptune in particular was practically off the charts—3 billion miles from the nearest Walgreen's, on average. "No one was seriously contemplating flights to the outer planets," Flandro explained. It was hard enough to keep a Mariner alive long enough to buzz Venus. With this new task, he felt he'd been pushed aside, studying vaporous missions requiring technology and equipment that hadn't even been seriously discussed. Nevertheless, the paycheck was helpful. So, familiar with the mechanics of such trips, young Gary scratched his crew cut and agreed to give it a whirl.

He already knew it would probably have to be done by leapfrogging from one planet to the next. The basic physical mechanism for that boiled down to an energy *exchange*, as it were, because the boost given to the spacecraft would come from a consequential degradation of the host planet's orbit. Flandro learned the nuts and bolts of it all while attending a 1959 short course at UCLA taught by Peenemünde refugee Krafft Ehricke. If such techniques were applied correctly, a spacecraft could be sped up or slowed down, depending on the goal. "I used Ehricke's works in all of my mission analysis efforts at JPL," Flandro said.

It's not that you couldn't get a rocket to someplace like Neptune without gravity's help—you'd just have to wait the forty years or so for it to arrive. "Suppose we've decided that we've got to go to Neptune," remarked Flandro. "And I go to the government, and I try to sell them on the idea of doing this mission. And I have to tell them it's going to take forty years to get there. What do you think my chances are of getting any money?"

Work began in a decidedly low-tech way. Flandro Scotch-taped two sheets of graph paper together, plotting time on one axis. Then he took a reachable copy of *Her Majesty's Nautical Almanac* off the shelf and used it to mark planetary longitude on the other axis. Now the analysis began. He needed actual time periods when the flights could reasonably be made. That meant plotting where the big gas giants were going and doing an appraisal of their proximities to one another. Orbital mechanics are like clockwork gears and ballroom dancing. Circles and ellipses.

On that summer day, tracing the arcs of the planets like he was, Gary Flandro had cause to set down the pencils and review what he'd done. The lines on his taped-up pages conveyed an outrageous tale. Grinding their way through space, four of the outer planets—Jupiter, Saturn, Uranus, and Neptune—would, in the next fourteen years, begin aligning themselves on one side of the Sun! It was not quite a perfect line, but definitely close enough for planets.

So what would gravity do? Already Flandro knew the answer: gravity would become his friend.

Flandro licked his thumb and extended it out at the far wall. They would align *closely enough* that if a spacecraft were launched toward Jupiter, the gravity of such a huge planet would fling the craft onward to Saturn. The line was *close enough* that the same thing would happen at Saturn—a big, zippy kick in the pants sending the craft screaming on to Uranus. From there, Neptune would be *close enough* to reach with another spurt. Using the accelerative nature of gravity, Flandro could get to Neptune only seven and a half years after launch. The hair went up on his arms.

Unwittingly, he had also just discovered what would become JPL's redemption.

If I can just get myself to Jupiter, Gary Flandro began thinking, *I could get myself to just about anywhere in the solar system.* He traced a finger from the alignment on back to the y-axis, checking for the year: 1977. If they got it together and launched then, he could see Neptune before he retired. He could witness it personally, along with the entire world. Discovering a seven-and-a-half-year trip to Neptune, he'd come far since that day in the JPL cafeteria when he had asked von Braun to autograph his book.

For Christmas in 1940, Gary Flandro's mom gave him an astronomy book. His six-year-old eyes marveled at the drawings of the planets and the spacecraft that someday might visit them. At seven Flandro spent a dollar for a hobby-store telescope kit that supposedly equaled the power of Galileo's. His dad helped assemble it, planting a few seeds in the process. A long string of home brewed telescopes followed. "In my teens, I spent many fruitless hours looking for evidence of the canals on Mars," he admitted.

One drawing in that astronomy book always stood out. It showed the planets lined up together in a sort of roll call. The image found a dark little parking space in the back of his head for twenty-five years—until this very day in Building 180, when his cerebral filing system recalled the illustration and brought it forward once again. Nobody would need a drawing like that for much longer. Flandro might not have been the first to imagine what photographs of Neptune would look like, but he was the first to know they'd actually be coming.

Gary Flandro had just given birth to Voyager.

Budgets, meetings, spacecraft design, computer programming, good weather, birthdays, holidays, sick days, or snow days—impetuous, flippant planets would wait for nothing. All four were already on their way together. JPL had twelve years to circle some very, *very* expensive and complicated wagons in order to pull off what could easily be regarded as pure fantasy. So at that exact moment in time, a fraction of an instant after Gary's revelation, the clock started counting down.

Next afternoon, Flandro walked as if his flight was boarding. Those in high positions needed to hear, to be convinced of this outer-planet gob smacker. They needed to buy into it, and Flandro was now a crusading entrepreneur with something to sell. Rightfully, his first customer would be Joe Cutting.

Cutting's real first name was Elliot, but his dad always used a nickname based on Joe-Joe the Dog-Faced Boy. It stuck into adulthood, and everyone used it.

Cutting loved Flandro's idea. The pair slapped together a presentation, then barnstormed Homer Joe's office and assailed him with slides. JPL chief scientist Homer Joe Stewart happened to be the same Homer Joe who'd complicated James Van Allen's life years earlier by helping select the Vanguard rocket to be America's first in space.

Cutting and Flandro clacked images through the slide projector, carefully describing the alignment, basic probe trajectory, and possible big-picture science opportunities. Stewart embraced the idea like a new parent. He told the men what a great find he thought it was, a real treasure. He loved Flandro's graph. The slides really sold him. A few problems hung off the edges, though, leaving him a teensy bit skeptical—problems like getting through the snippy asteroid belt running between Mars and Jupiter. And what about the electronics—eight years in space? Things coming out of the Lab barely made it ten months. The communications would also stretch what they knew—could a useable radio signal even make it as far as Neptune? It wasn't possible with what they had sitting on the shelf right then, but people could start looking into the situation. And yeah, the people—hmm, that was a tricky one. Would the manpower exist for what they needed, or would the Mars program take a hit? The more they talked, the more complications arose.

Even so, Stewart had to agree that four planets for the price of one was simply magnificent. To Flandro and Cutting, he mentioned some work done in the fifties by an Italian named Gaetano Crocco, who studied round-trip flights past Mars and Venus, which then returned to Earth. Crocco termed his mission the Grand Tour. It never flew—perhaps something else worthy of the title now *could*. And under the circumstances, Stewart could think of no better name. *What did they think?* All agreed. Stewart dismissed the men to begin crafting a news release.

Now they needed detailed trajectory calculations, more precisely approximating the journey from one planet to the next. The task demanded intensive computing runs. "I tried to find out what software we had," remembered Flandro. "In those days, it was pretty sparse." Poking through the meager options, he chose a trajectory program done up by Lockheed in Sunnyvale—the same Lockheed that had received a little Easter basket from Victor Clarke three years before. Already, Flandro had eyes on a fairly specific time period. Within this, he needed to evaluate launch dates, flight times, and course corrections. So in the days following, he set up tasks for the computer.

And Flandro knew his place. Students and even outright JPL employees didn't run the computer. Just the programmer did; one of them was a really nice guy named Don Snyder. Every day, he worked with Flandro to set up punch card decks for the necessary computing runs, which then began with regularity. The next morning (if Flandro was lucky), a stack of printouts would be on the floor outside his door. He'd take the numbers and make plots from them, seeking to tease out the best flight paths.

There was way too much to try to comprehend all at once. It required compartmentalization. The road would be built one mile at a time. Dividing the job, Flandro first created a bunch of Earth–Jupiter scenarios to see where they fell. Then he ran the Jupiter–Saturn scenarios as his next little section of roadway. This process went on for two months, until Flandro had graphed his entire course. On the improbable chance that JPL didn't want all four planets, he also ran a few Earth–Jupiter–Pluto scenarios, some Earth–Jupiter–Uranus scenarios, a couple Earth–Jupiter–Neptune scenarios, and a handful of Earth–Jupiter–Saturn scenarios. Even basic contingency plans might well save the day if trouble ran the primary mission aground. By the time his initial foray wrapped, Flandro possessed six basic mission profiles. If they got approval, work could begin on payload size and on just how big a rocket they'd need to shoot it all off. Hundreds could soon be working from the ideas of one.

Almost everybody at JPL read Stewart's news release, which failed to include the name of Gary Flandro. "Homer Joe's name was of course prominently mentioned," Flandro said. "But that's okay. I don't think he meant to try to take credit . . . he was the spokesman for the Laboratory—perfectly okay." The ball started rolling. It hit the press, little articles appearing here and there in the newspapers. *Homer Joe's got an idea; Homer Joe's figured out how to get to Neptune.* "My name was not mentioned," said Flandro. "Homer Joe Stewart was named the originator of this mission—which is of course not how it went. But I didn't think a thing about it. I was just a student, you know? Why should I be concerned about this kind of thing?"

Many scientists found themselves all lathery about the possibility of discovering majestic new things at these previously untouchable planets. Things totally beyond their reach until now. It was big time. People like Van Allen might want on board. Flandro had actually seen Van Allen at JPL a few times since *Explorer 1.* "I would never have dared to go talk with him," Flandro said, "because I was just a kid."

Not everybody within JPL reacted positively. Flandro took time to hand-shake his way through the frontline engineers, speaking with guidance and electrical folks, the mechanical wizards, the communications people. He was campaigning for the Grand Tour. Somebody told him, "This is ridiculous—we can't build a spacecraft that will last over two years, let alone eight years!" But Gary Flandro didn't know any better, so he kept pressing the issue.

Heaps of trajectory work sat under Gary Flandro's belt. He'd found a road to Neptune—and that was the problem. All his efforts dealt with the flight path itself. Nobody knew for sure what kind of car to use on the trip. Flandro needed help. One new hire looked promising—Roger Bourke, though totally green, had just recently signed on as a member of the Advanced Projects Group. Somebody nudged him in Flandro's direction, concluding that Bourke's lengthy schooling in aeronautics and astronautics might apply itself well toward resolving some of the vehicular details.

If Flandro was still an infant, Bourke was a newborn. He'd come to JPL directly from Stanford and had only been at the Lab for a year. "Education, no experience," Bourke conceded.

In Flandro, Roger Bourke found an ideal complement. The trajectory man could stay focused on mapmaking, while Bourke tackled the hardware and overall mission goals. Bourke's day-to-day work resolved itself in a couple of key topics. First came addressing the high-level spacecraft considerations—defining equipment and strategies capable of successfully navigating Flandro's twisting highway. Then came the collateral issues like Earth-to-ship communications and more visibly defined science objectives.

Bourke's inexperience lent itself well to these particular tasks, freeing him of the inside-the-box, tried-and-true methods that a veteran might have quickly leaned on. For Bourke there *was* no box. "It never occurred to me," he later suggested of his 1965 mindset, "that you *couldn't* build a spacecraft that would last for twelve years or would run on nuclear power." To him, the thorniest issue remained that of navigation: could they still properly aim something when it was billions of miles out? His naïveté spoke again, *Well, how hard could it be?*

Over a dozen months went by. Bourke and Flandro assembled an official-sounding memo entitled "Comments on Proposed Grand Tour Mission Study," dated October 1966, which they then let drift on up the chain. In it, Bourke had typed up two and a half pages detailing the men's opinions on such issues as guidance, extreme flight times, long-distance communication, and the possibility of lifting their creation on von Braun's skyscraping Saturn V. Prerequisites would factor heavily into success. The pair mentioned vast educational benefits to flying a precursor multiplanet mission in advance of the whopper. They singled out Joe Cutting's Venus-to-Mercury sojourn as the perfect test bed.

They also farsightedly projected the implications of this Grand Tour. The memo demonstrates a clear understanding of not only their own project's big picture but also of the major forces at work beyond their humble canyon walls. They even got a bit philosophical. Part of it reads, "It is interesting to speculate on what can happen during the course of a mission of this duration. The government's administration will turn over at least once and the Congress several times. The space budget and interest could do anything. The Laboratory's whole direction may change: we could launch while devoted to space exploration and finish while interested in oceanography. At the end of the mission, it's unlikely that any significant fraction of the personnel on the project will have been associated with it at the outset."

But the overriding conclusion? Could it be done? Hell yes. Somewhere between Flandro's discovery and this memo, the Grand Tour metamorphosed from crazy idea to outright possibility. It existed only in a memo, but the ultimate space journey was already underway. They had eleven years to design it, line up money, pick the experiments, build the craft, finish testing, and get it off the ground without it blowing up. Then the hard part could begin.

Eleven years and counting.

16. Didn't They Get It?

Many people do not buy Mike's view, so he is upset.

Joe Cutting

As more and more tantalizing Grand Tour details appeared in the professional literature, Mike Minovitch became increasingly alarmed by a distinct pattern: none of it mentioned him. There was nothing about gravity thrust, nothing about his wondrous computer program or discovery of the unprecedented four-planet opportunity. Week after despondent week, he'd pore through the industry magazines and other professional scientific journals, feeling morbidly sick to his stomach. One article said Victor Clarke came up with the idea. Another mentioned Homer Joe Stewart as the Grand Tour's originator. What the devil was going on here?

After combing through the available possibilities, Minovitch began seriously contemplating a dirty fact—nobody at JPL wanted to properly credit him. They'd first reacted to his work by decrying it as a violation of physics. But when the realization of what Minovitch had in his hands finally registered, the mood abruptly changed. Clarke, Stewart, and all the others? Evidently, in Minovitch's eyes, they wanted it all for themselves and weren't legitimately interested in sharing the propulsion breakthrough of the century with a part-time grad student.

Already Minovitch had confronted Clarke about blatantly stealing his program; that was back in '63 when *Aviation Week and Space Technology* had reported an Earth-Venus-Mars scheme supposedly designed by Lockheed. Minovitch called it plagiarism—an ugly word in any context. But soon thereafter, Minovitch felt Clarke more or less blew him off and with a gassy sigh retorted, *sorry*, suggesting that nothing could be done about it.

In the late sixties and early seventies, a buoyant Michael Minovitch in-

28. Mike Minovitch, 1971. (Courtesy of Michael Minovitch)

quired to publications like *Science* magazine, *Astronautics and Aeronautics*, and *Scientific American*. Gamely, he foresaw the publishing of his lengthy compendium entitled "Gravity Thrust Space Trajectories" in the more popular, general-interest literature. "I have an obligation to astronautics," Minovitch pleaded in one cover letter, "to make sure this development is accurately documented and reported"—going on to suggest that "any further delay will contribute to a continuing distortion which I have observed in the literature." In the mailing to *Science*, he even tucked in a couple of photographs "for consideration as possible cover illustrations," as he put it. Minovitch implored the editor of *Scientific American* to "publish this article which will give me claim to my ideas and outside recognition which I do not have at this time." Undoubtedly, Minovitch found vigor from a singular internal JPL memo directed at Joe Cutting by a fellow named Bill Stavro. Of

Minovitch's 1961 work, Stavro concluded that "As far I can tell, this *is* the original idea and theory behind *all* gravity assist trajectories."

Every publication declined. Repeatedly. "JPL had so much power over the professional American aerospace journals," Minovitch later asserted, accusatorily implying that the Lab surreptitiously tried to withhold his name from the press and therefore obfuscated the true origin of gravity-propulsion research. These rejections "had the effect," as he put it, "of keeping the American people, and the world, uninformed regarding the major breakthrough in space travel represented by my invention." And so with each round-trip to the mailbox, with yet another rejection, that yucky, distended feeling in his tummy grew.

After waiting for many years—ten, in fact, if dated back to his original holy grail gravity paper of 1961—Minovitch felt the only remaining option was to take matters into his own hands. In a move that might be construed as arrogant, he applied for his own award. Formally, it's known as the NASA Exceptional Scientific Achievement Medal, which, to this day, is still given for groundbreaking technological advances. And it really is a medal. Underneath a ribbon of smartly contrasting blues, a bronze hand balances a ringed sphere on one fingertip. In anticipation of its bestowment, Minovitch even wrote his own citation, which began, "For the invention and development of the concept of gravity thrust space flight." And in a seeming attempt to unite his work with the impending Grand Tour, the sentence picked up with, "which form the basis for the multiplanet missions planned for the Seventies." It had to be approved by the head of the Systems Division, at the time C. R. "Johnny" Gates, who signed off on the paperwork and sent it up the chain. All the trajectory folks resided under the umbrella of the Systems Division.

To Minovitch's great surprise, the award that dribbled on back bore little resemblance to the original. For starters, no longer was it for the Exceptional Scientific Achievement Medal. Instead, the revised citation indicated Minovitch was now being issued the NASA Exceptional Service Medal, which he understood to be something of a downgrade and definitely a slam.

The change in wording read, "For the first intensive development of the concept of gravity thrust trajectories for interplanetary space flight."

His mouth went to chalk. That wasn't supposed to be it at all!

"It is a fraud," Minovitch pointedly insisted, "that had the effect of stealing the invention from me rather than acknowledging me as the inventor."

Sooty flames of betrayal exsiccated Minovitch from the inside as he looked crossways at Roger Bourke, his immediate supervisor, who was standing there with an award Minovitch was very much not interested in receiving. Bourke had moved up the ladder into management. "Aren't supervisors supposed to have the basic integrity to look after the interests of the members in the Group?" Minovitch rhetorically asked years later.

"Not Roger Bourke."

Nearly fed up by this point, Mike Minovitch boiled over when, in 1974, he saw a paper by Norris Hetherington of the University of Kansas slated for publication in an upcoming issue of *Aerospace Historian*. Hetherington's article on the historical origins of gravity-thrust research contained oodles of material provided by Victor Clarke. In Minovitch's eyes, it seriously overrepresented Clarke's contributions and was plumb full of lies.

Resolutely, Minovitch composed a strong letter to Clarke, outlining the improprieties and flagging Clarke on his wrongs. "I want to impress upon you the seriousness with which I view the situation," Minovitch cautioned, referring to the "gross distortions of truth." He indicated that the paper's unedited publication would constitute "a point of no return," offering Minovitch "no alternative but to take the issue into court in the form of a heavy law suit."

That same day, another similarly worded letter went out to Bill Pickering. "I have been extremely patient for several years," intoned Minovitch, "hoping that certain supervisors on your staff would recognize that they can not distort history to suit their own needs." Assuredly it contained language Pickering didn't typically encounter—things like "the situation can only be rectified by a law suit." The then JPL director responded a month and a half later.

He rejected the idea of changes to Hetherington's article, suggesting that interference "would be improper and contrary to the principles of academic freedom." The man is a scholar, Pickering suggested, who must decide on the validity of his own content.

That Minovitch contributed greatly to space exploration was, in Pickering's mind, of no doubt. "We at JPL hold your pioneering work in high regard," he mentioned, acknowledging the importance of gravity propulsion and Minovitch's presence in general. The director pointed out how other Lab veterans frequently referenced his work and that he'd recommended Minovitch for the NASA Exceptional Service Medal, which had duly been presented.

Moving beyond that, Pickering prefaced his next comments by reminding Minovitch that the events in question happened some years before and, as might be expected, "you and various JPL employees have somewhat differing recollections." According to Pickering's own review of the situation, JPL employees had never deliberately changed the facts for Hetherington or for anyone else. Vic Clarke may have made some broad generalizations about who did what and when, but Pickering assured Minovitch that Clarke would be clarifying his remarks to Hetherington in short order. Clarke penitently did so in a two-page letter dated that very same day.

Near the end of his response, Pickering affirmed, "We want you to receive full and proper credit for your work." And he didn't even keep such a pious commitment private. Pickering's letter was copied to a page-long list of aerospace heavy hitters, including Kurt Debus, George Low, Carl Sagan, Wernher von Braun, and James Van Allen—just so everybody knew what was what. Then the JPL director got back to his in-box, and that seemed to be pretty much the end of it.

So if Bill Pickering wanted Mike Minovitch to receive full and proper credit for his work, then why in the world wasn't Minovitch seeing any? Why wasn't Pickering getting it for him?

17. The Death and the Funeral

What do you think if I go home
and lie down? I'm not feeling well.

Sergei Korolev, to his friend during an assembly, mid-1965

At a Soviet Academy of Sciences meeting late in 1965, nuclear physicist Andrei Sakharov bumped into Sergei Korolev. Often, scientists and engineers mix with one another about as well as Siamese fighting fish, but these guys knew each other already. And Sakharov felt concern. Recently he'd learned of the United States handily placing something like nineteen tons of payload into orbit. Although his facts were off by a measure, the Americans' two-man Gemini program *was*, as of late, noticeably and distressingly upstaging whatever grand spectacles Korolev had managed to bring about. He was getting trounced. Blottoed. Pummeled like the chump in a rigged boxing match. So here now stood Sakharov, baiting his friend, knowing full well how no Soviet booster could lift the same amount of weight. *How worried should they be*, he wanted to know. *As a nation?*

A bulbous fifty-eight-year-old Korolev put his arm around the other man and said, "Don't worry, we'll have our day yet."

Korolev's eyes might have given it away with their exhaustion and despair. He'd been working around the clock, overextending himself, driving past driven—and that was after giving up on half his plans. Long spun off to Lavochkin, the pilotless lunar and planetary endeavors had sprouted new roots and fully exited his short-term consciousness. Theoretically, the subtraction left him plenty of energy for the new generation of headline-filling manned exploits. But if it wasn't the politics draining him, it was the budget. And if it wasn't *that*, it was the impossibly demanding, unyielding, unrealistic schedule that had been formed that way only to stamp footprints in the lunar dust before any Americans did. That race kept S.P. plodding on, toes dragging in a half-composed zombie state. He was constantly reviewing schematics, checking parts, yelling about something, or moving from one meeting to the next and from one crazy assignment to another. Somehow he was supposed to rendezvous Vostoks and cosmonauts in space, build a gargantuan N-1 moon rocket with its thirty first-stage engines, then test it so that it could send cosmonauts to loop around the Moon and orbit. And despite the steep reduction in Vladimir Chelomey's stock, Korolev wasn't finished grappling with competition: Valentin Glushko's irritating ways still represented a threat to the chief designer's controlling mindset and methods. "Glushko thinks that he is the chief successor and descendant of Tsiolkovsky!" yabbered Korolev to an associate. "And that we are making tin cans!"

All of it was too much, maliciously snipping little pieces from his soul—*clip-clip*—and casting them off like fingernail trimmings. "Everything exasperated him," interjected his wife, Nina. "Even his slippers were in the wrong place."

She'd get letters from Tyura-Tam written by a very different hubby than the one who'd brazenly marched out of the death camps: "I am somehow unusually deeply tired," he wrote to her. "It is especially heavy and hard." He leaned on tranquilizers to help things along. "I am in a constant state of utter exhaustion and stress," the letters continued, "but I can under no conditions show that these things are getting to me. I am holding myself together using all the strength at my command." Down at the organ level, his overtorqued heart was beginning to give out—slipping gears, weakening him, and dragging the spirit along with. Ten days last year had already been spent in the hospital while doctors furtively combed through his system without any idea of where to begin or even who he was. They found everything in wicked condition—heart flutters, inflamed gallbladder. Mikhail Tikhonravov overheard the hallway whispers and made a comment in his diary: "Things bad with the heart." Commanding the endless strings of meetings and planning sessions as well as troubleshooting and quality control, Korolev continued his steady plunge.

"Sergei Pavlovich would sometimes come home at wit's end," remembered Nina. "He could be so emotionally torn, so exhausted."

The day after Christmas 1965, Sergei and Nina went to visit the cosmonaut training center for a basic little pep rally. "We're preparing the launch of the

Soyuz," Korolev enlightened the assembled group. "We are also working on a space station. . . . We are also working on effecting an unmanned soft lunar landing and conducting research in outer space. You'll learn more about the work once you become involved in it." Overall, the cosmonauts weren't thrilled with his show. Just more promises, promises and nothing definite.

He didn't really feel so hot. Some element of his insides seemed out of sorts, like a ball and chain lassoed to his intestines. At the very end of the year, Korolev finally told his wife, "I can't continue to work like this, you understand!" and then went to see a doctor. Sergei's physical turned up a nasty situation: polyps on his rectum, bleeding and painful and in need of swift removal. The hospital—more of a segregated, elite clinic for high-level state officials and party members—blocked off time for the routine procedure.

"I never remember him to be sick," professed an office worker, Antonina Zlotnikova. Korolev was due to enter the hospital on January 5. More than once, he'd told Antonina that he'd die at his desk. "Even on January 4 he worked late," she said.

On the fifth, Korolev checked himself in, wife Nina coming to visit every day and speak with the doctors. He didn't always understand what they were saying. They tested his hearing and found it to be pretty much shot. Most every doctor failed to understand how in the world he could hear anything at all. A spatula heard things better. During all those years of rocket engine tests, Korolev had probably stood too close to the noise and roaring; now he might just as well pack his ears with mud. The hospital dispensed a hearing aid to counteract the destructive effects, yet Korolev never bothered to use it. He went into surgery on the eleventh, at eight in the morning. The twelfth was his birthday.

The assigned surgeon, no less than the minister of health himself, entered the room to begin. This operation was considered simple enough for another patient's to be scheduled immediately afterward. It went downhill early on. The polyps came out endoscopically; then Korolev's arteries let loose an onslaught of blood that failed to slow. They lanced into his abdomen in an offensive to halt the bleeding and—*gasp*—stumbled upon a massive tumor. It was one of those came-around-the-corner surprises. "Very big, like two fists," recalled a nurse who had been in the room that day. The sizeable mass had already wormed itself deep into his intestine and pelvic wall and metastasized for the long haul. Conventional wisdom suggests Korolev wouldn't have lasted more than a few months even *with* surgery.

But in the operating room that day, the surgical team gallantly forged on. Painstakingly, they'd have to disassemble Korolev's rectum to get it all out. "This took a long time," Korolev's daughter Natasha later summarized. They kept him out cold with nitrous oxide, a fantastically poor choice for such a lengthy procedure. But it was never supposed to go that long. "My father had an anesthetic mask on for eight hours," Natasha said. Other options existed, but only for those lacking the specific injuries Korolev had suffered long ago. "They should have put some kind of tube into his lungs," explained Natasha, "but his jaws had been broken in prison." Her stoic father had never revealed the malady to his physicians.

The nitrous kept running out—the hospital only stocked miniature tanks with twenty-minute loads. They injected muscle relaxants to keep him under. But how might that affect the decrepit heart? Nobody had ever bothered to run an electrocardiogram ahead of time, which didn't exactly bode well for a patient with cardiac insufficiencies. Frenzied shouts came from the operating room—a desperate call for the only other qualified surgeon in the area. *Where the hell was he?* The doctor was on vacation; somebody had to track him down. Korolev was down then up again. The nitrous mask came off when the cocktail of injected drugs took effect and then went back on as he repeatedly faltered. Rollercoaster vitals. Unable to get a tube down his throat, they delicately punctured the neck and went in that way. The ticking clock ticked louder. The other doctor finally materialized, snapping on gloves to gallop in for the final round. At long last they breathlessly finished cleaning out the chief designer and stitched his torso back together, then left the room to mop their brows. Half an hour later Korolev's pulse stopped, bringing the men racing back in to get him going again. Nobody could.

Valentin Glushko was in a meeting when the phone rang. Without emotion he listened to the news, then hung up and turned himself back to the group.

"Sergei Pavlovich is no longer with us," he said blankly. "Now where did we leave off?"

Premier Leonid Brezhnev decided to be one of the pallbearers. It was he who reportedly consented to let *Pravda* go and send the guy out in style, finally revealing Soviet chief designer Sergei Pavlovich Korolev's true name to the world on January 16—faithful party member, husband, and father. The newspaper article included a formally posed shot of Korolev wearing his medals. For a man who was never supposed to be *seen* let alone *identified*, the foresight to ever arrange such a picture in the first place had to be applauded.

On January 18, draped in flowers, his coffin first sat high on a pedestal in the House of Unions, lying in state as various design bureau engineers and scientists, cosmonauts, and workers diligently filed through in offering their last to Sergei Pavlovich, whose cremated remains lay nestled inside. Classical music softly played out over the scene. Many sent wreaths to protectively surround the coffin, wreaths often draped in banners congratulating "the outstanding Soviet scientist, twice Hero of Socialist Labor, Lenin Prize winner, the Academician Sergei Pavlovich Korolev." The Council of Ministers sent one, along with the cosmonauts, the air force, and the Ministry of Defense. In some form, they all owed debts to Sergei Pavlovich.

Rotating every three minutes, the honor guard was comprised of fellow engineers and cosmonauts alike. Leonid Sedov took a turn. And then at noon they all realigned for the procession to Red Square and the Kremlin Wall. As the convoy snaked along through Moscow, thousands crowded the Square to catch a glimpse of this hallowed man who had forever stayed in the back—for Sputnik, Gagarin, for them all—and brought glory to a nation perpetually out of step from the rest. He'd transformed their rough and rusted homeland from primitive agrarian tradition into one of only two world superpowers. He was the no longer anonymous chief designer. Today was finally his day.

They reached the Kremlin Wall Necropolis. The speakers took their turns, Council of Ministers chairman Smirnov eulogizing Korolev's "inexhaustible energy." He spoke of the man's organizational skills, his forward-thinking progress, and how only "the labor of collectives" could finger the kind of success Korolev had experienced many times over. "Farewell, my good friend and comrade!" Smirnov proclaimed. "I declare the funeral devoted to the memory of Sergei Pavlovich Korolev to be open!" Mstislav Keldysh took his turn as well, preaching Korolev's "burning faith in Tsiolkovsky's ideas on interplanetary flight."

Then the crowds quieted as Earth's first exoatmospheric individual humbly took the podium. "The name of Sergei Pavlovich is linked," began a som-

29. Korolev's big day—his funeral. Pallbearer Brezhnev stands immediately behind rightmost portrait holder. (Author's collection)

ber Yuri Gagarin, "with a whole epoch in the history of mankind—the first flights of the artificial Earth satellites, the first flights to the Moon and to the planets, the first flights by human beings into space, and the first emergence of a human being into free space." Gagarin's last pronouncement referenced the pioneering spacewalk of Alexei Leonov during *Voskhod 2* nearly one year earlier. Without a doubt, it had been the absolute pinnacle of Korolev's achievements. He didn't live to see another cosmonaut launched.

When Gagarin finished eulogizing, Korolev's remains were gently seated in a niche of the wall's necropolis. Within a few short years Gagarin's own remains would join him—the victim of a plane crash.

Other speakers at the funeral didn't recall the same cozy feelings about Korolev as Gagarin and Smirnov. They felt his shoes would be easy to fill, that there were many like him to be found. They would just crank out another. Their pronouncements cast a palpable awkwardness on the proceedings.

"This is not true," confided Nikolai Kamanin to his journal, safely preferring to rebuke the men in private. Kamanin headed cosmonaut training at Star City, Russia, during the times of Gagarin and Leonov. And history has only reinforced his conclusions: "It was he who was the chief designer of spacecraft, not only in post, but in essence as well. I will always place un-limited value on his talent. I knew features of his character which were not the best, but they cannot hide the magnitude of the figure of our chief de-signer. His name should be before the names of all our cosmonauts. I am deeply convinced that it will be so."

Sergei Pavlovich almost got to see Mother Russia's new generation of be-loved space hardware properly operate. It'd been a long slog up to the first pictures from the Moon's surface. Although coming closer and closer to success, the tally counted eleven straight failures before Lavochkin took over and haughtily nailed it on their first try. Korolev's group designed the thing, but Georgiy Babakin made it live up to the advertising. With one unassuming metal ball, his team of wizards effectively (but not easily) out-shined everything JPL's Ranger would ever accomplish.

On February 3, 1966, an oddball assembly of shapes progressively ap-proached the Moon's equator. For all the world, it most resembled a huge pop can balancing a silver ball high on one end with a metal coffee filter slung underneath. But for Georgiy Babakin, it more resembled Sergei Korolev's mechta. Barely sixty miles up, its tiny light-catcher sensed the Moon's pres-ence, arousing a small braking rocket to ignite and begin shedding the pace—the coffee filter's job. Backpedaling down from tens of thousands of miles an hour, there was no orange glow, no giant flaming burst. In space, rocket engines offer none of the glam and flare they do in the atmosphere back on Earth, but the effect is entirely the same.

Slow, slower. It came in gradually now, downshifting—the braking rock-et's task was complete with three hundred feet between it and victory. Slower still, further braking was now taken over by a quintuplet of small rockets. Steady as she goes at twenty miles an hour—real turtle speeds. With fifteen feet to go, at a speed leisurely enough for Aeronutronic to scream *unfair*, a long, hinged arm unfolded out from the bottom. Grazing the surface, it provoked a reaction. Instantly the two hundred–pound ball uncorked from the top, thrown clear of the rest and down to the dust, its impact softened by two hemispherical air bags that had inflated on descent. The can and the filter piled it in, crashed and dead, but that was the plan all along. The ball was fine. It bounced a few times and then peacefully came to rest on the Ocean of Storms. They'd done it. They were down. And astronomer

Tommy Gold was dadgum *wrong*; the lunar surface contained no endless feet of dust to instantly swallow a visitor. You *could* walk on it.

Say hello to *Luna 9*.

Just a few minutes later came time to stoke up the ball's innards. Weighted with an offset center of gravity, the heavier bottom had settled—allowing four triangular petals on the top half to draw a few watts of battery power and quietly swing open. Four radio antennae gamely extended skyward. Then, from the center of *Luna 9* pedestaled up a small, lightweight turret— the camera. A *television* camera, unfettered from limitations imposed by a finite length of film stock.

What did the Moon look like today? Babakin's team found out fifteen minutes later, when a poorly exposed image floated home. It looked like streaky charcoal shapes. The Sun was just rising on the Moon. But it was still too low—they needed patience.

Sergei Korolev might have seen the task brought off by his *Luna 8* that past December, but the airbags had caught on a bracket and tore open. It was the latest in a demoralizing run of duds covering almost every fault imaginable: the wrong commands went up, improperly mounted sensors lost sight of Earth, the autopilot failed. At one point, the Kremlin suggested ditching the whole program and had to be talked back into it.

Babakin's *Luna 9* winner was the result of meticulous reengineering and painfully stringent quality control. The electronics were improved to work normally at 180 degrees Fahrenheit. Because moving parts can be disastrously sandbagged in a vacuum, *Luna 9*'s contained almost none. And look—they hit a home run with the opening at bat. To most everyone's surprise but theirs, a first-time spacecraft gamer had upstaged the stalwart Jet Propulsion Lab.

"New Prominent Achievement of Soviet Science and Technology," announced *Pravda* on February 4.

An uncomfortably long day following the landing, when the Sun was ideally positioned, requests again beamed up for the modest *Luna 9* to generate more images. At the time, the Soviets favored a camera design wholly unlike the ones used today. Created strictly for panoramic landscapes, the delicately machined camera weighed in at less than three pounds. Its simple imager took the form of a slit, measuring the brightness of a thin vertical Moon slice before mechanically advancing to capture the next. It took more than an hour and a half, but by gradually working its way around, the magnesium-bodied camera could image a full 360 degrees.

The telemetry feed—seventeen photographs in all—was strung together, broadcast from *Luna 9*, and then promptly captured en route to Moscow by the Jodrell Bank Telescope. The people of Jodrell unscrambled the signal, decrypted the pictures, and slid them out to British tabloids a full day before they hit the pages of *Pravda*.

Jodrell public affairs officer Reginald Lascelles had been the one to notice the distinct tones of a telephoto fax machine coming through the dish—the same tones used to wire news pictures across the ocean. Lascelles used to work at a newspaper and knew the sounds well. *What could they be?* Jodrell wasn't in the news business and didn't have a fax receiver.

He'd raised a flag to the others, who laid their hands on a suitable receiver from the *Daily Express* over in Manchester and lugged it back to Jodrell. They taped the six-minute burst of sound and, to their absolute shock, realized that the photos had not come from overseas but from 240,000 miles away in space. A few of the discarded spacecraft remnants could even be seen.

Then the pictures went out to the tabloids. Not only did Jodrell founder Bernard Lovell catch hell for apolitically doing so, he also managed to completely misrepresent the correct shape of the images. They were too squeezed together, imparting a camel hump in the Moon's middle. Lovell hadn't understood how the camera worked and had gotten his proportions totally out of whack. It led to a nasty conversation between him and the Soviet Academy of Sciences.

Over a sum of eight hours, four panoramas went out from *Luna 9* before its batteries could do no more. After the third round came in, Lavochkin realized their craft had settled a few inches—offering a perspective slightly offset from the first two. They lined up the two angles and made three-dimensional pictures, useful in gauging the distance between features.

The world took awhile to let go. "It is the first time the Moon has been observed at surface level," reported the BBC in its coverage. Where JPL's Ranger advertised a resolution in the yards, *Luna 9* had it in the inches. "The craft doesn't seem to have gone in deeply at all," suggested British astronomer Dr. Raymond Lyttleton on BBC Radio the day the landscapes were published. "There's some remarkable support underneath. This is new. There was no way of knowing this before." If only Sergei Korolev could have seen the surface.

Not wishing to be scooped again, Lavochkin retooled their broadcasting setup in order to defend itself against future piracy. They employed the screen play, operating a complicated pattern of transmission and reception meant only to confuse the British facility. It took weeks to configure. As a Lavochkin worker recalled, "We were able to send back information in two bands . . . playing cat and mouse with Jodrell Bank." And sixty days later, when *Luna 10* flew into the Moon's orbit, their scheme worked. "We successfully completed our nearly round-the-clock work to send back the images, and breathed a sigh of relief," the technician continued, thrilled at stumping the Brits.

"It was as if a great weight had been lifted from our shoulders."

18. One Hundred Percent Failure

By God, my piece of equipment will not fail.

William Pickering, in a letter to Ed Cortright, February 20, 1963

Before NASA ever existed, a man named Ed Cortright had been toiling at Lewis Research in Ohio on such arcane yet essential aerodynamic concepts as inlet and nozzle shapes. By the end of 1958 Abe Silverstein had toted Cortright over to the new NASA, positioning him to oversee the Lunar and Planetary Programs by conjuring new goals and missions while simultaneously weaning the old ones from ARPA. Even though his research at Lewis commanded a decidedly micrometric level of attention, Edgar M. Cortright possessed a keen sense of the big picture. Additionally, he'd been the one to haul Oran Nicks on board NASA with all of Oran's hard-edged ways in tow.

While settling into the new digs, Cortright began daydreaming about some kind of lunar machine that wouldn't just snap a few pictures while hitting the Moon but would *orbit* for extended photo sessions. Heck, if it were built right, it could accomplish a near-complete mapping of the lunar surface while hosting, in tandem, a suite of long-running experiments. And then came the big kicker: tucked underneath, a petite lander would unhook and drop to soft touchdown. Badda bing, two machines from one rocket. The lander could have a camera on it and maybe a soil scooper or something, running off solar panels. Wouldn't that kind of thing make more sense than piling it in like Ranger? Scientists had been asking for comprehensive lunar surveying missions almost from the beginning; maybe something could be developed to make their dreams come true. By 1959 NASA called it "Surveyor Orbiter." JPL officially picked it up in the spring of 1960, where it became yet another distraction from the task of Ranger.

Four busy years downstream, in late 1964, a cathedral of angst sprung up around the embryonic Surveyor Orbiter. It was too much spacecraft; JPL simply couldn't get a handle on it. But Oran Nicks felt he knew a way to make peace with the whole sludgy mess—divide it into two separate machines. Make one the Orbiter and one the Surveyor—meaning just a lander—as two distinct programs. "I can't take any credit for that at all," Cortright said of the idea, "but I agreed with it once I saw the arguments." Nicks recommended leaving the lander with JPL and giving the orbiter to someone divorced from the current brouhaha in Pasadena. And by now, it was all Cortright could do just to get up and start the day. Every time a Ranger craft failed, Cortright was one of the NASA men up in the hot seat. "As it got worse and worse, we'd spend weeks in front of congressional committees explaining it," he said. "I had to do most of that. Telling them all these millions of dollars [had] gone down the drain."

What a hassle. "The tone was one of constant work." Only forty-one years old, Cortright was well accustomed to hard decisions and even harder effort in his chosen industry. "I think I worked between sixty and seventy hours a week," he recalled of the NASA atmosphere. "In Washington, it's so hard to get anything done. You have to work twice as hard as the ordinary person just to get the same thing done, because you have to get through the bureaucracy and all the crap that goes on up in that place."

Oran Nicks mentioned to Homer Newell how the Langley Research Center just over the way in Hampton, Virginia, might be an ideal foster family to run the Orbiter project. Newell received the idea with some concern, as Langley had never driven an on-line flight assignment outright from start to finish and wasn't really configured to do so in the first place. Like Ames and Lewis, Langley focused on core aeronautical research; in their case it went way the heck back to the First World War. But JPL continually fatigued NASA headquarters. If it wasn't the attitude, it was the systematic incompetence and daily finger-pointing that was on the verge of becoming policy. A shell-shocked Ed Cortright came away from just about every JPL meeting feeling someplace else could do better. Should they switch?

JPL had already done so much developmental work on the orbiter . . . a little more information would be helpful. So Chuck Sonett dispatched a man to compile some intel, and his findings nailed the lid shut. JPL cameras lacked the necessary longevity for an orbiter. They couldn't be straightforwardly adapted for a lander either.

"They had optics that were sort of unbelievably complicated," Sonett explained. "It would have taken twenty years of development to bring them up to a standard!"

Then JPL communicated to Oran Nicks how they planned—not *requested*—to spend one and a half million dollars of their budget revisiting the original Surveyor Orbiter concept. Nicks was aghast. Immediately dropping into his seat, he scribbled out a reply informing them of the multitudinous orbiter studies already under way. Nicks also used a few inches of typewriter ribbon to remind JPL of the deepening well of problems *already* on their plate with Surveyor—with or without an orbiter.

Some wanted to cancel the project. Charles Sonett vividly remembers NASA wanting to rip JPL's plug from the wall and chuck them in the dumpster like a broken waffle iron: "There were people who wanted to pull the whole contract out from underneath JPL," he said of the Surveyor dilemmas. And beyond that, others wanted to cut off Pasadena entirely. "Get them out of the space business," as Sonett put it, in a tough love kind of way that suggested he was not overly concerned about what the Lab might subsequently find to do.

But then who could develop these new machines? Langley was as good a place as any to start, so Oran Nicks unearthed the original dog-eared Surveyor Orbiter plans and sent them over to the prospective freshmen. Would they be interested in resuscitating a project?

The idea intrigued many at Langley, though nobody said yes right away. Center director Floyd Thompson knew a measure of thoughtful consideration was definitely called for. So Langley ran themselves a little feasibility study. Management discussions began, enveloping appraisals of the task at hand—plus honest self-reflection over whether they'd really be a suitable match. The work was organized, logical, and thorough.

Watching them sensibly proceed made Nicks and Cortright and Newell sit up and smile and secretly wish for the diminutive facility to give the thumbs up. They finally did; two months later Langley's Clinton Brown came back and said how much they'd *love* to run an orbiter project. Cortright gave it to them. Nicks purred. NASA called it Lunar Orbiter, fed it to Apollo like Ranger, and blasted another hole in JPL's tower of immortality. NASA would get their orbital pictures; one of them was even supposed to be of a landed Surveyor craft.

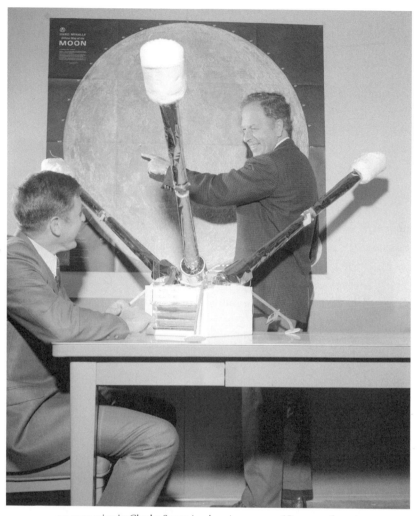

30. NASA scientist Charles Sonett in a happier moment. (Courtesy of NASA)

Pasadena's moon machine had received a preemptive and very necessary amputation. Now the project encompassed a modest tri-legged lander to affirm soil mechanics and lunar-module design. That was about the extent of it.

Lunar Orbiter—the no-nonsense title aptly described Langley's approach. They ran it with the efficiency of a McDonald's by first establishing project offices in their commandeered sixteen-foot wind tunnel. Actual flight operations were to play out across the country, using JPL's mission control center—if the Lab felt like cooperating. Some at JPL had already issued disclaimers that any and all trajectory problems *were not* and *would never be* JPL's responsibility. Langley would only be renting out the facilities.

Langley operated under the first NASA contract tailored on the psychological principle of conditioning. Their agreement included cost, delivery, and technical performance incentives. More dollars flowed for staying ahead of the schedule. Penalties kicked in for missed deadlines and budget overruns. These harsh terms and explicit rules nevertheless defined a known and therefore more comfortable atmosphere to work in.

Langley knew their place. They were ace managers but had never built lunar hardware. So Langley shopped around. Of the five proposals to come in from third-party contractors, Seattle's Boeing had the most complicated design at the highest cost. They won rather easily and promptly assigned 1,800 people to the task—over half the number at all of JPL. Boeing's $60 million proposal cost so much money compared to the others that Newell and Cortright got themselves hauled before Congress to explain why in the world they'd chosen whom they did. *Why not STL and all their great development work?* Congress wanted to know. The men explained that STL had proposed a high-speed camera system with onboard liquid film developing, and it teetered on the edge of fantasy. *Why not Hughes Aircraft?* demanded the congressmen. Because their ship didn't generate enough power, Cortright patiently explained. Besides, the Hughes camera would've been brand new.

The NASA men defended their choice. Boeing recently had built their Kent Facility specifically to test space hardware in real-world environments—shake tables, vacuum chambers . . . the whole enchilada. Sure, Boeing lost to Grumman in the bid for the lunar module, but they accumulated a library of experience in doing so. They had a bulletproof solar panel design that prevented cracking or buckling within a five hundred–degree temperature spread. Their spacecraft design lifted Mariner's proven chassis wholesale. On top of it they intended to mount a sun sensor from Bendix that was already flight qualified and reliable. For propulsion, Boeing planned to use a rocket engine already in advanced development for Apollo. Cortright waved his hands like a television pitchman: *But wait, Representative Karth, there's more!*

Boeing had a great relationship with Eastman Kodak, which had been chosen to supply the large-format camera and Bimat developer that was currently being proven in real time over everyone's heads on Corona. Sergei Korolev would have given his left arm for the Bimat system. Its tiny setup did away with liquid chemicals and their problematic use in weightlessness. Instead, Kodak made a sandwich. Freshly exposed film frames got a strip of dry chemicals pressed on top and, in one step, were developed and fixed. Moments later, the strip peeled away as the negatives loaded onto a take-up reel. Fresh chemicals every time!

Boeing's machine was also built for work. Cortright claimed that "one-half million times the area coverage of a Ranger TV impactor" could be handled by just one Orbiter. And if they governed this the right way, Cortright added, all these glommed-on Ranger missions could be dropped. That would have sounded fine to Bud Schurmeier, not yet aware of the new plans and thigh deep in his to-do list as it was.

Back at JPL, *Ranger 6* idly lay with its belly open in the Pasadena operating rooms when Schurmeier heard the Apollo landing gear design had been frozen in stone. NASA decided its critical structure without one single close-up lunar image or blip of seismometer data. Schurmeier's unfulfilled program was now obsolete—the Manned Spacecraft Office no longer cared much where on the Moon's surface Ranger hit or even if it did. They'd moved on. The goal of placing men on the Moon already commanded 50 percent of NASA's budget and kept growing.

By the time it all ended on February 2, Bud Schurmeier figured he just might get his own ass fired, too. He was in JPL's control room. A watertight *Ranger 6* had concluded its trip to the Moon and just minutes ago vaporized in the Sea of Tranquility. The television cameras never turned on, and no pictures were ever transmitted. In the proverbial JPL backroom waited scientists holding magnifying glasses, poised over large tables, warmed up and ready for images and now wondering why they'd even bothered to show up. It was after one thirty in the morning. Schurmeier sat motionless, coming to grips with what happened.

Then Jim Burke, who was standing right next to him, smiled and said, "Well, you have got one more stage of this thing working!"

One minute before, the normally cool-headed Bill Pickering had stood in the midst of it all and uncharacteristically cackled at people nearby. "I never want to go through an experience like this again! *Never!!*"

A minute before *that*, the overhead PA system had lit up with a weird yet sultry female voice: "Spray on Avon Cologne Mist, and walk in fragrant beauty." Apparently, the radio receiver had picked up a bit of stray signal on the way in. The Avon invitation blared through JPL control and floated around von Karman Auditorium, which was crowded standing room only with reporters from everywhere in the nation. Avon was perhaps the best news they'd gotten.

JPL employee Walt Downhower had been asked to call *Ranger 6*'s descent on the PA for all those reporters mobbing the auditorium. He put down the microphone and left, avoiding the press conference. Driving home in the small hours, Downhower clicked on his car radio. The air around him filled with a replay of his own deadpan commentary: "Ten minutes to impact . . . We still have no indication of full power video." And Downhower clicked it off in shame. He'd never do that job again.

Another minute before the Avon ad, *Ranger 6* had hit. The impact was noted after fifteen minutes of tension, attempted problem-solving, mute horror, and clenched-teeth, pell-mell, mad-dash topsy-turvy, because *Ranger 6*'s bank of cameras had failed to give any indication whatsoever that they'd started working.

Just two days before, endless rounds of brow mopping coursed through the Lab as *Ranger 6* unfolded itself and left Earth orbit with nary a hitch. These were happy moments. Spot-on trajectory. And at the time, with *Ranger 6* really and truly on its way, nobody seemed too uneasy about that curious anomaly just a couple of minutes into ascent: still buttoned up in the Agena, *Ranger 6*'s cameras had turned themselves on just after staging and then went off again eight minutes later.

"Everything looked like it was gonna work great," Schurmeier defended of his impetus to go buy a case of champagne after launch. "It had been such a long, rocky road up to that point." He'd filled a washtub with bottles and ice and had stashed the whole trophy in the back of his van for the blowout celebration party. But after impact he drove home low, waking early to visit the garage. Now it was just him and the tub and a whole lot of silence.

The ice was melted, and most of the labels had soaked off and floated up on top. It looked like garbage. Schurmeier sighed and fished everything out of the water and methodically fit the labels back on the bottles. Then he got into his orange dune buggy and veered over to the Lab. Other Ranger workers were already filing in, early on this Sunday morning. Nobody had called anyone else; they all just showed up. "Yeah, we're ready and willing," one of them told Schurmeier. They commandeered a room and began the failure analysis.

The next week, an industry publication called *Missiles and Rockets* carried an article about the Ranger program by its editor, William Coughlin. Coughlin called *Ranger 6* a "100 percent failure."

By March 17 all the discovery had wrapped, and fingers began pointing once again. NASA's review board accused RCA and JPL of gutter-ball engineering, showing an abysmal lack of television-electronics redundancy. Nicks, Cortright, and Newell got their hands on the report and read it as a group. In general, they were disgusted with the reviewers. So what if there wasn't total redundancy in the camera system? Plenty of other systems ran without a spare. The attitude control system didn't have a backup. And bad engineering? Where? As an example, the report specifically mentioned unvented electronics boxes. Newell said they'd never had a problem doing it that way. He wrote up a final report and slanted its tone in the direction of common sense. Sure, it concluded, we could beef up oversight even more, but don't sink the whole project with this kind of talk.

The major thing to address was that goofy postlaunch abnormality. On the side of Agena, a lunchbox-sized door covered an array of terminals used in the ground tests. Even though *Ranger 6* was all tucked inside, its vital signs, cameras, and batteries could still be accessed by flipping open the door and jacking into the ports. When the *Ranger 6* Atlas staged, a superheated burst of plasma shot up inside the panel; in that state, matter is an electrical conductor. It jumped several pins to make a number of unintentional connections. None of the jolts actually destroyed the working parts, but they did make *Ranger 6*'s sequencer rotate a few clicks. "The most amazing thing about the incident," Schurmeier recalled, "was that the stepping relay in the TV system stepped the *exact* number of times to end up in the *off* position again."

A try could be made for *Ranger 7* during a five-day launch period less than eleven weeks away. Anything past those dates would have to hibernate, though—two Mariners waited on deck for the fall Mars launch period, saturating the Cape's Complex 12 facilities. Like Schurmeier needed any heat.

Back in Highstown, New Jersey, twenty-five JPL people moved into RCA's plant and vowed to remain until the company handed off a fail-safe camera system. RCA cancelled all its car pools, which had been taking people home at four thirty, and started running three shifts seven days a week. If something was late, it wouldn't be from them.

On July 28 Bud Schurmeier watched his *Ranger 7* fly. It was a Tuesday morning. Over the next forty-eight hours, Schurmeier held his breath as the company began arriving; scientists meandered into the JPL flight operations facility and took up residence in the adjacent experimenters' room. Then Homer Newell, Oran Nicks, and Ed Cortright alighted. Reporters from every major newspaper and magazine were camped out again in von Karman. Impact was scheduled for early Friday at 6:25 in the morning. Nobody slept much on Thursday night.

At seven minutes after six o'clock, Walt Downhower's replacement got up on the PA. "Full video power, strong and clear!" belted out George Nichols. His voice colorfully resonated, in high contrast with Downhower's monotone. Von Karman erupted in applause. Above the din, Gordon Kautz called out, "It's too soon!" But the other bank of cameras had warmed and were already sending back good video. It was working. The homecoming pep rally in Von Karman built even louder with cheers and shouts. Bud Schurmeier glued himself to the readouts. Announcer Nichols enthusiastically called it like a sudden-death foot race.

"All recorders at Goldstone are go!"

"Video signals still continue excellent!"

"Everything is 'go,' as it has been since launch!"

"Two minutes, all systems operating!" He was panting now, heart racing, at the top of his lungs.

"Pictures being received at Goldstone!"

The crowd lost it.

"One minute to impact . . . excellent, excellent . . . signals to the end. *Impact!*" And *Ranger 7* surrendered itself to the Sea of Clouds, obliterating its painstakingly engineered form over the very spot it had just sent pic-

31. Late for a party: Homer Newell on the phone with President Johnson.
(Courtesy of NASA/JPL-Caltech)

tures of, banishing demons from the heads of Schurmeier and Burke alike,
the chirping telemetry fallen silent.

A standing ovation awaited Homer Newell, Bill Pickering, and Bud Schurmeier
as they lined up outside von Karman for the press conference. But somebody interrupted their progress; the flashing hold light of a call was waiting in the office next door.

Newell lifted the receiver. "Hello?"

Thirty-five miles north of Barstow, Goldstone Station had 35mm film of
the Moon's surface from as close as four miles—recorded from the video
stream merely two and a half seconds before *Ranger 7* hit. From Goldstone

the film went to the Hollywood airport, where it was driven through the early morning hours to Consolidated Film Industries. In 1964 alone the company took home three Academy Awards for Technical Achievement in Filmmaking. It was the natural place to service *Ranger 7*'s $34.7 million film, and like everybody else, Homer Newell ticked off the moments until it came back. But right at that second, Newell was on a phone call he didn't want.

"On behalf of the whole country," the president's voice began flowing through the receiver to Newell's distracted ear, "I want to congratulate you and those associated with you in NASA and the Jet Propulsion Laboratory and in the industrial laboratories."

Newell smiled weakly. He was supposed to walk through the door next to him and give a press conference. Right now! Over a hundred people waited, right on the other side of that door. Would this take long?

Johnson scriptedly blathered on. "All of you have contributed the skills to make this *Ranger 7* flight the great success that it is." Newell looked at Pickering. And then he heard Johnson say the wrong words. "This is a basic step forward in our orderly program to assemble the scientific knowledge necessary for man's trip to the Moon."

Scientific? Don't let Harold Urey hear that!

While *Ranger 7*'s images were developed and printed, guards stood outside Consolidated Film Industries with loaded handguns. Six hours later the work was done and placed in a car, which threaded a race line at high speed to JPL. By then, the formal press conference was long over. A newsman had asked Bill Pickering about the Lab's future in *Ranger 7*'s wake. Pickering smirked, "I think it's improved." A few scattered laughs. Newell, and then Bud Schurmeier, offered some quick words. They all wanted out because *Ranger 7*'s film was due to show up any minute.

When the reporters began packing, the trio hopped up and briskly strode toward the experimenters' room—but not before Pickering scooped up Jim Burke to come along.

"All I can say is that it was typical of him," Burke recalled, "in the midst of all the publicity and hoo-ha of the day, to call me in for a quiet moment." And indeed, no one spoke as Bud Schurmeier finally uncorked his champagne while Pickering opened a large manila envelope from the delivery courier and spread out the Moon for all to see.

19 June 1963
JET PROPULSION LABORATORY INTEROFFICE MEMO
TO: All Concerned
FROM: J. N. James
SUBJECT: Significance of 19 June 1963

Today, the 19th of June, the USSR spacecraft, Mars I, made its encounter with Mars—dead as a doornail. You are receiving an issue of the memorandum because you are one of the individuals who by your daily actions and efforts can make Mariner C the first spacecraft to take measurements on the planet Mars.

It's as tough a job as we could pick. The Soviets have made at least seven launches to Venus and Mars, none of which have succeeded. You have tried twice—to Venus—and succeeded with Mariner II. On the basis of those statistics you are better than they are.

I believe we have a first rate design in the Mariner C. We have plenty of talent assigned to all Project areas, and the schedule isn't too bad. So it is pretty much up to each of us to make every day count.

The Soviets will launch again for Mars in 1964 but they will have some company.

J. N. James

Even though he'd been the very guy to write it, Jack James needed the pep talk like everybody else. *Mariner 3* had to be flawless; the Lab's reputation was on the line. James had heard rumors that some people inside NASA wanted to cut them off. And considering their *Mariner 2* conquest, expectations naturally ran high on the eve of America's first attempts at the red planet. While still on the ground, the ship would be referred to as *Mariner C*. Once in space, it would become *Mariner 3*. For Mars or for Venus, they would use the name Mariner.

Oran Nicks went to see Jack James one day, and it was not because of technological migraines like heat sterilization or trajectory analysis. Nicks, rather, had a bone to pick with James's plan to decorate the exterior of Mariner's skin. Already James had mocked up a panel with what he had in mind. Nicks turned it around in his hands. It was the seal of the United States of America, lightly embossed on a cover panel for the electronics. James had nothing but good intentions, *this is an American spacecraft, right?*

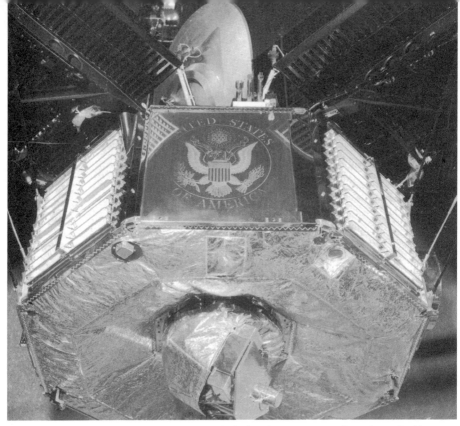

32. Mariner's decorated exterior. What was all the fuss about? (Courtesy of NASA/JPL-Caltech)

Immediately Nicks expressed a funny feeling about the idea. *Kind of showboaty, don't you think?* To James, this reaction was a bit unexpected and the sort that contributed to the perception of Nicks as something of a heartless scrooge.

James decided to push it a bit. *Didn't the Russians already crash a pennant on the Moon? What's the big deal?* Nicks failed to appreciate the comparison. "I wanted no part in that disgusting game," he remarked of the idea of "competing symbols." His persistent focal point remained *mission results*, not untoward attention on something as gimmicky as artwork. What next, the Goodyear logo? "My greatest concern was that critics would misinterpret this symbol as a lack of seriousness on our part," he said.

Oran Nicks decided to approach the issue from a more-pragmatic angle. *We're depending on this bare metal to shine and help disperse the heat,* Nicks suggested. *Won't the embossing affect that?* He knew it was a reach—both new Mariners featured heaps of thermal control measures: insulating blankets, surface treatments, and louvers that opened if the temperature went too high. James wasn't looking to do up Mariner in plum paisleys. Including a little bit of America wouldn't be a problem, would it?

James already had an answer for Oran Nicks. *Why not run a few heat tests to see?* "It was fastened by only a few screws," Nicks remembered. "The final decision on whether to use it could be made at the last minute." If the design failed to pass muster, it could easily be swapped for a blank one. Nicks really had no choice but to agree. He pointed at the Mariner program manager. "I further insisted on a low-profile, no-publicity approach for the addition," Nicks recalled of the conversation. Jack James ran the tests. The U.S. seal aced every one, so James had a pair of flight versions made up and distributed to the assemblers.

The first one did not make it to Mars. On a dark Thursday in November 1965, the ascending craft fell victim to a lack of materials testing—not in the spaceship itself, but in its womb. Surrounding *Mariner 3* was a brand-new shroud of a nose cone built from aluminum honeycomb at the core with fiberglass layered on top—a new and properly lightweight Agena fairing.

Nearly any fiberglass-car owner can philosophize at length on how tricky the stuff is to work with. Modern lay-up procedures draw out all the bubbles, but in the sixties, people were still trying to get a handle on how best to sandwich it together. Haphazard workmanship easily permitted the entrapment of tiny air bubbles under the surface. Lots of them.

Testing Mariner's nose cone occurred in a variety of environments. It was tested at extremes of temperature. It was tested in a vacuum. Nobody tested both at the same time, and just over five minutes into powered flight, everyone found out what that test would have shown. Reaching altitude, trapped gas bubbles imploded the nose cone. It didn't destroy *Mariner 3*; rather, the collapsing nose instantly shrink-wrapped itself around the spacecraft like a wet blanket. The payload stack held together during climb, but once above the atmosphere, *Mariner 3* and its 138,000 individual parts and seven delicate experiments couldn't shake the covering even after the explosive bolts fired to separate the probe. *Mariner 3* powered up anyway, just like it was designed to, immediately beginning to drain the singular and runty battery because the solar panels couldn't unfold. The pitiful thing was doomed, offering Jack James a moment to reflect on the four short weeks remaining for him to get *Mariner 4* off the ground before Mars went away.

The circle of responsibility ran around almost the entire continental United States. Lewis Research in Cleveland had designed the nose fundamentals, spending at least a year on this one design alone. Lockheed actually built it, while JPL unwrapped and installed it. The blame-storming sessions began.

Just a few days after *Mariner 3*'s failure, delegates from each outfit assembled in one place to run a few tests. Right before they got started, a tightly wound Lockheed executive fished a quarter from his pocket and climbed inside the flight spare nose, tapping everywhere in attempts to sound out a flaw.

"Tapping on a composite structure is kind of a standard way of testing for voids," remarked Bud Schurmeier of the incident. He didn't think it was a crazy thing to do at all.

They pulled the executive out and carted the nose into a vacuum chamber, sucked out the air, and turned up the heat. Everybody settled back to see what might happen, and that's when it went off like a bomb. Bingo, the problem had been re-created. Lockheed worked with JPL and Lewis to build something else. They threw out the fiberglass idea and went with magnesium, being only slightly heavier. Only three weeks left until Mars went away.

Douglas Rickard smiled at the breakneck turnaround and flawless launch. NASA entrusted Rickard and a handful of his fellow Australians with operation of the Deep Space Network's tracking facility in remote Woomera. A computing and radio engineer, Rickard shared responsibility for *Mariner 4* operations while the probe lay in range of Woomera Station and its two-hundred-foot dish.

Nestled in south-central Australia, the small town embraced community, fellowship, and celebration. The night sky beheld a clarity typically found only in outer space. And in February 1959 JPL had come through and taken a liking to a naturally formed low-lying area near the Island Lagoon dry lakebed. They stuck in a tracking station and eighty-five foot dish—their first outside the United States.

Woomera owed pretty much everything to the station and extensive rocket test-range facilities, which made good use of the endless scrubby miles. The town's name almost became Red Sands, following on the name of White Sands in New Mexico. But after reconsideration, a change was felt—something a bit more local, perhaps. In Aboriginal language, "woomera" means

33. "A beautiful, though forbidding place." Woomera's Island Lagoon Station, 1961, fifteen miles from Woomera proper. (Scan by Jan Delgado, courtesy of Colin Mackellar)

"spear-throwing device." The name stuck, and by the early sixties the town itself had swelled to six thousand inhabitants, who came from everywhere in the British Empire. Australians mixed with Londoners, who mixed with South Africans. NASA standardized on Greenwich Mean Time, or GMT, to keep everyone spread over the world all on the same clock. It was also known as Zulu time, leading some South Africans to complain about the term and its reflection on native tribes. NASA memos thus went out reminding everyone to use the politically correct GMT.

The outpost featured multiple electronic links with Pasadena, including telemetry data, voice, and Teletype lines. Woomerites routinely "chatted" with their counterparts at JPL using the Teletypes—especially between missions. On Christmas Eve in Pasadena, the Teletypes suddenly began chattering out random "X"s and "O"s. It was a transmission from Woomera. As more paper spooled through the printer, those at JPL noted a pattern emerging—that of Santa's sled being pulled by six kangaroos. During another, slightly embarrassing moment, Woomera technicians were forced to inform JPL of delays in spacecraft tracking because local parakeets had eaten off part of the antenna cover.

Oran Nicks, Jack James, and *Mariner 4* already owed much to Doug Rickard. Two months into the seven and a half–month journey, Rickard casually happened to notice how the probe's radio link seemed to consistently measure six decibels lower than the spec. Granted, it was not all that much of a discrepancy. But by the time of Mars encounter, the spec called for just four decibels above the noise floor—where radio signals become indistinguishable from natural background static. If things didn't change, *Mariner 4* would go mute.

Rickard got himself on the phone with the communications people at JPL—who failed to appreciate the significance of four minus six equaling disaster. Thankfully, someone in Pasadena graciously double-checked their math and found that Rickard was right. Even though *Mariner 4* now flew tens of millions of miles from Earth, its reprogrammable design meant that changes could be done up and radioed to the craft. And they had two months to do it. The hearing aid patch worked.

"Thank goodness for programmable computers," Rickard later offered. "And a nosy Aussie."

Not only a radio engineer but a radio buff, Doug Rickard no doubt enjoyed his stint at Woomera. During quiet times at home, he crafted musical instruments like banjos and guitars. Blue-and-white buses operated a free local transport service. The local *Gibber Gabber* newspaper kept everyone abreast of the gossip and goings-on. All ages hung out at the Coffee Lounge to eat a few sweets, listen to the jukebox, have a Coke, and play a little pinball.

Like most every Woomera building, the movie theater had a wooden floor. During shows, kids would finish drinking their Cokes and then irritate other moviegoers by loudly rolling the empty bottles all the way across the floor. Then, after the lights came up, mad gymnastics ensued to round up as many bottles as possible in order to claim the deposits. After the theater burned down, children perused the burnt remains and took only empty pop bottles. Every year the Christmas party began when a helicopter showed up overhead with a guy dressed like Father Christmas leaning out to wave as the helicopter landed on the school's running track. Some children grew up thinking Father Christmas flew his helicopter everywhere in the world.

Mariner 4's expedition was designed to be anything but ordinary and simple. Whipping past Mars, the ship would fly through a prime, close-up location—a sweet spot—for all of five minutes and fire off video stills with its camera. The distance home was too great to relay them in real time, so *Mariner 4* would lay down the shots to an onboard tape recorder holding enough length for about twenty-two images. It was a two-track model, set up to record on one track going down and then reverse direction filling up track two on its way back. Nearly all this action would be on the planet's farside and therefore beyond radio contact. One reason for doing it this way was basically to conduct a free science experiment; as *Mariner 4* peeled around behind Mars, its outbound transmissions would sweep through the fringes of the Martian atmosphere to provide measurements of atmospheric density and height.

By today's standards, *Mariner 4*'s picture resolution measured as archaic as its data rate. Solely black-and-white, the image potential was only two hundred–by–two hundred pixels—galloping home on a radio wave at eight bits per second. At that pace, lightly woven into *Mariner 4*'s basic telemetry, sending all twenty-one images back would consume months.

Many times during this flight, Woomera remained the sole facility in contact, as it would be during *Mariner 4*'s first information relay after emerging from the back side. And a total coincidence of timing meant Woomera would enjoy *Mariner 4* completely to themselves for eight solid hours at the beginning of picture transmission.

But JPL had built only one machine to display the images. Guess where it wasn't? Rickard and his mates felt a bit perturbed that the Lab wanted them to record everything on tape and then ship it off to Pasadena without even having a look-see. All the excitement would be across the Pacific Ocean.

By July 14, 1965, *Mariner 4*'s planetary science instruments had all been switched on. At the JPL Space Flight Operations Facility, crews stayed close to the raw data readouts—long strings of numbers transmitting binary figures. Sometimes it began 111111, which meant engineering data would be coming next, and sometimes it began 0000111011100101, which meant science data would follow. Once the JPL computers noticed either of these patterns, they'd lock on and transform the gushing river of binaries into something usable—approximately 23 *million* individual measurements.

This wasn't exactly a buttery-smooth process. Sometimes the number strings contained gaps, and the computer didn't stitch them together. Or the tracking station might lock onto the wrong signal and invert everything. Every single bit coming in meant zip until one of these patterns was detected, so people hovered dry-eyed over the machines, in a state of continual data monitoring to spot the pattern and help the computer lock on. It was a game of concentration they had to win, the grand prize being $20 million worth of spacecraft data.

Woomera stood by. Picture taking was scheduled to commence in twenty-four hours. Wasn't there anything that could be done for a quick peek? "Being nosey buggers as we were," Rickard later said, "we just had to do something about it." As *Mariner 4* raced within six thousand miles of Mars, it fired off shots with its video camera and laid them down as numbers on the spool of tape. The craft then hurled itself around the planet's rear end, breaking contact with Earth, right according to plan. It was supposed to emerge on the other side in about an hour. Later, after *Mariner 4*'s mission had ended, a tabloid newspaper explained the sixty-minute communications gap as being caused by UFOs.

Down in Woomera, somebody laid their hands on a colossal sheet of graph paper, marking off two hundred squares in each direction. Each square was then further subdivided into four smaller squares. Before long *Mariner 4*'s precious telemetry began arriving, and the Woomera tape recorder hummed with life, meticulously recording every bit for later shipment to JPL. Unbeknownst to the Lab, eight Australians hunched over a large grid with sharpened pencils at the ready. Another stood at the radio console and began calling off numbers. The values ran zero to sixty-three. Each number essentially corresponded to a differing shade of gray—zero meaning plain white and sixty-three meaning black. At the rate of one a second, the console man called out his numbers; starting from the top left, the Woomerites began shading in their squares. 63. 63. 63. 63. It went on for half the picture. 63. 63. 63. Nothing came in but black for over four solid hours. The men kept shading and shading, debating whether or not the spaceship had taken a dive like so many others.

Then it continued: 63. 63. 47. 44. 39. 32. Okay, they were getting values now, and the mood in the room improved. Row after row, the shaded little squares built to define a curving, arcing edge. They had the edge of Mars. It was just down near the bottom of the image, at least on this first one. 26. 29. 24. 24. More shading. From just a few feet back, the human eye joined all the squares into a respectable image of Martian craters. Craters? They kept going. One of them got on the line to Pasadena, reporting the successful arrival of data. Unable to resist, he also began describing the Martian crater patterns. Mars had no canals, no life they could see—just craters. And there also was what seemed to be a cloud up there high in the atmosphere.

Doug Rickard said, "The Americans were furious and told us to stop using the comms circuits for jokes." The Woomera guy on the phone didn't mention what they were up to—only that the team had been able to decode the first image. "But," Rickard went on, "as we continued to describe the different craters as they appeared, they finally came to realize that we were not joking."

The Woomera clan had indeed seen a cloud, although they didn't quite know it yet. Jack James and the rest of JPL—and then the world—saw it, too, quickly consigning the high-altitude smear to that of an optics error. One person at JPL reckoned, "We assumed it was a flaw in the camera lens." And in weeks to follow, a long string of tests ran up and down through JPL striving to prove just such a thing. They overloaded the spare television system, trying to create overexposure or fogging. Some tests deliberately induced noise and artifacts in the pictures. They even commanded *Mariner 4* itself to bark off a few more shots—of nothing, actually, since Mars was long past. As test after test ended by disproving every single theory, one after the other, a new pattern of thinking took hold.

Mars had weather!

On August 10, 1966, the first Lunar Orbiter went out. And Langley threw what amounted to a perfect game. Before Orbiter began, two independent studies concluded that the probability of two successes in five flights ran maybe 80 percent. Inside of twelve months, Langley went five for five. Proposed Surveyor and Apollo landing sites were verified. Every objective was met. Their schedule ran over by only sixty days.

The orbiters worked amazingly well. These spacecraft could shoot oblique pictures of the Moon, seeing the surface from the kinds of angles manned landing crews would as they approached. The high-resolution Kodak photographs were quickly processed for use by Apollo training, who took them to the flight simulators. Now, bona fide surface images could be projected in front of the simulator windows, advanced frame by frame to depict the approach.

Such impressive marks chiseled away at JPL's reputation, providing Cor-

34. Comforting success: presenting a *Mariner 4* image of the Martian surface to the Oval Office, 1964. *From left*: Bill Pickering, Oran Nicks, Jack James, Lyndon Johnson, and Jack Webb. (Courtesy of NASA/JPL-Caltech)

tright and Nicks, as well as Homer Newell and Chuck Sonett, a brief glimpse of worlds beyond Pasadena. Clearly, it wasn't a totally fair comparison. Mission parameters differed, sometimes greatly. With a ship in orbit, Langley always had the option of slowing down to analyze and resolve any intermittent troubles while an Orbiter gently circled the Moon. Time was on their side. On occasion, revised instructions were pieced together and sent up. Ranger's suicidal flight profile left no room for error. Neither did Mariner's one-shots. And by August 1967, with *Lunar Orbiter 5* away and crooning perfectly, it was time to be done and call it a success.

Floyd Thompson and the others at Langley carefully filed away the Lunar Orbiter drawings and quietly began work on a new project named Viking. And where Harold Urey got his pictures, JPL got an expulsion. Yet another congressional-subcommittee review had formulated harsh words for JPL. While acknowledging the complexity of running lunar and planetary programs at JPL, "major improvement actions take place primarily as a result of failures." After the last Surveyor spacecraft finally fell silent, the Jet Propulsion Laboratory would savor no overall responsibility for unmanned exploratory missions. This would not change for years—not until 1977 when the grandest journey of all would begin.

19. Three-Problem Shipley

I think it's a great pity that the engineers responsible
for this spacecraft are not better known.
They should be on postage stamps.

Carl Sagan

Space mission designers face endless questions. It kind of goes with the territory. Half the time, they're aiming for a goal nobody's ever tackled before. And when it came to this whole "Grand Tour" to the outer planets—well, that stuff was just totally off the map.

Unreserved brain space fills quickly, as discomforting facts pile on top of complications. *By the time anything gets to Neptune, our Sun is just a tiny orange ball. Solar panels won't work, but the spacecraft can't run off an extension cord. And the distance—3 billion miles—holy cow. Will the radios go that far? What about the time delay? Does anyone have experience stringing up a billion-mile phone line? Will a camera still work after years in the numbing environment?*

Yeah, and what about those pictures? Only a few years before, photography wasn't even on the menu. Scientists tended to hold their noses in the air if the subject came up. The overall scientific value of photographs was right down there with boogers.

But what they kept forgetting is that photography unites the world. It's a commonality running through all our veins. Dismissing overworked prose and fancy adjectives, images get straight to the point. Everyone relates, as was proved by Ranger, Luna, Mariner. Unmanned spacecraft might not have their flyboy jock heroes in the nifty suits—but, oh, could they snap the most breathtaking of pictures. This machine definitely had to have a camera.

Of course, the cool shots become possible only if the spacecraft makes it out near the intended targets. Lists of problems grow distressingly huge long before Neptune. First, the travel hazards need to be addressed, such as the asteroid belt past Mars. Why put in all the effort if the ship can't even make it down the road?

If the trip then worked out, the mission itself would become everybody's next point of contention. And in Jupiter's particular neck of the woods, environmental hazards eclipse those of Venus and Mars by several orders of magnitude. *How much of a problem is Jupiter's intense radiation? What will it do to the electronics?* From there, questions regarding the science experiments begin to crop up: *How close can we get to the moons of Jupiter and still get good scientific data from the planet itself? Can we control the ship precisely enough for that? Then, how close to Saturn's rings? And the scientific data there?*

All in all, there were too many questions for two guys and a cause. No matter how hard Roger Bourke and Gary Flandro worked the situation, the pair kept getting steamrollered by the project's sheer scale. As Flandro succinctly positioned it, "We were to build a machine capable of flying for ten years without making a mistake—using the electronics of the time!"

Questions and uncertainties piled higher and higher; a mountain of unknowns threatened to avalanche Pasadena and bury all of them like Vesuvius and Pompeii. Curiously enough, though, everything boiled down to a simple chicken-and-egg problem. Nobody could design a spaceship for Neptune because they didn't know what to use—no knowledge base existed. There was no stockpile of recipes from which to draw—no tricks. That is to say, the absolute nuts-and-bolts technology required for an eight-year Neptune flight simply was not present. What they needed was some kind of research phase. Wisely, JPL thus sanctioned a formal "technology effort," as they internally referred to it, christening the new endeavor with a suitably yucky bureaucratic acronym like good government agencies always do. They called it TOPS.

In June 1968 NASA threw some dollars at JPL to fork into the work of TOPS. The money was caught by a fellow named William Shipley, who accepted the role of point man and signaled the opening kickoff by labeling a clean, new three-ring notebook and placing it on a mostly empty shelf in his office. This is where the work would live. Shipley's assignment was to play Victor Frankenstein. If they could create a wunderkind spaceship to travel to the outer planets, what would it be like and what would it require?

35. During a rare moment of inaction, William Shipley pauses for his 1970 formal JPL portrait. (Courtesy of NASA/JPL-Caltech)

Shipley's presence had been gracing the Lab's concrete walkways almost forever. "I arrived at JPL the same week President Eisenhower announced that the country was going to launch a satellite," he recalled. "My first assignment at JPL was to compute the temperature of a satellite in a two hundred–mile orbit." Solving that math problem contributed to *Explorer 1*'s success. Ten years downstream, he linked up with representatives from the relevant JPL engineering divisions and began meeting to see what they could figure out.

At the top of almost everyone's list was how to juice the ship—power, *not* propulsion. Propulsion is what gets you there, and in this case it'd be the launch rocket, gravity, and a few onboard thrusters for little corrections. Power refers to the electrical supply for the spacecraft itself, and every single other unmanned craft that had been launched up to that point used either chemical batteries or solar panels. But if they really got to Neptune, the

Sun would be smaller than a pinhead. Solar panels the size of tennis courts wouldn't do jack. Their chosen solution, then—really the *only* solution—rested with a comparatively unproven use of radioactive plutonium. Understanding this assigns comforting meaning to the TOPS acronym: Thermoelectric Outer Planet Spacecraft.

Shipley added a page to his loose-leaf binder and noted their choice. The premise of it was simple enough. It wasn't about the *radioactivity* of plutonium; it was all about the *heat*. While plutonium is spitting off enough radiation to kill humans—even telemarketers—or pretty much anything else in short order, it's also creating heat—hundreds of degrees of it. With enough plutonium, the heat lasts for decades.

Their creation wouldn't run off of the heat directly. A touch of rigging is necessary to channel the heat and make it do something useful. Two little strips of dissimilar metals are sandwiched back-to-back. This is called a thermocouple. Searing heat is ducted past one of the metals, while the other hangs its tail out in frigidly cold space. Set up like this, an amazing thing happens. Working in concert, the metals automatically reconcile the temperature differential between them into a smidgen of electricity, and *bang*, there's your energy source. Two strips of metal. Hot and cold.

The operative form of this power source resembles modern art, and in America its creation is strictly the domain of the Atomic Energy Commission. Metal plutonium is bonded with ceramic materials to create something resembling an off-color ping-pong ball. A thin metal shell is formed around it, and then the entire ball is wrapped in a high-tech yarn woven of graphite. This graphite layer is built up to about half an inch thick and is there for the specific purpose of holding the ball together should the launch rocket disintegrate.

The balls get stacked inside a coffee can–sized enclosure that is also made of graphite. There are six layers with four balls on each. Outside the can, obnoxiously big metal fins ring the edge to aid in shedding hundreds of surplus degrees. Plutonium creates enough heat that most of it can't be processed by the thermocouples and is bled off into space.

Each can is good for around forty usable watts of power—enough to run a very dim light bulb. But they can be strung together for more output. Once complete, the appliance lacks any sort of "on" button. It's always running.

When it comes to overall efficiency, however, the setup is abominably poor—down on the order of 10 percent. And over the course of many years, the intense and unyielding radiation exposure will damage the spacecraft's onboard electronics. That can be mitigated by mounting the cans out on a swing arm away from the ship, but the hundreds of thermocouples don't enjoy that luxury. They degrade to the point where enough have stopped working and too little power is being generated to keep anything alive. You have to start out with more energy than you'll need at the bitter end.

A coffee can with fins—there it was. A neat little package ready for the hardware-store shelves. Bill Shipley would have to line up some assistance with the AEC—really there was no other way to do it. The TOPS group therefore registered their planned use of this radioisotope thermoelectric generator (RTG).

Four RTGs would probably be enough to run the self test and repair (STAR) computer over at UCLA. There, a Lithuanian-born computer scientist named Algirdas Avizienis was seven years into a little pet project. Avizienis liked computers. They made him feel happy. And figuring he might like them better if they worked a tad more reliably, he set off on a quest to build a computer from scratch that could fix itself, using seed money from JPL. STAR computers were, at the time, entirely theoretical, but the concept was hardly new. For years on end, they'd been the subject of paper studies with the idea that some day one of the gizmos might actually get built and used in something for real. Its underlying concept rivaled artificial intelligence. The STAR computer would run a suite of isolate programs capable of operating a spacecraft in a completely autonomous fashion. It would possess enough horse sense to recognize when it was under the weather—or in the process of becoming so—and be able to nurse any threatening illness back to solid health.

Avizienis was just thinking longitudinally. When he started, Ranger and Mariner were barely past their design phases and ratcheting through assembly. Eventual mission success would lead to longer missions, he rationalized, requiring computing brains capable of staying in it for the whole twelve rounds. The need for a STAR computer on Flandro's Grand Tour predicated itself on the obscenely long radio times between Earth and the outer planets. Nail-biting hours would tally up between ground transmission, reception by the ship, and then confirmation back on terra firma. Nobody would be watching its back in real time. As Bud Schurmeier explained, "By the time you got telemetry from the spacecraft and analyzed it and sent a command back to fix it, then there'd obviously been a disaster."

Gobs of approaches existed, all with pros and cons. Today's space shuttle computer, for example, is actually a council of five. All are identical; any single one can drive the bird. And when conflict arises, they all vote. The scheme is intelligent enough that the majority can actually override the dissenters and ignore them or turn them off altogether. They have no constitution.

The main problems with this are *weight* and *power*. Weight budgets are immovable brick walls relating directly to booster lift capability. Designers kill themselves trying to shave ounces. America's petulant scientific community would never stand for fewer experiments going up because all the weight went to the controls. It was the same deal with power. So as Avizienis pondered all the combinations, he arrived at this principle of what he called selective redundancy. Instead of multiple computers, he envisioned a *single* computer with multiple *spare* components. Sort of like having a whole string of extra keyboards attached to a home personal computer. If one blanked out, the computer would recognize what happened and rotate in the fresh piece. The spares could remain unpowered and dormant, meaning the whole STAR consumed only as much electricity as a regular system would without any spare parts. The setup demanded a greater weight allowance for the computing brain, but not as much as bolting in a whole extra system.

Bill Shipley duly jotted all this in his nondescript notebook, which already had sprouted an impressive line of descendants. "I had a bookshelf full of three-ring binders that were problems," he explained years later. Shipley killed trees by the acre, consuming defenseless reams of paper in working through the difficulties. When a new problem came along, he'd open up another binder section and enter it. "I used to tell people that was my $20 million bookshelf," he said, "because that represented about $20 million worth of problems."

Each day brought lively challenges and perspective. Any one of a zillion given quandaries never got resolved by the end of the workday, but oftentimes a suggestion or two had been made regarding the next direction to proceed. An innocent night of sleep might change all that back again. Shipley would come in the next morning and look at team member Ray Heacock and say, "I thought about that; I don't think that was a good idea."

"Yeah, it doesn't sound very good to me this morning, either," Heacock often responded. They'd be back to square one again on that issue and enter more scribbling in the binders.

Over time, as the group's design and thought processes matured, Bill Shipley began to close out his notebooks one by one. Of the TOPS experience, he offered a tip on basic strategy: "I always tried to keep it to the point where I had two, at most three, big problems going at once. I really worked hard to get one out of the way if I saw another one coming. If you get more than about three big problems going at once, people decide you really haven't got control of the situation."

A couple of the big hurdles had now transitioned into an environment of greater understanding. They knew what to run it off of and how to make it think. With muscle and brains out of the way, discussions turned to the skin. Enough was already known about Jupiter's strange ways to appreciate how the colossal orb gave off a lot of radiation. Nobody knew for sure how much. Unfortunately, the years of ground-based observation had generated a suite of ballpark values hardly worthy of celebration. "High" wasn't the word to describe them. Try "extremely severe." The planet absolutely reeked of radiation. Depending on the tour's approach path and the length of time the spacecraft marinated in Jupiter's rancid emissions, the whole dang-blasted thing could give out entirely unless armored to the teeth in heavy lead shielding. Only much, much more precise detail would allow Shipley, Heacock, and the others to properly establish the requisite amount of protection. If only they knew just how bad it really was or wasn't out there.

Harder numbers were on the way. Up the coast in Santa Clara County, just north of Mountain View at Moffett Field, the bashful yet flourishing NASA Ames Research Center quietly continued preparations for their upcoming *Pioneers 10* and *11* flights: a pair of identical and extraordinarily simple unmanned spacecraft to reconnoiter Jupiter in 1973. The flight plan called for flybys only. If things went well enough, one or both might also try plodding on to Saturn.

They'd go as a logical extension of Ames's existing bed of groundbreaking space research. The center's tally of flights was only up to four, but each was so perfectly executed—and by a total space newbie at that. They got started when, some years back, Ames engineer Charlie Hall had bent the ear of Ed Cortright with regard to a modest bloodline of solar probes:

Let's fly out by the Sun, Charlie had proposed, *and discover something new.* Cortright was found to be surprisingly amenable and perhaps even eager to line up a touch of JPL competition. Ranger malaise was pushing everyone to wit's end.

Ames came in as a total breath of fresh air: under budget, unpretentious, and unfathomably successful. Their machines were ingeniously capable and focused entirely on scientific return. A whole quartet of them! And so with *Pioneers 6, 7, 8,* and *9* still in perfect working health and returning valuable solar data, approval came through for these new, dual Jupiter flights: *Pioneer 10* and *Pioneer 11.*

Back down in Pasadena, the TOPS group felt Ames's forthcoming excursions lent ideally to understanding the Grand Tour's needs. Explained Bud Schurmeier of JPL's Pioneer attitude, "That was a precursor to tell us whether we could get through the asteroid belt and what the radiation environment was at Jupiter." In an act that could be perceived as selfish, JPL lobbied NASA to have Pioneer fly the tour's intended course, no matter Ames's original plan. Understandably, Ames failed to appreciate the inglorious idea of becoming cannon fodder. JPL wanted them to just go and test the waters?

"You might say that's how we kind of looked at it," offered Schurmeier.

TOPS, by this time, needed to wrap its research and get on with a mission proposal so they would at least have a chance of making the 1977 launch date. Loose ends abounded—things like radio contact. Any increases in the ship's antenna size would nicely multiply its transmission capabilities. But the absolute top-end dish size was already known, as restricted by the booster. Shipley's group planned to circumvent that using a big collapsible antenna, folded up umbrella-style for launch and then snapping open once in space.

Time ran out while they were still working through the antenna details—along with a hundred other things. Shipley commented, "We didn't complete all that we wanted to, but . . . we had got a very good understanding of the system and how it ought to work." And he set about using the harvest from TOPS to cook up a proper spacecraft and mission.

By the time Shipley finished his draft, the flight manifest rivaled Pan Am. In 1977 twin launches would be sent toward Jupiter, Saturn, and Pluto. They had figured out a way to get to Pluto. A second round would be launched two years later attempting Jupiter, Uranus, and Neptune. Four ships in total could entirely cover the bases. These were grand plans indeed. "We kept coming up with cost estimates that were three quarters to a billion dollars," Shipley remarked of the Grand Tour proposal. The paperwork went on through to NASA. All he had to do was get the approval signature.

With absolutely no fanfare whatsoever, *Explorer 1* reentered Earth's atmosphere on the last day of March in 1970 and burned into dust somewhere over Easter Island. Gravity will slowly decay any spacecraft's orbit unless it carries the ability to push itself back up. The little ship eked out 58,376 orbits before succumbing.

James Van Allen received some touching mail that week:

April 1, 1970
From the University of California at Los Alamos National Lab
Dear Jim,
I was sorry to hear of the demise of your good friend Explorer I on March 31st. Enclosed is a clipping about it from the Albuquerque Journal.
Sincerely,

Edward W. Hones, Jr.

Then this:

April 6, 1970
From Ernst Stuhlinger
To James Van Allen

Dear Jim,
On April 1st, at a dinner party held at JPL in connection with an OSSA senior council meeting, Jesse Mitchell brought out a toast to Explorer I, which on the previous day had terminated its successful life of a little over twelve years.

Bill Pickering and Werhner von Braun were there to reply briefly, and we all thought of you also as the third member of the Big Three of January 31st, 1958. Best Personal Regards,

Ernst Stuhlinger

The unassuming Iowa boy never felt overly sentimental about the little guy who forever pushed him out of academic obscurity. Honest. "I was very proud of it, of course," he said in retrospect, with a smile of accomplishment. "But I knew it was going to come down; it had long since stopped transmitting, of course. It was just sort of an inert object by then."

However, even James Van Allen was subject to occasional whims of humanity, which required lots of cajoling for him to admit. One whim did surface. Occasionally, after the Explorers began flying, he'd be out driving I-80 past Coralville some evening and spontaneously take his eyes skyward off the road in a brief glance. He'd do this just on the off chance a wee little shape of reflected light way up there might resemble something he'd once cradled in his own two hands. Never once did he see that light, drawing up a thin sadness. "But you know, they were invisible," he admitted—almost as an excuse and snapping back to pragmatism. He waved it off, "Entirely in your head."

Already by the time *Apollo 11* landed in July of 1969, the Grand Tour had a problem known as the Space Shuttle. Even with flights planned through *Apollo 20* into the early seventies, NASA had already contemplated the end. They were also contemplating how their nice, rosy future lay nestled in another emotionally heart-capturing program like the Moon race. And obviously, the next important effort would also have to be manned, because nobody held parades for beeping balls of science experiments.

Storm clouds broadened into 1971. NASA and JPL reps sat in Congress one day on irritatingly hard furniture, learning more than they cared to about waning public support for America's space programs in general. Moonwalking boys in their gorgeous white suits were no longer a license to print money. And Apollo's fiscal ending was not going to free up billions for unmanned spaceflight. The out-of-control situation in Southeast Asia continually dominated the front pages. In April two thousand Vietnam veterans rallied Washington DC; some of them chucked their medals on the Capitol steps. The *Pentagon Papers* were beginning to appear in the *New York Times*. Forty people were killed during the armed retaking of Attica State Prison in New York. Not even the stunning Phyllis George and her triumphant crowning as Miss America of '72 could regroup a wholly fractured nation. *Apollo 17*'s imminent return from the Moon in 1972 would bring Kennedy's manned landing program to a premature close. NASA had their priceless rocks; there was no reason to keep going back. So with regard to the Grand Tour—bad news, JPL: space budgets would be tightening up snugger than a corset. The only space hardware getting big money would contain seats and windows and Tang.

Who cared about the outer planets anyway? Was *Uranus* ever in the newspaper—like, on the front page? Did anybody even know the right way to pronounce it? Who cared about Neptune when you could do something like eat a turkey dinner in space? That's what people really wanted to see. Who cared about a computer in a box, and so what if it could fly between Saturn's rings? The computer wasn't even coming back. To win support, it needed aviator glasses and razor stubble. James Van Allen opined, "The idea of the Grand Tour became a sort of dirty word, so to speak, at NASA headquarters."

Moving forward, NASA proposed building a space station high in Earth orbit as a launching base for manned missions to Mars. Constructing all that required an ongoing transport system of materials and men—ergo, the Shuttle program. It received a thumbs-up from Congress. By the time of the Grand Tour proposal, the Shuttle had already been chewing through funds for three long years.

One thing was definitely for sure: unmanned programs lined the bottom of the barrel. Later JPL director Bruce Murray summed up the attitude nicely: "JPL was on everybody's hit list awaiting the next big NASA cut." And nobody in Washington DC cared much about flying to Neptune any more. "They concluded that the Grand Tour was simply too expensive," Roger Bourke explained, "and they wanted us to cut it back to something that was more feasible financially."

Bud Schurmeier agreed. "The Shuttle, in fact, proved to be our biggest competitor for available dollars. It wasn't that the agency didn't want to explore the outer planets; it just wanted something less costly and ambitious." But words of warning had also come from the scientific community. "Scientists are used to making an experiment and finding out and then changing it and doing something more," Schurmeier went on to say. "The scientists were concerned that you design it now and then it would be ten years later before it was launched and they would get the data. They were afraid that a lot of the instruments would be out of date."

In January of '72 Bill Shipley dragged himself back to Pasadena from Washington DC on yet another red-eye flight. NASA administrator Jim Fletcher had called a little meeting out east and delivered a real no-shitter. He said straight-out, "If you can design something that you can get the Space Science Board to approve, then we will propose it." What Fletcher really meant was, "Drop your budget."

"That was an emotional setback," offered Roger Bourke. And one day soon thereafter, Shipley commandeered a meeting room with Bourke and Homer Joe Stewart and Bud Schurmeier and practically anybody else around, so they could all crack their heads wide open and figure out what to do next.

Was there anything—perhaps a bit more limited in scope—that still took advantage of the magical planetary alignment? One half of the problem was the budget, but the other was their chrome-plated, hot rod dream machine. What they'd come up with was not only expensive almost beyond justification, it was also ungodly overcomplicated. So much bleeding-edge gadgetry littered the specs that even if they did forage the money, the ship might not make it through development and testing before the final launch period closed in 1979. That was also the same year that Shuttle was supposed to begin flying its "completely reusable" missions. Time, gentlemen, to discard ballast.

They took out the STAR. It promised redundancy at the expense of simplicity, and by all rights, it would suck dollars like a Hoover until all the bugs shook out. And one thing had always bothered them about its architecture: chain of command. How did they know the computer was making the right decisions? Who tests the tester? So it came out.

What options were left? What could they base the ship on? Somebody asked, *What's the most reliable thing we have going right now?* But he already knew the answer. Mariner. By far, its legacy had proven most successful. And this funneled the discussion toward an obvious deduction: why not build from its time-tested design?

People started talking really fast, and the temperature in the room went up. What if they were able to make the Mariner computer smaller by putting in a way to change its programming during flight? They could then use two fully redundant minicomputers able to watch each other's back on the heavy tasks of navigation and camera pointing, while sharing the less-important jobs?

Shipley brought the room to order. *Look*, he intoned, *we still have to keep a very sound program plan. Why are we doing this work? To explore.* Separate from this engineering team, a whole JPL-based Science Working Group had been gently upholding the project's responsibility for creating a viable, purpose-driven mission.

For two weeks this nucleus of brainpower worked day and night, like butchers with too much fat on their best cuts. Every member was relentlessly driven by the external factors of science and budgetary constraints—all the while remaining cognizant of the deadline.

Something tumbled from those weeks—a repackaged plan of flights that was not exactly worthy of the name Grand Tour. This was more of a weekend tour. The new project even took on a completely inglorious name: Mariner-Jupiter-Saturn 1977 (or MJS'77 for short). Largely based on Mariner architecture, the retooled flights would launch only in 1977, targeting Jupiter and Saturn. That was it. The really expensive stuff—the long-term electronics, the fancy folding antenna—were mostly for the final stops.

Three viable-option packages sat ready for NASA browsing. On the top end lay JPL's ideal version, featuring a rich complement of science experiments and plenty of redundancy in the ship itself. Down at the anything-is-better-than-nothing level sat their outline for a bare-bones reconnaissance of Jupiter and Saturn with emphasis on returning broad brushstrokes of information. Splitting the difference between these two yielded a third, middle-of-the-road option. Hopefully, NASA would bite on any one of them.

All options slashed the flight roster down to three ships—one in each year of the main launch period. Very few of the precious Titan IV boosters remained, and NASA didn't want to buy any more. Even though there were still seven years before it would supposedly be ready to fly, NASA had already decreed that all future unmanned probes would launch from the Space Shuttle cargo bay. "Well," Bud Schurmeier remembered of that NASA decision, "it didn't make any sense right when they did it!"

Somebody drove out to the warehouse and did inventory: only two Titans were actually left. So JPL reduced their roster further to just two ships now. If the first one conked out or hugged an asteroid, the second might deliver the goods. One backup would be enough. But what if the first ship gave them everything right off the bat? What could they do with the second? Roger Bourke knew the same thing that everybody at the Lab knew—Uranus and Neptune were going to be there whether Jim Fletcher liked it or not. What about building in a little extra under the hood of MJS? Maybe they could finagle some kind of escape clause. If one ship made Saturn with no trouble and the other was close behind, a bit of backroom table talk might unwrap a few greenbacks for Uranus or Neptune.

Finally, in May 1972, JPL got a phone call from Washington DC. NASA said they'd take the middle-of-the-road option. "There was some relief when the word came down," recalled Bud Schurmeier. "We had put a lot of effort into making it happen, and that was the first big memorable moment in the project's life."

The billion-dollar Grand Tour received an unjust name change and funding revision. The allocation was down to $150 million, though some extra money would be forthcoming to beef up the scientific possibilities. Unfortunately, the scope was clear—no official support beyond Saturn.

Sitting back in his large, comfy chair to watch all the goings-on reclined an aging man, now operating a renowned bed of teaching and research from a top-floor corner office in a building with his name on the outside. James Van Allen recalled, "You know, we all sort of laughed to ourselves and said the planets are going to be there anyways. So no matter what you call it, the opportunities are going to exist." And he was right, whether Congress liked it or not. So the magicians of Pasadena inaudibly kept their minds and options open on the off chance that their creation might yet be able to go the distance. Just in case.

20. Pete and Al's Little Field Trip

We've not only placed man's eyes on the moon in the form of
a TV camera, but now we've also put his arms and hands there.

Ben Milwitzky

Space probes almost never come back to Earth—at least, not in one piece. So for someone who worked on a program, saying goodbye sort of goes with the territory. All through design and fabrication, through assembly and test, and through the writing of glitzy press releases, everyone on the project keeps this fundamental truth in mind. Just like watching any movie about the *Titanic*, they all know how it'll end up.

Within the hallowed walls of the Smithsonian Institution's Air and Space Museum, one prized artifact on display is the complete—and *flown*—television camera from *Surveyor 3*. It spent thirty-one months on the Moon before being returned in an act that might be considered Apollo's tithe for all the funds, manpower, and attention President Kennedy's little adventure stole from unmanned explorative programs of the sixties.

July 20, 1969, was quite a day for human adventure. Neil Armstrong piloted himself and Buzz Aldrin to a gentle landing on the Moon's Sea of Tranquility. Their *Apollo 11* mission collectively signified the fulfillment of Kennedy's challenge, the winning of the space race against Russia, and the brief coming together of humanity. If only S. P. Korolev could have seen it.

In the final moments of descent, Armstrong realized his landing ship's computer was automatically and unwittingly taking them right into a big crater filled with boulders that could be considered "sharp enough." By then, *Apollo 11*'s crew were well down inside the dead man's zone—too high to cut the motor and plink down on the surface but too low to abort. Armstrong coolly flipped the craft into manual control and dropped two gloved hands on the joysticks, taking his chances on sixty seconds of hooch left in the gas tanks. The men drifted downrange, skimming boulder tops. Standing side by side in the cramped lander, Armstrong worked his controls while Aldrin called out altitudes, which dwindled as quickly as the fuel. Finally, Armstrong located an uncluttered, flat place to land, where he set the craft down so gently that its shock absorbers barely had to compress.

But there was now one problem: nobody knew the exact location. Several top government officials, who seemingly could find the thorn in any rosebush, made it known to all involved that the next landing better be square on target. Right there in mission control on the day of *Apollo 11*'s victory, Sam Phillips decreed, "On the next mission, I want a pinpoint landing."

Phillips was the Apollo program director. Near him that day stood one Charles "Pete" Conrad, who overheard the whole snotty remark. Conrad paid more attention than most, as he was in line to command America's *second* lunar landing. Pete Conrad knew then that he would have to park it on a dime.

So along came *Apollo 12*'s lunar lander, *Intrepid*, dropping from orbit toward the Moon's Ocean of Storms in late November of 1969 with Conrad at the helm. To his immediate right stood Alan Bean, a fellow naval aviator and, perhaps more importantly, a longtime friend. Bean functioned as the mission's lunar module pilot, which was something of a misnomer. Lunar module pilots didn't actually pilot the lunar module. The commander did.

The pair steered toward a rendezvous that had been suggested months earlier by Jack Sevier, a planning coordinator buried somewhere inside NASA. His stunningly bright idea purported to land *Apollo 12* at a location noted for more than basic geological interest or craters with oblique names. It held something a bit more tangible for the average human. At the same time, it would also fulfill Phillips's precise targeting requirement and remain beneficial to the program—what a combination. Sevier recommended landing near Ed Cortright's meek, little *Surveyor 3* and actually visiting it. They'd have to land within walking distance, which meant *Intrepid* couldn't be farther than two thousand linear feet from its quarry.

Surveyor 3 lived in the belly of the Snowman. Like stellar constellations, patterns of lunar craters ended up with unofficial names to aid in navigation. Typically, they became relevant only from above. Imagining a snowman amid all those craters wasn't obvious from every angle, but from the direction that Conrad and Bean would descend, *Apollo 12*'s principal landing site looked more or less like a classic three-ball snowman. Conrad christened that middle section Surveyor Crater, and on Wednesday, November 19, 1969, he set down the lunar module right between it and the Snowman's "head" crater. Bull's-eye.

One day after landing, with a successful three and a half–hour moon walk already behind them, the flight plan next called for heading to *Surveyor 3*. In spite of the mission's growing list of triumphs—flawless transit, pinpoint landing, extensive science package deployment—neither Conrad nor Bean would consider their tasks fully complete until they had finally visited the ghost ship that now lay just a short walk from them.

Every Surveyor looked a lot like a stripped-down toy version of the manned Apollo landers. In particular, they bore quite similar designs when it came to the landing gear and shock absorbers. Both had to land only one time. For shocks, each utilized simple, crushable aluminum honeycomb. It could take the thump of landing without leaking hydraulic fluid, which was heavier and altogether completely untested in a vacuum.

Years before, when the overall calendar was first getting hammered out, the spindly little bug was scheduled to land *way* before Apollo so that the shock design could be verified. Caldwell Johnson thought it over in September 1961 while driving to Langley for a meeting with some other NASA types. The rest of the attendees were trying to get a handle on Surveyor's overall purpose and mission.

Johnson worked for Langley as a draftsman, and his engineering drawings were pure artwork. Now he was supposed to draw the design for a lunar landing gear, but the starting point remained evasive. Nobody knew what it should look like, because they weren't entirely sure what the Moon itself was like. Langley engineer Max Faget sure had his doubts. Before the September meeting, Johnson had been standing in Faget's office, where they both kept repeating the same thing. Any passerby would have wondered what was happening behind the door, as the men kept shouting at each other, "It's got to be like Arizona!" It sounded almost like a playground game. "The moon has just *got* to be like Arizona!" The Langley meeting that Johnson now steered himself toward would ideally resolve this issue. All he had to go on so far wasn't much beyond this nutty hunch about Arizona, so maybe the Surveyor folks could help.

When Johnson walked in to join the group, he looked around. Right away, the people from NASA's Lunar and Planetary Programs asked how they and Surveyor could most effectively support Apollo. Someone at the meeting later described the scene: "They were sitting there with their pens poised, ready to write our answer down."

That's when Caldwell Johnson decided to speak up. He said, "Crash into the Moon and smash all to hell, and then at least we'll know we don't have several meters of dust on it."

By the time Conrad and Bean set *Intrepid* down, *Surveyor 3* had been cooling its heels on the Ocean of Storms for over two years. Its low, tripod-framed body alighted on April 20, 1967, while half of NASA was still reeling from a fire that had killed the *Apollo 1* crew during a ground test. *Surveyor 3* ran beautifully for the next fifteen days. Like its two predecessors, the machine sported a camera; inside those two weeks, its lens recorded some six thousand high-quality pictures of the surrounding terrain. And *Surveyor 3* was the first of its breed to feature a sampling scoop able to scissor out and dig into the dirt—much like a child's beachfront toy. *Hey, Max and Caldwell, you wanna know what the surface is like? Here you go!*

Then the lunar night fell on May 4, plunging camera and scoop and everything else into unparalleled blackness. The circumstances left *Surveyor 3* unable to charge its batteries using the dual solar panels mounted overhead on a telescoping boom.

Sunrise finally came two weeks later and cast light across a dead spacecraft. Everyone on the ground knew that all along. It was part of the flight plan—the part that goes on the last page. So that's where *Surveyor 3* sat, untouched by human hands.

That is, until Conrad and Bean came a-visiting.

Unlike virtually every other unmanned program, the first one out did what it was supposed to—in spite of the rocket's brand-new upper stage *and* the new spacecraft. It even succeeded despite all the fighting among JPL, who managed it; Hughes Aircraft, who built it; and NASA, who begrudgingly paid for it. *Surveyor 1* blasted off on May 30, 1966, touching down on June 2. Even though the objectives were few and altogether basic, they rated high on the difficulty meter. All anybody wanted was for the thing to successfully get out of Earth orbit intact, stay in radio contact during the voyage, and land—preferably upright. *Surveyor 1* brought it all home, vindicating the design, the booster, and the team—every single one of the goods. The trip was a piece of cake.

Selling the program wasn't. Presumably, Surveyor's top priority was to validate Apollo's landing gear designs and choice of locations. Max Faget figured the whole program merely wasted NASA money and effort.

His opinions emerged as early as the spring of 1959, when the young engineer was asked by Keith Glennan to make a small-group presentation on the feasibility of manned lunar landings. Faget laid it out for his audience, and nothing he said included probes. First would come the scouts—men in a craft sent out just to loop the Moon once and come right back. Even with one single loop, they'd have plenty of time to peer down through telescopes and explore locations. Following that would be the manned orbiting mission. The next trip would orbit the prime landing spot to confirm its suitability, and then land after a few revolutions. And at that moment, Wernher von Braun raised his hand.

He said, "Max, you're overlooking all that we're going to learn in the Surveyor program."

The group reviewed their proposed machine. At the time, Surveyor's flight plan didn't involve orbiting; the intention was to zip in directly from Earth and execute a pillow-soft landing. But Faget couldn't see it. He thought the idea was bunk. All he could see was a hurtling chunk of tinfoil, careening through space at better than ten thousand feet per second. It would need one hell of a rocket to slow the pace for controlled descent. And that rocket would have to fire at about a hundred miles above the surface? Automatically? Forget about it. With that, Faget offered, "It would be a pretty bad day if when you lit up the rockets, they didn't light."

In attendance, William Pickering realized how far beyond Max Faget's cerebral horizon they'd already progressed. "We'll have done this a number of times!" he called out. "We're already working on the techniques!" The JPL director began talking as if this was his very last opportunity to explain. He talked about the guidance. He talked about the approach pattern. He talked about the retro-rocket. It was already at an advanced stage of design; they had most of the requirements figured out.

The blizzard of facts began to let up. And all Faget could offer was, "Well, gee whiz, you know . . ." He wasn't sure what to say. In his eyes, a *lot* still had to be figured out. Lately, nobody was having much success reaching the Moon at all.

By 1962, with Kennedy's difficult challenge on the table, things naturally began heating up. Congressman Joe Karth came to visit the Manned Spacecraft Center in Houston. The place had just opened. Max Faget was on the front lines of engineering and design and was one of those responsible for coming up with how to build a working moon ship. Over lunch, Karth offhandedly asked him about what they expected to reap from Surveyor by the time all of the probes finished landing.

Undoubtedly, Karth had to have been aware of Surveyor's developmental problems. Running almost in parallel with them were the Centaur booster's problems. At least von Braun's Alabama geniuses at the Marshall Space Flight Center held responsibility for that one. Centaur was planned as a new, high-energy upper stage for Atlas, and its advertised capabilities would be needed to soft-land a probe on the Moon. No Centaur, no Surveyor. But everyone at Marshall came to work each day sandbagged with the Saturn booster for Apollo. It was always top priority. Even worse, Centaur's specs fluttered wildly about like a mosquito in the corner. It kept drifting out of the crosshairs, leading JPL and Hughes to constantly monkey with Surveyor's design as they tried to keep up. NASA was about to pull Centaur from Marshall's overworked fingers and give the whole dang deal to their Lewis Research Center in Cleveland.

And with all this surely in mind, Karth now asked Faget, "What kind of a problem would it amount to if the Surveyor program failed?"

In a flat tone of voice, Faget matter-of-factly informed the congressman that they had already factored Surveyor out of any decision making. Whether the program succeeded or crashed and burned made not a lick of difference. "That wouldn't be any bad problem," Faget told him. "We can do it without those guys." Karth was suitably perplexed. He'd brought up Surveyor in part to justify all that moola his constituents were laying out. Apollo didn't even want it?

Seemingly oblivious to the delicate situation, Faget kept right on jabbering away. "We've got a great, big, wide landing gear, and we just can't afford to be vulnerable to the loss of that program. We'd go ahead anyway."

Faget expanded on even this response, telling Karth that he and his boys had already made plans to get the same information on their own. Rough shapes had been worked out for a chunky survey module capable of staying a week or more in lunar orbit and staffed by bona fide humans. They planned on flying low and slow above the surface, reconnoitering prime spots just like Faget had said back at that uncomfortable 1959 meeting. Engineering drawings were already in progress, laying out penetrometers that could be dropped from orbit down to the surface. It was all figured out. "It would have been a good program," Faget later remarked of the plan. "But it didn't happen."

What did happen, though, is Jack Webb got Bob Gilruth on the phone in a great big fat hurry when he heard about it the next morning. To Gilruth, Webb rattled off how Joe Karth had called and given him hell. Had somebody cut the phone lines between NASA's manned and unmanned divisions? Weren't you people talking to each other? As soon as Gilruth put the receiver down, he ordered Max Faget to his office for his side of the story.

"I gotta tell you, Max," Gilruth began, "you really blew it." Gilruth impressed upon Faget the importance of keeping mum if questions came up about independence or partnerships or allegiance or any of that stuff. Zip it shut. That was a direct order from Webb.

Geologist Eugene Shoemaker was heavily entrenched in this Surveyor venture, and the man had suffered before. "We were all shell-shocked from Ranger," he explained. So when *Surveyor 1* made it down, all Shoemaker could say was, "My God! It landed!" Oran Nicks couldn't believe it either.

Some of the first congratulatory words to Nicks came from Max Faget. *I didn't believe you could bring off an unmanned landing,* Max said over the phone to him. *Especially on the first try.* A part of Nicks reveled in such turnabout, and he soaked up the praise. They had one sitting on the Moon, upright and in one piece!

Surveyor 1's first hour was spent running through systems tests, but everyone chomped their bottom lips wanting to switch on the television camera. Finally, they sent commands to fire it up, and what slowly came back was the image everyone had been waiting for: a close-up of one footpad. It was the first of some ten thousand supreme-quality television pictures *Surveyor 1* would return until the lunar nightfall descended on June 14. And not every picture was of the footpad—the camera sat on motors able to spin it around 360 degrees and pedestal up or down like a periscope. Great images kept coming in, making the *Luna 9* pictures look like mush. "Hell," Shoemaker later offered, "I wouldn't have given you a 10 percent chance that *Surveyor 1* was going to land. I think we were all aware that it was going to be a matter of luck."

Was the design already this rock solid?

That question received a definitive "No" about four months later. An accommodating Centaur upper-stage kicked *Surveyor 2* out of Earth orbit right on schedule, but sixteen hours later the spacecraft itself hit a snag. During the single flight-path adjustment, one of *Surveyor 2*'s miniscule attitude-control motors had to work properly for no more than ten seconds. The motor didn't light up, but its two siblings did, throwing the ship's alignment out of whack and losing the mission. No method existed for getting it back. But then came a third Surveyor, rocketing away on April 17, 1967.

As Pete Conrad stepped away from *Intrepid*'s ladder, he called out, "Boy, you'll never believe it. Guess what I see sitting on the side of the crater!"

His partner responded, "The old Surveyor, right?" Al Bean was still up there in the lander, peeking out the half-height door from a position on hands and knees.

"The old Surveyor," Conrad repeated. "Yes, sir." He laughed. "Does that look neat! It can't be any further than six hundred feet from here. How about that!" His distance estimate was as spot-on as his landing, due in part to the crew's extensive time around a model of the robot spacecraft. Conrad later said, "We knew what size it was, because we worked with an exact mock-up."

Surveyor 3 had actually landed three full times. At initial touchdown its attitude-control engines kept firing away, kicking the ship back up. Sixty feet downrange it bounced a second time and once again took to the skies. The Moon-to-Earth radio delay is about a second and a half, and when flight controllers recognized the problem, they instantly spit up a command for the engine to knock it off, which it did while at the top of its skyward arc. The spacecraft flew another thirty feet and slammed the crater's edge a third time, sliding twelve inches down before finally coming to rest. In spite of the triple play, *Surveyor 3*'s aluminum-honeycomb footpads still did the trick.

36. Moon walker–to–be Al Bean faces a Surveyor mock-up during his training. Slant of floor matches *Surveyor 3* crater. (Courtesy of NASA)

To play it safe, the moon walkers approached from high ground, coming in from the lip of the crater and sort of spiraling down in. The pair drew right up alongside, taking pictures as they shuffled in.

It was *Surveyor 3* all right. Anybody doubting that these things ever went the distance only had to come spend a quick minute in Conrad and Bean's million-dollar boots. Standing before the men was a perfectly utilitarian design. Three legs held the working parts. A central boom rose up maybe three feet to support two panels filled with hundreds of solar cells. All totaled, they gifted eighty-five watts of life to the machine—a record-setting amount for the time.

Like it was supposed to, the craft's retropack had been cast off the bottom just before final approach, crashing miles away in parts unknown. It was a discarded marvel of engineering and thrust-to-weight ratios. Boasting the highest-performing solid-fuel motor yet constructed—it had worked like a champ to slow the weary traveler down to manageable landing speeds. This was one of the magic pieces that even Max Faget hadn't been able to imagine a decade before. The rocket complemented Surveyor's other brand-new and rather exclusive enhancements, such as the delicate positional-adjustment rockets and a sophisticated landing radar. Even though both were moon ships, the dissimilarities between Surveyor and Ranger put them worlds apart—light-years, even. Surveyor was lunar exploration come of age.

Al Bean looked at his target. He said, "Hey, we got a nice, brown Surveyor here, Houston." *Surveyor 3* hadn't changed colors; as it turned out, the spacecraft was covered with dust. Conrad ran his gloved finger along the television-camera mirror, tracing a line. It seemed to have a fine, uniform coating. The astronauts spent a bit of time getting close-up shots of the footpads and of the trenches left by the scoop. Such areas had enjoyed repeated photography during *Surveyor 3*'s operational life. Now, earthbound scientists remained frightfully interested in what sorts of differences might be visible thirty-one months later.

The scoop hung on *Surveyor 3* in such a way as to be easily visible by the camera at all times. It dangled at the end of a scissor-arm capable of extending over three feet. The small bucket could and did tamp-tamp itself on the soil, craft six-inch-deep trenches, and portage diggings up closer to the camera lens for scrutiny. Only four inches long and two wide, the scoop banged, scraped, dug, and trenched on the surface for a cumulative total of nearly eighteen and a half hours and functioned as a primary reason for targeting *Intrepid* at this particular Surveyor—versus one of its cousins.

Thirty-one months earlier, the scoop had been driven by one Floyd Roberson, who relied on the ship's television camera to know what he was doing. Operation had not been entirely straightforward. The camera lens pointed straight up and relied on an angled, motorized mirror to alter its view of the terrain. Every move looked the exact reverse of what was actually happening. So Roberson, during JPL practice sessions with mock arm in a sandbox, watched all his work in mirror and learned how to think backward. On top of this was the second and a half delay between moving the controls and having the spacecraft respond. A discomforting experience not too far removed from drunk driving. The tedious process required patience, but endless training made Roberson a pro. After completing NASA's carefully scripted tasks of using *Surveyor 3* to knock on the surface and dig trenches, it was time for some real sandbox fun. Floyd spent ninety minutes maneuvering inch-by-inch toward a small white rock, waiting after each step for the televised image to refresh and confirm his location. The scoop's jaws finally closed around the rock, picked it up, and put on the Doberman squeeze. A hundred pounds of pressure clamped down on the sample, which failed to yield in any way. Geologists in the room giggled and oohed at this new gadget. And naturally, Ronald Scott was thrilled with how well it seemed to work.

Scott was the guy who'd actually designed the device. For many years his love had been soil in general: what different kinds were made of, how it moves and shifts, what happens when it freezes and thaws, how to test it. It was his thing. He contributed much toward the world's understanding of what is actually happening at the dirt level during your run-of-the-mill earthquake or landslide. When the Baldwin Hills Dam failed on an otherwise peaceful Saturday in 1963, Scott materialized on the scene to analyze the mechanics of why 250 million gallons of water spilled over an earthen dike into suburban Los Angeles. And not long afterward the Scottish native and Caltech engineering professor responded to a unique challenge: develop a suite of preferable methods for the examination of *lunar* soil.

Pete Conrad leaned in as they scrutinized the landing footpads. A studious Al Bean radioed, "That aft honeycomb shock absorber struck the dirt and looks like it took some of the shock. Other than that, the front one didn't appear to do that. Stereo there. Sure isn't going to slide down the hill though, that's for sure."

The two snapped tourist pictures of each other. First, Pete stood next to *Surveyor 3*; then they traded places. Next up was surgery—time to cut parts off of *Surveyor 3* for return. Bean handed Pete Conrad a set of bolt cutters that had been lugged along for a quarter-million miles. "Here comes your cutter, babes," he told his partner.

They'd rehearsed the actions a zillion times in training with a full-scale Surveyor model. Conrad later said, "We would practice with it and even cut the appropriate tubing; and then they would replace the tubing, and we'd do it over again. I don't think we cut the scoop off, but we practiced with those big bolt cutters. And the big thing was that we had to know what was the right thing to cut, because that thing still had propellant in it and, for all they knew, it was still pressurized and everything."

Benjamin Milwitzky functioned as Surveyor program manager on the NASA side. By this time, he'd been working in concert with Oran Nicks for several years. Prior to NASA's formation, Milwitzky cut his teeth at Langley as a landing gear expert—standing out as a detail man, a technical man, a man who went out and did what he said he'd do. And to Milwitzky, troubleshooting routinely involved a sort of research approach—one that struck the hardheaded Nicks as a bothersome, long-winded method of resolving problems. But Nicks could hardly dispute the guy's results. He was a landing gear man; that's why they'd chosen *him* to head up this project so distinctively focused on qualifying a landing gear design.

Gentle Ben Milwitzky's first major Pasadena-flavored hassle revolved around JPL's desire to run this project in-house from top to bottom. That is, they'd work up the blueprints, craft every test model, produce NASA's end product, and fly the missions. But Surveyor's formal approval as a lunar program had come in the midst of Ranger's downward slide. Why give it to JPL? And what about Mariner? It was in the Lab's pipeline at the same time. How on earth could JPL handle more than *those*? How could anybody?

Despite Mariner's relative smoothness in development, it, too, had been beset by procedural deficiencies that threatened the overall mission success. "It is a good thing we do our own selecting here," went one letter from a principal scientist. JPL had furnished them with electrical components that proved to be "completely, electrically dead." Another letter from a science investigator began, "The JPL timing for release of scientific data is not acceptable." This escalated to, "the alarming situation which seems to be shaping up at JPL," regarding the Lab's handling of raw data from the spacecraft. Subsequent to this, the University of Chicago, which was providing a cosmic ray telescope for an upcoming Mariner, went so far as to send a man to personally inspect JPL's readiness. "Mr. Gordon Lentz of our laboratory," a letter explained, "indicates that it is not likely that anybody is available for assignment to these tasks in the near future." And things concluded on a sour note. "If the situation as I present it is correct, it is certainly unacceptable from the point of view of the PIs. What will be done about it?" "pi" referred to "principal investigator"—a term for the scientist in charge of an experiment.

The Iowa physics department was fighting their own battle with a hard-nosed jpl. "We do not concur with the figures you give for labor," read one letter to Van Allen's department, calling Iowa's accounting practices into question. Overall, the document was more than a little accusatory. Beyond that, the Lab's delinquency in forwarding qualified parts meant Iowa couldn't maintain their promised schedule. "Almost all original deliveries of components are overdue," Iowa pleaded. At one point, jpl axed Van Allen's contract in a strongly worded telex that read, in part, "Immediately stop all work, terminate subcontracts, and place no further orders." Iowa's defense began, "We have not received a sufficient number of parts to build," which of course were supposed to be coming from jpl. "Please forward parts immediately." The contract eventually got switched back on. But when Iowa delivered the flight hardware, they found out that jpl technicians liked to crack it open for no real reason. "It has come to my attention that sui experiments are to be disassembled by jpl personnel after satisfactorily completing the final inspection by a jpl inspector at sui," read another telex. During a jpl test, the unit was contaminated with oil and needed to go back to the university for cleaning and recalibration. Such problems failed to nurture a warm cocoa-and-snuggles feeling between jpl and the very scientists it needed to justify these horrendously expensive missions.

Yet the Lab's cumulative experience simply couldn't be ignored. Hardly anybody did this kind of thing. jpl possessed the brainpower and the facilities. They had the infrastructure. Surveyor would be too big and complicated for so many other places.

What to do then?

In a maneuver designed to placate the issues, nasa insisted on an outside contractor for Surveyor. jpl could design the thing and be responsible for calculating its trajectories and supporting the mission. They'd hold responsibility for delivering each Surveyor to the lunar surface—but come hell or bad bananas someplace else was going to build the hardware—a third party. On *that*, nasa wasn't budging. Predictably, jpl responded to the idea in a less than enthusiastic fashion but ultimately agreed to the terms. Their tentative (and laughably underestimated) budget got approved. It specified money for seventeen gleaming new Surveyors, which ended up being not nearly enough for the seven that actually flew.

The Lab put a guy named Eugene Giberson in charge of running the Surveyor effort at JPL, along with finding someone to build it. In January 1961 Hughes Aircraft ended up winning the contract after bowling over four other aerospace competitors. Oran Nicks felt Hughes had everyone beat on the technical aspects, which he and Ben Milwitzky felt to be most important. Hughes had no other spacecraft programs going. This would be the first.

And as Ed Cortright summarized it, "JPL and Hughes had all they could do, to do the Surveyor."

Hughes would earn their money in the landing. A chasm of proficiency separated Ranger's Neanderthal balsa-wood impactor from this new and comparatively dignified arrival on shock-absorbing legs. They would finish out an entirely controlled descent—requiring a lightweight retro-rocket to initially slow the craft, plus three swiveling vernier engines to nudge Surveyor's position as it doddered in on terminal approach. The scheme flat out hinged on cues from a dual landing radar quite similar to what real, live humans would embrace when *they* finally showed up at the Moon.

During approach, Surveyor's first radar would simply ping its target, actuating the solid-fuel retro about sixty miles up. The ship would be moving so quickly that the very brief time lag between initial command and motor ignition allowed it to drop eleven heart-grabbing miles. Less than a minute later, *clunk*—the retromotor would be discarded. From there, Surveyor's second and more exacting radar would continuously monitor not only the craft's distance from the ground but also its velocity—vertically *and* horizontally—feeding tweaks to the autopilot. It could both decelerate and steer. And people thought parallel parking was rough. If the parts did their jobs, the spacecraft would end up fourteen feet off the ground when the engines died for good, allowing all 650 pounds to land at seven miles an hour with practically no sideways motion. Seven miles an hour is a respectable burst speed for the average thirteen-year-old.

That's how it was *supposed* to work anyway. Each component, along with the procedure for its use, was all brand-new and featured no precedent. "We were starting essentially from scratch," remarked Hughes engineer Leo Stoolman. At the project's outset, Oran Nicks harbored supreme confidence in Hughes to conquer such unfathomable technology. Quickly though, he came to wonder how seriously this vendor appreciated Surveyor's overall engineering challenges. If the drop tests were any indication, this was going to be tough.

The prospects of the drop tests started off sounding great. A round of simulated landings would be made while trying to approximate lunar conditions as closely as possible. Naturally, some compromises would have to be made. Six times as much gravitational force on Earth would skew the results. An atmosphere meant aerodynamic forces would affect Surveyor on Earth but not on the Moon. And rocket engines themselves behave differently enough in an atmosphere. Testing went forward nonetheless with these inconsistencies supposedly in mind. Nicks flew down to White Sands to check it out.

First, proof test models were to be slung from cranes and dangled over the barren ground. In this fashion, the engines and complicated landing radar could undergo a few workouts. The real tests came later on, when giant balloons would tow Surveyor up 1,500 feet and then let go. The entire descent sequence could then be run through with the spacecraft performing a genuine landing on the New Mexico desert. Nicks had set a requirement that Hughes pull it off three times in a row before he'd call the tests completely successful.

But Oran Nicks discovered that a lot of the machinery in the test model wasn't even real flight hardware and that Hughes wasn't going to follow the same protocols they'd use during actual missions. Nobody at Hughes was taking the test seriously. What could realistically be gained from drop testing if the tests themselves were so different?

The first full-release test in April 1964 went horribly wrong. Surveyor delicately hung under its gigantic balloon 1,500 feet up. Crummy weather skulked through the test area and quickly morphed into a thunderstorm. Electrical disturbances in the air prematurely triggered Surveyor's release, and the wad of expensive parts and engines obliterated itself on the desert floor. They had another go at it in October, just before *Mariner 3* lifted off. Another letdown—five of them, in fact—on this one spacecraft alone. Five major components had completely failed.

"We pretty much underestimated the magnitude of this job," one of the Hughes executives admitted at the time. "It is a much bigger job than we originally anticipated."

Oran Nicks emitted a low, primal noise. Simultaneously he juggled this situation with that of Lunar Orbiter and Boeing. Their contract wasn't totally sorted, and plenty of influential congressmen still publicly questioned the wisdom of choosing the most expensive candidate in the first place. *Mariner 3* was sitting there on the pad in Florida. Another NASA quarterly review always hung on the horizon. Commercial air travel benefited greatly from Oran Nicks's schedule: up to Seattle for Boeing, over to Cleveland for Lewis and the Centaur, reverse to Pasadena for JPL, then back across the country to Langley.

It didn't end.

Then, JPL decided they'd had it with Reaction Motors. Functioning as a division of Thiokol, the manufacturer had not been able to deliver satisfactorily working vernier engines—the ones to maneuver Surveyor during that nail-biting terminal descent. Encountering numerous technical and managerial problems is nothing new to aerospace firms, but JPL apparently considered Reaction Motors beyond hope. Nicks learned, after the fact, how JPL had dumped Reaction Motors and instead reached for Space Technology Labs. Oran Nicks found nothing amusing in JPL's seemingly desperate actions and went investigating. Alighting first at STL, Nicks came out with mixed feelings. The outfit definitely seemed up to the task but was going to require understandably large amounts of time that JPL and NASA simply didn't have.

He left STL and traveled to the ousted provider, which did much to help him understand JPL's drastic action. After taking a look around, Nicks proclaimed the Reaction Motors facility "a bucket shop," which no doubt would have enraged the bucket-manufacturing community to assaultive lengths. The filthy place contained little equipment and even less of the actual hardware. Furthermore, the company's test results put them squarely into remedial rocket making. Nicks went for a little chat with the contractor's senior management, who were still reeling from the decision. *Sure,* they told him, *we've got problems. But who doesn't? We're creating something from nothing. Don't you think it's a bit harsh to pull the rug out from under us?*

A terrified Reaction Motors leveraged whatever they could to win back JPL's affection. They pledged to fix the motor problems. They said they would use their own money to do it. They said the terms of the old contract could be thrown out; they'd renegotiate. And above all, management claimed that they genuinely believed in Surveyor and wanted to continue participating.

Interestingly, JPL agreed to cancel the new deal with STL and reinstate a thankful Reaction Motors. Less than four months had gone by since Pasadena had told them to forget about it. A congressional House Oversight Committee would summarily report, "termination of the RMD contract seems to have had a salutary effect. Evidently, technical and management problems were solved in rather short order when the contractor realized what was at stake and that the government was willing to cancel his contract."

Nicks had a review of his own going on. A quick synthesis of its findings with the congressional report indicated that he and Ben Milwitzky really needed to act. The pair took their recommendations to Ed Cortright, who backed the idea. So then it was off to JPL for another round of uncomfortable discussions. Sitting in conference, the three NASA men patiently explained how they felt JPL required a fundamental sort of change, one that might prevent these bothersome Ranger-like evils from persistently recurring. They wanted a new position created at the Lab for a kind of glorified general manager—somebody well versed in all the icky details that JPL didn't like and wasn't any good at, things like contract negotiation and top-end management. This position would act as a ramrod to push projects through. Several months later, the Caltech board of directors finally caved. They even had a name, strongly recommending that Bill Pickering hire retired major general Alvin Luedecke as deputy director. He was leaving the Atomic Energy Commission and could start right away. Bill Pickering crossed his arms and knew full well what was going on before him. What NASA really wanted was a babysitter—some kind of on-site taskmaster riding herd over all the JPL tomfoolery. And Luedecke would be this man.

He strolled in on August 1, 1964, to hang his hat for the first time. August 1 happened to be a Saturday, but weekends didn't matter much to people like Alvin Luedecke. He immediately began logging seven-day weeks, remaining at the Lab for as many as sixteen hours at a stretch. Upon his departure after thirty-six months of employment, Luedecke tallied approximately seven normal work years and defied almost every expectation as to what his presence would contribute. Oran Nicks was in love.

In concert with NASA, Luedecke quickly assessed the faltering Surveyor program and agreed that the time had already passed to freshen up its leadership. For starters, everybody wanted someone besides Eugene Giberson driving the show from Pasadena's end. Giberson gulped—he was better than this Nicksian treatment indicated; in fact, he would recover to command other JPL projects. Robert Parks ended up at Surveyor's helm. And what did Parks's coworkers make of the switch?

"They thought I had taken on an impossible task," he said.

With both NASA and Caltech behind him, Luedecke now turned his full attention upon Hughes Aircraft, who as of late had been causing more trouble than an insubordinate toddler. So many errors had occurred, with most of them so largely attributable to basic human negligence, that Oran Nicks decided humiliation could well be a fitting addition to his game plan. The idea came to him from his time back in the army days at the rifle range. If somebody missed their target completely, a red flag was held up to indicate the lack of any score. "Maggie's Drawers," they were called. Oran Nicks was inspired.

A few weeks later, Hughes vice president John Richardson opened up a package from NASA headquarters bearing his name on the address label. Inside was a note on top of something red. The note was from Nicks, cheerfully explaining how the enclosed crimson pennant could be flown from the company's main flagpole every single time a significant error was committed. Going even further was the recommendation that signs be posted in the work areas of those involved, calling attention to whatever blunders had been made. Going even further beyond *that*, Nicks paid for the whole thing himself: flag, box, and shipping.

Major General Luedecke employed a contrasting approach. He reviewed the Hughes contract and felt overcome by sympathetic concern. Much of the company's obligations lay scattered in discontiguously written pieces. Numerous changes had been made to Surveyor's design—forty-six modifications across eighty change orders, to put a number on it. It wasn't the contractor's fault; changes happen in any sort of project. You almost can't build a doghouse without making changes. Here though, the situation burst into outright confusion. In all this time, the contract had never been updated to keep pace with the project's evolution, so nobody knew for sure who held ultimate responsibility for what. Maybe the formalities just needed some ironing out. Bravely, Alvin Luedecke strapped on waders and stepped into this bureaucratic quagmire. Already, it was late in the game: *Surveyor I* had been long accepted, shipped, and folded up into the Centaur. But appreciating that late was always better than never, Luedecke hand delivered the revised contract to Hughes on May 30, 1966—the very day of *Surveyor 1*'s launch. Both events signaled a turnaround in relations between all involved. Maybe the program could succeed.

Conrad and Bean began their dissection. A few pieces of cable were clipped loose and dropped into a special container for return to Earth. Al Bean told his commander, "Why don't cha . . . Why don't cha . . . Why don't you give that a cut right there, Pete?"

Conrad said, "All right."

"Give them a couple of pieces."

And then came the biggest prize of all: that television camera. The camera was such a big deal, in part, because it was more easily recoverable, unlike some piece of electronics deep inside the ship. Obtusely it stuck to one side, almost phallic, making for a simpler target. And this plum picking was also a complete assembly. Conrad and Bean wouldn't be bringing back just a *piece* of an experiment; they'd have the whole darn thing. It was relatively small and manageable. It would fit into one of their rock boxes. The camera also contained a diverse assortment of parts: metal and glass, electronics and mechanical gears. There was a little bit of everything right there.

Conrad directed his partner, "Okay, whip around the other side of that scoop, and let's get that camera."

The two spent a couple of minutes getting into position. Then Conrad opened up the cutting jaws and got to work. Bean offered a play-by-play: "All right. Wait, I got to open your bag! Stop! Okay. Cut. That a boy. Good cut."

Conrad kept going at it. "Ah!"

"Good cut," Bean offered again. Then another cut. "Good one, Pete."

Conrad was just about there. "Okay, two more tubes on that TV camera, and that baby's ours."

After another bout of effort, Bean said, "Let me get a grip on it now."

Conrad said, "Okay." And then the camera was free. "That's ours!" hurrahed Conrad. He went off into gales of laughter.

Underneath his commander's cackling, Bean called out, "We got her!"

"Beautiful!"

Conrad laughed while Bean continued talking. "It'll fit right in that sack. Hey, is this ever lighter than that one in Houston!" The two men stowed their prize in a backpack that zippered shut. Conrad was able to break off the sampling scoop. He'd been coached ahead of time by Surveyor engineers who went over the model on Earth with him and pointed out some weak spots that might break the easiest. The whole scoop assembly, with dirt still inside, went into another bag.

When *Surveyor 3*'s remnants arrived safely back on Earth, the camera was gingerly delivered to Hughes Aircraft, who assessed it in surgical clean room conditions. They started with the big picture—noting color changes, micrometeorite pits, and so on. Overall, it exhibited signs of weathering, even though there was no weather up there to speak of. Next came the arduous process of taking the thing apart, and that's when the technicians made a startling discovery. Sandwiched between two aluminum circuit boards, and nestled deeply within the camera's insulating foam, lived a cluster of Earth bacteria. For two and a half years it lay dormant, surviving in a complete vacuum ever since the moment *Surveyor 3*'s Centaur booster had put the machine above Earth's atmosphere. And then the stuff had come back to life sometime after returning. Delicately, deftly, workers removed samples via a small hole in one of the boards.

Pete Conrad later had this to offer regarding the event: "There's a report on that. I always thought the most significant thing that we ever found on the whole goddamn Moon was that little bacteria who came back and lived, and nobody ever said shit about it."

The bacteria nobody apparently ever said shit about was unmasked as alpha-hemolytic *Streptococcus mitis*. Essentially, this is the long way to describe strep throat. If Jim Burke's Rangers had not wilted so visibly in the presence of heat sterilization, *Surveyor 3*'s bacteria may well never have had a chance to stow away on board. But of course, oodles more was known about our airless and virtually dead Moon by the time *Luna 9* and *Surveyor 1* landed. In the presence of such knowledge the regulations had been amply slackened. Even so, the problems caused by sterilization have led to much more reliable hardware.

Since 1969 time has shed more light on the stowaway bacteria. The very real possibility exists that *Surveyor 3*'s strep throat wasn't just resilient as all get-out to hold its breath for almost three years. Rather, the bacteria may well have not jumped on board until some point *after* the camera returned to Earth. But if that was the case, how come only this one particular strain survived? Wouldn't a broader cross section of germs be present in a camera that simply picked up a bug after coming home? And wouldn't they be more spread out?

21. Irradiated Plans

I know in his heart of hearts he would not have wanted
to give up the job as project manager of Voyager.

John Casani, on the man he replaced

"I personally believe that he is the best project manager JPL ever had," explained Ray Heacock, who managed the development and testing of the MJS'77 hardware. "He is very people- and team-oriented. He picked individuals one by one, on the basis of forming a real team and team concept. His management style also fostered and strengthened the team."

Another high-level JPL member put it like this: "Schurmeier was really the architect of the project."

"He's a very important guy. If you had to pick one person, the one project manager, the one leader to hang the Voyager trophy on, it would be Schurmeier," added Bruce Murray, who took over as JPL director from Bill Pickering in 1975. "I can't overestimate how much of a role Schurmeier had in making this thing work."

In such glowing terms, they spoke *not* of wealthy and prominent St. Paul, Minnesota, citizen E. J. Schurmeier—who, in 1897, advocated pardoning the surviving members of the bank-robbing Jesse James–Cole Younger gang. Rather, Heacock and Murray spoke of Bud Schurmeier. Yeah, *that* guy—Ranger inheritor and old JPL hand, who was now newly anointed as project manager and general head honcho of . . . well, they hadn't yet changed its name to "Voyager." Officially the project was still referred to as Mariner-Jupiter-Saturn 1977, a moniker about as attractive as quintuple bypass.

Whatever they called it, at least somebody good held the reins. "It wasn't that I had all of the smarts to know everything," Schurmeier was quick to concede in his typical aw-shucks fashion. "But somebody has to listen to all the pros and cons and then make the judgment that this is the way we should go." That person would be he.

Schurmeier's real first name was Harris. So what earned him the alias? "I'm still not really kind of sure," he offered, in something of a rare moment of uncertainty. "My older brother was named after my father, Gustav Benjamin, so he was called 'Ben.' And then I was named after my uncle Harris Macintosh. But I don't know; somehow I got to be called 'Bud.'" He'd married a woman named Bettye Jo Parris, but everybody called *her* "BJ." "She was a gal that worked at Caltech in the wind tunnel there," he explained of the circumstances leading to their introduction. "That was in the old days, before they had electronic computers, when everything was calculated on a Friden." Schurmeier worked there, too. "So I saw her a lot and eventually got to ask her out for a party, and one thing led to another!"

"I had known Bud very well," continued Bruce Murray. "I realized he was an extraordinary person. He had set up the Systems Engineering Division at the very beginning, but he also had more lateral vision; he could deal with a variety of people and institutions and sort them out much better than most."

The project was Schurmeier's to run as he saw fit. Equipped with wispy golden sideburns and a prematurely receding hairline, Schurmeier unassumingly walked down the monochrome hall, past his secretary, and on into his office—a real office, with a door and window. "In those days that's the way it was," he explained. "It didn't get to the cubicle design until later." And now Bud sat down at his institutionally metallic desk with the fifth- or sixth-hand filing cabinets to begin commanding a truncated project.

As a matter of record, Schurmeier always maintained that not much new really went into enabling a mission extension to Uranus or Neptune. "We did that as we went along," he contended, dismissing any suggestions that the group outright conspired to sneak extra yummies into the basket for the long haul. According to Schurmeier, whenever it was realistic the engineers merely swapped in upgraded equipment, which conveniently enough just happened to be more suited for long-duration flight.

"At the time of the MJS proposal, we planned to come back with a dual launch in 1979 to Jupiter, Uranus, and Neptune," he elaborated. "NASA turned it down because they did not want to purchase any more Titan-Centaurs. All the money was going into the Shuttle."

37. This detail from an MJS'77 group picture shows Ed Stone at left with Bud Schurmeier front and center. In the back row on the right is Ray Heacock. (Courtesy of NASA/JPL-Caltech)

Just one guy behind a desk could not put a spacecraft at Neptune, or Jupiter, or even on the flippin' launch pad. Schurmeier's first big obstacle was to formally open shop by compiling the MJS'77 team proper and somehow get them all facing in the same direction. He said, "I had known all the guys from previous missions and knew their strengths and weaknesses." And many of them had been in on the original Grand Tour proposal. So like a seasoned football coach, Bud Schurmeier slotted his players.

Ray Heacock was in for sure, at JPL since 1959—as good of a hardware man as they came. Jowly with a shock of black slicked-back hair, he'd put in his time on Ranger and already had forgotten more about the particulars of TOPS than most would ever know. Overall big-picture responsibility for spacecraft design and testing went to him.

Schurmeier's palm next rested on the angular head of Ed Stone—Caltech professor of physics. The thin, big-eared Stone would act as project scientist, becoming a central sounding board and clearinghouse for all the academic types chosen to populate the spacecraft with experiments. When they came parading through with concerns, conflicts, and consternation, all could look to Stone as their man on the inside—wearing their colors and thinking like they did and in a position to spin around and do something about it.

Then there was Ralph Miles. He'd function as mission design manager—a deceivingly nebulous title. Miles's responsibility entailed nothing less than mapping a precise course through the solar system's planets and moons—a highway predicated on science objectives and gravitational influences. Properly leapfrogging from one heavenly body to the next meant nailing a series of precisely defined corridors, or gates, like a downhill skier going *sluice-sluice-sluice* through the black field of moons and planets. Angling the wrong way past Jupiter *even slightly* meant missing the exit for Saturn, and so on down the line.

Miles's work didn't end there. Science plans dictated fuel, navigation, instrument pointing, and telemetry requirements. How much fuel could be used to correct for each planetary encounter? How much extra should go in reserve? Which experiment got what time on the dish? How would JPL determine where, exactly, the ship was at any point? All of that was in the dominion of Miles, interfacing with Ed Stone as his gateway to the scientists.

Other titles filtered onto the roster: project assurance manager, spacecraft team chief, imaging representative. Then there were the systems managers—the task leaders. Alongside them and definitely not underneath stood the thousands of bright JPL minds in endless permutations of supporting roles: draftsmen, machinists, technicians, accountants, security guards, and receptionists.

Humbly shuffling papers up in his little office sat Bud Schurmeier, capstone of the pyramid, refining the MJS'77 organization and overall strategy. His helpless desk endured a continual state of near avalanche. "I'm a pile man," Schurmeier explained of his filing system. "Yeah. Only reason that it ever got straightened up was the secretary." With the core team finalized, things shifted into high gear. They faced an absolute deadline. People started clocking in longer days than anybody already was—days and nights alike. For many of them, JPL became "wife number one" in the vernacular of one teammate. "I'd take a lot of work home," Schurmeier offered, "but I would try to keep it away from the family and dinner."

Gary Flandro wasn't on Bud's roster. PhD in hand by 1967, Flandro retreated to the University of Utah in Salt Lake City even before TOPS began, where he hung the title of assistant professor of Mechanical Engineering on his office door. He'd have to watch his baby's first steps from afar.

During their endless back-and-forth negotiations with NASA, JPL had promised up and down like good little boys and girls to hire a properly qualified highfalutin outside contractor to actually build the spacecraft. "Oh boy," Schurmeier had been lectured at one point, "you gotta get a contractor that knows how to build hardware instead of these young kids."

Yet Bud and his cohorts reviewed the evolving legacy of machinery they'd brought forth and quickly concluded that it might not be a bad idea at all to renege and make the thing in-house. Sure, Hughes Aircraft had eventually delivered a perfectly working Surveyor. But their cost-plus-a-fixed-fee contract drained five times the money originally budgeted.

JPL's Mariner line had aged well beyond puberty—now in its fifth design iteration and galloping like a thoroughbred. It had been built right there in Pasadena. After a spring launch, *Mariner 9* was currently running trouble-free laps around Mars like it was no trick at all. In yet another meeting, Bud Schurmeier argued, "Look. We're gonna use a lot of the Mariner design and hardware, and it can be a lot cheaper if we do this as an in-house project."

NASA administrators ultimately waved the suggestion through. "That was not a hard sell," explained Schurmeier. "Cost savings."

Overall planning now required the processing of Gary Flandro's original flight paths into something more responsive to tangible objectives—like photography and science returns—while remaining inside the mission constraints. Where do you want to be at Saturn? What do you want to look at? What do you want to accomplish with the rings? Ed Stone's harem of scientists wanted it all.

They also wanted to see moons. The thinking of the era held most all solar system moons to be analogous to our own—that is, gray, lifeless, and fundamentally less exciting than underpants on your birthday. But there was this one moon out at Saturn—an exception, conceivably the *only* exception, to the rule. Of all the worlds in our solar system beyond Earth, Saturn's moon Titan evoked particular interest. Hazy with atmosphere and potential lakes of watery hydrocarbons, the body had long been considered the most likely candidate to have something growing there. A Nerf-orange globe, it represented primeval Earth circa 4 billion BC.

JPL computers began lumbering through the possibilities for Jupiter–Saturn–Titan—JST, to use their internal shorthand. Every decent option had a gotcha. For a nice, close Titan snuggle, MJS'77 strategists would have to allow her to be disquietingly fondled by Saturn's gravity in a very specific fashion. An energy boost would thus be realized, maneuvering the craft delectably close to those rich orange clouds, which can't be penetrated by any telescope on the ground. But herein was the rub—after the Titan encounter, MJS'77 would unavoidably get flung upward, out of the flat plane that our planets travel in. She'd never be able to reach any other worlds—no Uranus or Neptune or anything else.

But all agreed Titan was something that nobody could ignore. Hanging out there 800 million miles from Earth, this singular hot toddy evoked such wonder and carried such promise that an entire spacecraft might well be rationalized just to check out the big T.

One ship would be gone.

But they had *two* flights, two bullets in the gun. What if the first shot went swimmingly well? What if their aim was perfect? If *MJS'77* remained operational past the original mission, JPL might try to punt for some overtime on the scoreboard clock.

"Our idea," mused Roger Bourke of the group's plan, "was that once the thing is pointed at Neptune, maybe the money will somehow materialize." Who could say no to Neptune, when a great ship had done no wrong and was already headed there?

With primary flight objectives covered, the actual road map now required creation. But Ralph Miles began having problems with these agonizing muscle spasms in his back. Even before that happened, the notion had crossed his mind to leave MJS'77 and pursue his own line of research—so Ralph bowed out altogether. In his place arrived Charley Kohlhase.

Ex-navy man, photographer, golfer, and backpacker, Charles Emile Kohlhase had long been a detail man and simultaneously credits and blames his father for that approach to life. "When I was young, my father was a perfectionist," Kohlhase explained of his background. "He was critical of me; he was very hard on me. He was an excellent cabinetmaker and jack of most trades. If I worked on a small box or something, if I didn't do it exactly right, he would criticize me, and he was constantly after me in that sense." And a shadow always lurked over the younger man's shoulder, even when Dad and his exacting methods weren't around.

In the areas of math and physics, however, the elder Kohlhase foundered.

38. Modern Renaissance man: Charley Kohlhase. Here, he's in the thick of an early 1960s meeting to coordinate lighting conditions at Mars during a Mariner flyby. (Courtesy of Charles Kohlhase)

When young Charley graduated high school in Chattanooga, Tennessee, and went to Georgia Tech, he evoked the ire of his father by jumping ship from mechanical engineering to physics. "Somehow I had the courage to rebel," Kohlhase explained, "even though he cut off all financial support after that time."

Oak Ridge offered the new physicist a job straight from Georgia Tech, but hydrogen bombs didn't get his blood moving like space exploration did. It meant lower pay than nuclear research, but Kohlhase decided to have a go all the same. That was 1959, and now here he was—ensconced in a numbered building, eating cafeteria food every day, trying to build a mission and lay out a course that no one else ever had. Or has since.

When all the possible flight combinations were brought together, they added up to nearly ten thousand possibilities. Somehow that had to boil down to a pretty low number—like *two*. Kohlhase and his merry band of designers and plotters thus began considering all the scientific and photographic requests, working to eliminate the more restrictive or undesirable flight paths. Nested somewhere in those ten thousand were the money flights.

The three possible years for launch included thirty-day periods in 1976, 1977, and 1978. Each lay about thirteen months apart. A trajectory specialist at the Lab named Paul Penzo wrote his own computer program, and with it they began sifting through the options. Their first opening in 1976 became less attractive as time went on. Jupiter would still be gaining on Saturn and, in fact, trailing so far behind its ringed neighbor that the required gravity assist would have been dangerously close to the planet and its intense radiation. Kohlhase discarded any launch opportunities during America's bicentennial. Two thousand options fell away.

Another possibility lay in a 1978 launch. But there, Jupiter would have already gained on Saturn, leaving Kohlhase and Penzo to bet the farm on a very, very slight gravity boost. A limp throw like that translated into a far passage of Jupiter, meaning the poorest reconnoitering of Jovian satellites. The group settled on a 1977 launch, calculating an ideal push from Jupiter and what promised to be an illuminating autocross right through its field of moons. With every science objective factored in from start to finish, the overall Neptune flight time grew to a dozen years.

Choices like these rippled downstream into everything. It cascaded into weight allowances for the experiments, the computer, metal framework tubes, radiation shielding, insulating blankets to wrap the bus. It impacted how much corrective fuel the ship would have to carry for in-flight adjustments. Kohlhase's team came up with a number for that last one, but it struck them as way too low. "Everybody said, 'Oh come on, it's got to be more than that,'" Kohlhase remarked. But when the work was rechecked, nobody could find anything wrong. They decided on a value, forwarded the information to Propulsion, and then tried to call it finished. Kohlhase went home and immediately started having uneasy dreams at night. His subconscious kept looping these chilling little movies in his head where they hadn't figured the errors correctly and the ship blew through all its fuel even before Saturn. Nothing shut the dreams off. He kept waking up, thinking to himself, "Oh God, how would I like to be responsible for overlooking something like that!"

Hardware problems escalated. The umbrella-style folding radio antenna had been a budget casualty, replaced with a cheaper and more-traditional rigid graphite-epoxy dish. But there was only one problem: "Nobody had ever built a graphite-epoxy antenna that size before," explained Bill Shipley.

It was maybe 300 percent bigger than any others. Shipley's revised budget covered five dishes, assuming a few casualties along the way. True to form, the first one out of the mold broke apart like a Chevy Vega. "We had a major disaster," he said. They started picking up the pieces.

"But we learned something," Bill Shipley optimistically began, thinking of the situation. "We learned it was really very easy to patch. And once we knew it was easy to patch, we didn't need five of them any more. So we cut the quantity down to three." Money saved on two unbuilt dishes rolled right back into the program. All three dishes flew; the last was pinched years later from a display model and used on a completely separate program.

Then in December 1973 the MJS'77 team heard some news to stop them cold. A certain little spaceship from Ames Research Center, a spindly skeleton thing named *Pioneer 10*, was at that very moment out cruising past Jupiter and on the phone with superbad news. The local weather forecast called for radiation levels far more deafening than what anybody figured—by at least a factor of ten. Hot 'n' toasty *Pioneer 10* wasn't even at closest approach. Said Charlie Hall, Pioneer project manager, "They were plotting energy versus distance to the planet, and that line was going straight up." The ship began generating false commands back onto itself. Pictures stopped coming. One experiment died. *Pioneer 10* appeared to be headed for total meltdown.

One guy said, "My God, we've had it."

As this reality pinged through Ames, Hall remarked that the place began resembling "a funeral parlor." A million GE Spacemakers had nothing on all that radiation from Jupiter soaking into their machine. In fact, *double* the amount of energy wafted from Jupiter than what it got from the Sun. At the huge planet's core existed some very odd behavior. And without question the still-unfolding saga introduced a series of queasy new wrinkles to the Jet Propulsion Laboratory, which didn't have any money left to iron them out. Pretty soon the budget wasn't going to have enough left for Schurmeier to hang around, either.

"That was a major impact on the program," explained Bill Shipley of this radiation discovery. He spoke in deceivingly understated terms. "It was not in the original program plan."

Darn that *Pioneer 10*. Its presence signified a conundrum of blessings mixed with opening-night upstaging. The ship owed its birthright to an obscure governmental outpost fastened onto the corner of a weathered air base down near San Francisco. Ames Research Center was the kind of no-nonsense facility that always moved in small steps and budgeted most of its time for basic aeronautical studies like aircraft flutter and reentry heating.

They also had something to prove. As the original Grand Tour proposal was being cogitated and dressed down and compromised, NASA had already announced its separate desire to entertain wide-scale proposals for asteroid belt and Jupiter-flyby missions—albeit on a much simpler scale than the tour's.

The asteroid belt? That very real band of rubble between Mars and Jupiter was a formidable geologic feature. There are those who believe it's what remains of a planet that formed—or maybe half-formed—long, long ago between Mars and Jupiter. But the planets were all too close to one another, leading to gravitational tugs-of-war in which this middle protoplanet lost and today is an asteroid belt. That's the theory anyway.

Although the total number probably reaches into the millions, the belt contains tens of thousands of larger, catalogued asteroids. "*They're* not the problem," is how James Van Allen put it. "Nobody ever had any apprehension about hitting one of *those* or being hit by one of *those*; they're so sparse. It's the pebbles and debris left over from these asteroids colliding with each other." Reams of dust and gravel—*that* was the problem.

Could a ship make it through? Only three years beforehand some of the greatest minds in planetary science had locked themselves in rooms and sweatily banged their fists on the table and decidedly proclaimed that it couldn't be done. PhDs hung in the room like cigarette smoke. With neckties loosened and veins popping, the researchers tore into their paperwork and flip charts and frenetically squabbled. Papers flew. Nobody stopped talking. They drank gallons of coffee, hurled brimstone and lightning bolts, and preached sermons of destruction. A spacecraft might take *months* getting through the region. If debris in the belt didn't get you, the radiation would.

Van Allen deadpanned, "Opinions sort of ranged very widely." Personally, he had no idea whether they'd make it or not. "But we'll go through it, by George, and find out!"

Ames stepped up to the plate and, in February 1969, landed a prestigious commission. However the asteroid belt formed, it might be a brick wall—or at the very least one hell of an obstacle course. Everybody needed better information. NASA's plan to get it revolved around simple craft tailored to reconnoiter the basic lay of this land and report what was found. And since Jupiter was next off the highway, why not go there too?

Calamitous Jupiter offers a local approximation of the natural forces ebbing in our universe at large. And the planet's numerous moons form a sort of miniature solar system of their own, revealing clues about how actual solar systems form. These kinds of rationales are what focused all the spotlights on Jupiter. Past there, NASA's intrasolar appetite more or less took a nosedive.

In 1971 Disney World and the Aswan Dam opened. Idi Amin gifted Uganda to himself. D. B. Cooper hijacked an airplane for money and parachuted out the back, never to be seen again. Angry Belgian farmers crashed an international economics conference using live cows. And in California, instruments and subassemblies began arriving from scattered providers throughout the world, gravitating together as 570 pounds of sculpture that interacted with divine cohesion. The U.S. government exchanged $75 million for 2.5 million man-hours and a dozen science experiments.

One of those experiments resulted from a cluster of sharp minds operating out of an unassuming, drab stone building on the corner of Dubuque and Jefferson streets in Iowa City. Visitors usually walked right by on their way to see the old state capitol building instead. Yet more was always happening in the other place—its dim brick lobby, no bigger than a parking stall, gave that away with its shelves full of rockoons and space probe models and international awards. At the opposite end of the minimuseum and the glassed-entry double doors, a single elevator whisked up to the seventh floor. Once it opened, you'd arrived. Right across from the elevator, and ten steps from the loo, was Room 701. Inside, past the cast-iron sink, bookshelves, and stacks of paper high enough to violate any fire code sat one man eager to learn what surprises the outer solar system might have in store. James Van Allen's cosmic ray instrument had been accepted for Pioneer.

Choosing spacecraft experiments is not a popularity contest. Van Allen never got in just because he was Van Allen. Rather, he faced the same protracted selection process as all the others. The same dry essays to complete and questions to address: In what ways is this experiment consistent with the mission's objectives? Discuss. What new investigations does it purport

39. *Pioneer 10* coming together, 1971. Peeking through a notch in the dish is *Pioneer 10*'s imager. Finned RTGs point toward us and slightly to the left. (Courtesy of NASA)

to make, perhaps of phenomena predicted but not yet observed? Please describe your instrument's power consumption and telemetry requirements.

By this time in the early seventies, the Iowa City contingent had tallied an impressive track record of successful, productive experiments on virtually every unmanned American program that had ever come and gone. Perhaps the long string of prestige and government contracts was enough to net Van Allen a new Ferrari 365? Hardly. "We never made any money of course," he indicated. "We were always a nonprofit organization, to say the least." They never had any money to insure the flight instruments against catastrophe. And typically, the contract between NASA and Iowa to supply them did not provide any sort of compensation to Van Allen at all. "So they got a terrific bargain from us, actually," he noted. None of this experimental work says anything of Van Allen's long days of teaching the existing course load, grading term papers, showing up at the faculty meetings to discuss who was going to get tenure, or staying late in the office when an undergrad had a question about General Astronomy. In light of such headaches, the competition, and the microscopic number of slots, an easy way out did always exist: inaction. *Don't* submit a proposal, *stay* at the university job, go into work every day, teach the classes, and go to the faculty meetings. Why not just stay in your own lab safe and secure and collect the paycheck?

Because it's in the blood, whose pressure always increased when an AO arrived. One day back in the late sixties, right after Pioneer had been approved, Van Allen had a little something extra waiting there in the mailbox for Room 701. Along with many others in the profession, he'd received a bulletin from NASA entitled "Announcement of Opportunity." They still exist today. When an AO comes in, the hearsay stops—that watercooler speculation about where NASA is planning to go next. Researchers ache for these sorts of things. Everyone has their shelved pet projects waiting for the day when NASA's next announcement suddenly makes *them* priority one. *Their* box might *fly*. And flown experiments can make careers. The AO details how many slots are open for what kinds of experiments, and typically the number of applications exceeds the number of slots by a factor of ten or more. It makes becoming a Tour de France champion look easy. In Pioneer's case, the mailman brought NASA seventy-five applications vying for thirteen openings. Seventy-five typewritten dreams.

Potential experimenters must read the AO and ruminate over some vexingly hard questions. Then they have to be capable of answering honestly. *How is my work served by flying this sort of experiment? Does it make sense to apply?* Professional advancement is one thing; relevance is another.

Once the snowstorm of incoming applications subsides, NASA review committees face their own internal rounds of questioning. Each proposal gets shunted into a category reflecting the meat of its proposal. *Does this cover what we aim to study? Is the experiment viable? Does it use existing technology or blue-sky stuff nobody's gotten to work?* Such elemental sorting makes the ultimate selection process less painful for all involved. Over the years, it's proven to be a methodical and slow-moving process—yet fair and universally accepted by all involved, including by the losers.

For *Pioneer 10*, the University of Iowa's physics department built and tested their experiment right smack on campus, like they always did. Then

Van Allen made arrangements for substitute lecturers while he visited Ames a couple times to check the fit of Iowa's box on the spacecraft chassis. Later on, during his formal delivery, Van Allen secretly committed a fundamental breach of protocol. *Pioneer 10* launched in March 1972, and Van Allen could forget about it for a while. He had other things bubbling.

Guess what? The belt proved to be a total MacGuffin.

Over a year later when the ship finally approached Jupiter, the physicist ably smooched Abigail and the children goodbye, said he'd call in a few days, and then headed west with his pipe and briefcase to Ames and the Holiday Inn in Mountain View for the purposes of attending the two-week-long encounter.

"It was great stuff. Those encounters were really terrific."

Almost gleefully, Van Allen recalled a very different atmosphere at Ames Center than at JPL. "We found Ames, actually, much more pleasant to work with, in particular because of this one man, Charlie Hall, who really had primary interest in accomplishing the scientific objectives"—instead of flying for flying's sake, which is what Van Allen considered JPL to be more focused on. Charlie Hall preferred good data any day of the week. "He believed in science."

For the encounter, Van Allen rounded up a few graduate students to come along and help mitigate the intense work periods. Ames had reconfigured its large Room 209 with desks all around the perimeter, and each of the scientific groups commandeered spots in very much a me-first kind of way. Iowa was down the hall in the conspicuously smaller Room 238, wedged into a corner with one bookcase over to the side. It all felt like a big science fair. "We had a twenty-four-hour patrol on the data, you know," Van Allen explained of the process. "Every couple of hours we'd get a fresh batch of printed-out data and make plots and diagrams." To accomplish that, they used nothing fancy: green graph paper and colored pencils, one color per detector. Passing through Jupiter's enormous magnetosphere took several days. Explained one of the students, "We taped subsequent graphs together until we had the whole encounter on ten feet of graph paper." Each morning about six o'clock, Van Allen parked his rental car in the Ames lot and took over from the graveyard team—but not before sorting through whatever had happened overnight.

Soon after, all the principal investigators filed into Charlie Hall's office for a daily general meeting on significant events of the last twenty-four hours. A devilishly effective man, Hall deliberately kept the room too warm and never permitted anyone to sit down. "People understood that that was the way he conducted meetings," Van Allen explained. "And his theory, which was really pretty good, was that if you don't sit down, you're going to be more brief. Get to the point!"

Several days into the Jupiter encounter, with *Pioneer 10* still actually gaining on the planet, Van Allen contemplated his data plots on the Jovian radiation belts. In his head, a picture of them slowly came into focus, and he sat there in a J. C. Penney sportcoat with his unlit pipe in one hand trying to figure out the best way to communicate this during the next day's news conference. In cross section the inner belt resembled an egg, while the outer one was more like a bloated acorn. With a snap of inspiration, he grabbed at the institutional rotary phone and called up Michelle Thomsen, still back in Iowa City. Thomsen was one of his grad students and due to fly in the next day.

"Michelle," he lightly began, "what I need to demonstrate this is a little ball about two inches in diameter to represent the planet." Thomsen held the phone with her shoulder and grabbed a notepad to jot it all down.

Van Allen continued expounding on his little scheme. "And, um, maybe you could paint on some design or sort of suggest the structure and the red spot and so forth on the planet." His grad student kept writing. Now Van Allen was getting to the good stuff. "And then I need to have a row of holes drilled . . . sort of a ring of holes around each pole. And then include some coat hangers." She kept on writing. "And see if the shop can mount it on a motor," Van Allen finished. He hung up and went back to work.

Talk about things falling in your lap. Time was short, but Thomsen was nothing if not resourceful. And she regarded the task not as drudgery but merely as one lively addition to her overall experience. "As a fairly new graduate student, I was thrilled to be in on such a historic moment," she explained. "The excitement, anticipation, and camaraderie of the encounter still give me pleasure to recall."

The next day, Thomsen rolled in and handed off Van Allen's little care package. "Here you go, Van." After you'd been around awhile, you got to call him Van. He dumped the box of parts out on the desk of his motel room. They spread to the bed and floor—Van Allen on his hands and knees, then

40. Van Allen meets up with Michelle Thomsen at the *Pioneer 10* Jupiter encounter. (Courtesy of the University of Iowa, Michelle Thomsen, and Rick Rairden)

back up over to the bed, sorting through them like Legos. It was almost time for the press conference. Dicing up the wire hangers, he reshaped them to match his data plots and fit the ends in the rings of holes. Then he picked up the whole shebang and left.

Whatever any of the other scientists commented on at that press conference, Van Allen quite easily upstaged them all by simply plugging in the little rig and spinning it round for all the world to see. "*That* stood out," he recalled with a broad smile. "That's still, you might say, *the* model of Jovian radiation belts." And after *Pioneer 10* ended, Van Allen remanded the thing back to his office where it lived in a spot of prominence.

"That was all an overnight operation!"

As serious as he generally took himself, Van Allen was, of course, not above moments like this—or like those moments of whimsy when he sought out *Explorer 1* in the sky. Nor was he above his actions during the final *Pioneer 10* assembly. On that particular day, some of the approximately thirteen experimenters gathered before the mostly complete ship holding their instruments, which were not yet installed. It was a photo session. "We were all standing around having our pictures taken, with these soft white gloves on so we wouldn't mar the surfaces," Van Allen explained. "We weren't supposed to handle 'em with our bare hands for fear of putting smudges on the gold plating on the boxes."

41. Showtime! Van Allen unveils his motorized model of the Jovian magnetosphere at a *Pioneer 10* press conference. At left is the apparently envious John Simpson. (Courtesy of the University of Iowa)

An Ames quality controller left the room for just a second. Van Allen whipped off his glove and deftly planted one bare finger tip on the underside of his gold-plated experiment. "I just did it for impulse, on kicks. I get my fingerprint on this, and we're going to Jupiter with my fingerprint on it!" He chuckled, rubbing his hands together.

And thirty years hence—as *Pioneer 10* prepared to breathe its last—Iowa's instrument would be the last one running.

And what of Pasadena's own radiation problem? In the spotless JPL assembly bay sat a mostly complete MJS'77 spacecraft, now in need of broad refitting with radiation-resistant parts. It called for one big ugly do over. But totally refurbishing the thing wasn't going to be possible; the integration was all locked down. So here and there, the machine would receive exotic covers made of tantalum—a bluish gray metal that is corrosion resistant and very hard. Using tantalum was one of those hypothetical, what-if exercises done way back in TOPS land, and somewhere along the way it got shelved. Nobody could remember what exactly they concluded. The whole thing had to be dusted off and revisited.

Parts ineligible for the special covers might benefit from swapping in radiation-hardened components. How do things get radiation hardened? The Lab didn't have to ask that question in a void. As Bud Schurmeier explained of the time period, "There was a lot of going on in the weapons business to harden stuff for nuclear blasts," and JPL could build off of this head start. Using something like lead was easy enough. Some of the individual assemblies could just be swaddled in little lead ponchos. Other parts could be pre-degraded with exposure to radiation—almost like getting a flu shot containing active flu cultures. The battle to harden *MJS'77* spread down into its nooks and crannies. Someone figured out that if the little circuit board lines were spaced just a bit farther apart, it would help. "So we ended up doing some of all those things," clarified Schurmeier.

The entirety of 1974 was principally spent beefing up the radiation shields. Schurmeier's final, agreed-upon guideline was a factor of two: that is, all spacecraft pieces should be rated to continue functioning even after twice the expected dosage of Jovian radiation. In some places, that meant some really pricey workmanship—like making the camera optics out of pure quartz so that the radiation wouldn't darken them. In sum, this issue cost the program an additional $13 million, and NASA thankfully covered them on that one.

It was all becoming a relentless march forward. Long days turned longer. Widespread e-mail and voice mail didn't yet exist for anyone to hide behind. Even so, nobody hid behind their secretaries either. "I don't think there was much of that that went on," Schurmeier alleged of today's evasive trends. With conveniences like videoconferencing still years off, both NASA and JPL counted on good ol' face time to keep themselves aligned. Often, high-level JPL staffers sat in on the last Pasadena meeting of the day only to weather a red-eye flight across three time zones to Washington and the first NASA meeting of the next day. Then they'd repeat the operation going back. They started living the project. One engineer woke up three times in a single night worrying about some obscure hardware issue. The next morning, he quiescently stood in the shower for thirty minutes tumbling the problem around in his head before realizing that he was still standing there in the shower.

The computer design was finalized. They programmed the ship to look after itself and remain on 100 percent autopilot for two entire weeks without so much as a peep from the ground. To do this, *MJS'77* would operate on a level of intelligence derived from STAR—kind of a STAR lite. If, for example, the ship realized that it was spinning too fast, it could halt the motion and wait for instructions in a safe mode. Or it could flip over to a backup computer. The scenarios went on forever. They had another command known as the Omen, because it ordered the computer to grab its ankles and prepare for some imminently catastrophic failure. All of this logic required only a very small amount of real estate. As finally configured, each spacecraft computer launched with four kilobytes of memory—as much as an Atari game cartridge. Even so, it was light-years above anything in space. *Pioneer 10* operated on practically no intelligence whatsoever. It was a real-time machine. Commands arrived by radio, and *Pioneer 10* acted on them and then forgot what they were. Much like a fourth grader, it lacked the ability to plan ahead.

One problem, though, with any arrangement of this MJS'77 computer logic is that nobody really knew *precisely* what normal operation entailed. They hadn't done this before. Mission flexibility thus bore unto itself a sharp learning curve that couldn't really be adequately explored until the ships were already gone. "You learn by flying the spacecraft," a team member explained. Since this hardware brain couldn't be modified after launch, its instructions were laid out to be as open-ended as possible. Even deep into the mission, the whole thing could be reprogrammed with a burst of new instructions from the ground.

Soldered onto circuit boards and tested at extremes of temperature and vibration, these hardened electronics became parts of larger parts, which in turn became elements of even larger parts. Meticulously they coagulated in the JPL high bay. At one point a round of spray painting was going on outside, and the hydrocarbons got sucked into the air circulators and ruined the detectors on a bunch of science experiments. They all had to be changed out.

Yet another hardware glitch reared its pimply butt. The smorgasbord of instruments contained a magnetometer. They'd been on unmanned flights almost from the beginning, gathering information about magnetic fields thrown off by the planets and moons. On Earth a compass needle points north, harmonizing quite nicely with more or less the "top" of our planet, but it's not the same deal everyplace else. Depending on how the molten inner layers of a celestial body shift about, the magnetic field differs. If a

spacecraft carries something to measure that field—like a magnetometer—
then its internal workings can be divined. Including one on *MJS'77* was only
to be expected.

But these devices are particularly susceptible to magnetic energy given
off by virtually anything. Larry Cahill flew magnetometers on some of
Iowa's rockoon flights and had trouble calibrating them inside the packed
confines of the physics building. Even his telephone could screw them up.
Cahill was forced to calibrate his instruments on the wide-open and decidedly nonmagnetic lawn outside.

Someone was going to have to figure out how to get *MJS'77*'s magnetometer
in a position to do the job without any false readings from the very equipment driving it. What they really wanted to do was somehow hang it off
the back of the ship, a nice forty feet away from everything else. But how
do you do that and still cram everything in the launch rocket?

Ideas went around meeting tables and lunch tables and drafting tables
and undoubtedly more than a few dinner tables. A big Slinky? Collapsible
rod? Scissor-jack thing? Huge spool of rope? Nothing was out of bounds.

Their cyclic discussions came back around to—of all things—the Slinky
idea, where it evolved into a workable solution. Solar system exploration
would forever owe a debt to the child's stair-hopping plaything, although
the final iteration more resembled one of those trick mixed-nut cans where
you take off the lid and spring-loaded snakes pop out. For the ship, a thin,
wobbly fiberglass boom did the trick—so frail it couldn't support its own
weight on Earth. It could be delicately coiled up into a small can about two
feet long and then extended. In testing, they hit a snag. The energy potential stored in those coils was enough to *thwip* the magnetometer right off
the end if it unwound too fast. An additional, complicated, and very much
one-time-use arrangement of cables, pulleys, and motors went in to control
the speed and torque. It was more weight and space that could have been
used for something else, but they had to have it.

The major spacecraft elements thus fell into place: One big long pole
stretched out the magnetometer. On a much shorter swing-out arm hung
the fabulously hot RTGs. And then directly opposite them, positioned as
far as possible from the radiation, a long boom held the remaining experiments—some of them on a geared, fully articulating mount at the end of

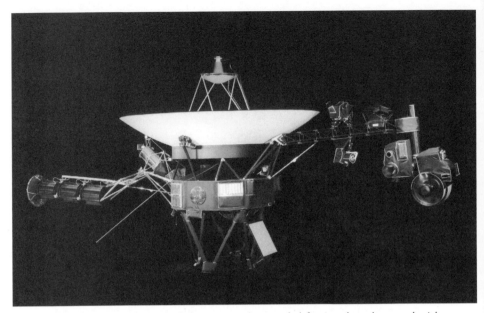

42. How the ship would appear in flight. RTGs extend away to the left; science boom hangs on the right. White square attached to lower struts serves as a calibration target for the cameras. Note the Golden Record, added after she became one of the *Voyagers*. (Courtesy of NASA/JPL-Caltech)

the arm. They called it the scan platform. All parts congregated on a ten-sided middle—the bus—enclosing computers, radio gear, little thrusters, and propellant tank. White antenna garnished the top.

Two years before launch, Lab director Bruce Murray approached Bud Schurmeier with an offer he couldn't refuse—or perhaps one that he was not supposed to. Murray was about to drop a bomb. With the Moon race over and NASA seemingly gearing down space exploration, the new JPL director had been thinking long and hard about future times.

Primarily, he felt the Lab shouldn't keep all its eggs in the same basket. They should branch out some, leveraging the on-campus talent for other uses. Alternative energy seemed the most obvious choice; the Middle East oil crisis was in full bloom. It had everybody lining up at the gas pumps and filling their houses with insulation. Said Murray, "I decided we would make a major assault to build a big civilian business for JPL in energy." And of all people he wanted Schurmeier to head it. "I pushed Schurmeier up," Murray offered. "He was too important to let him do something like run Voyager. I needed him for broader responsibilities."

"I suppose I could have refused," Schurmeier ruminated, caught totally off guard by the request. "I could see the Lab was gonna be in deep trouble here, 'cause the way NASA was doing things. So that's what probably swayed me to say 'okay,' although I hated to leave."

Somewhat apologetically, Murray narrated the regime change. "So I took him off the nicest project, Voyager, and put him on a very unrewarding job."

"It did bother me not to see it through," Schurmeier admitted of his departure from MJS'77, never having been able to finish with the Systems Division either.

The familiar face of John Casani thus materialized in Schurmeier's office to job share over the next three months and hopefully make for a smooth transition. Casani had previously been laboring over in Guidance and Control, providing support for the impending mission. "Things had been set up pretty well," he commented of their overall readiness. "We had the 'A' team. We had the best people in the Lab working on the project at that time. There was no question about that."

The new MJS'77 project manager was no spring chicken. "Casani has a very systematic approach in examining issues or problems," explained Thomas Gavin, a fellow JPL teammate who spent half a fretful decade wading through the murky peat bog of radiation hardening. "When you had to present a problem and the potential solution, Casani would very quickly work the discussion to the boundary of your understanding of the issue." Strategies like that were truly the hallmark of any successful project manager. "He always worked it with you," Gavin continued, "so that you were discovering the soft spots in your solution."

Casani's first order of business was telephone service. He changed the last four digits of his phone number to 6-5-7-8. And with that crossed off the list, he began shadowing the outgoing manager around in every corner and cubby of Schurmeier's work. "In three months he tried as hard as he could to show me all the ropes and Washington connections," the new man remembered of his overlap. "He was a great mentor and a great tutor and made a very easy transition for me." And what of Schurmeier's final days? "I think it was with really a heavy heart that he left the project," Casani offered. "I mentioned that to him several times in years gone by, but he just looked at me and smiled. He would never confirm or deny it."

What's with that phone number? Drummed out on a touch-tone keypad, the combination spelled *Mariner-Jupiter-Saturn-Uranus*. Like Roger Bourke, Casani was bound and determined to reach beyond the rings. "We wanted people to be thinking at least beyond Saturn," he explained.

Casani's next concern revolved around the name. He absolutely hated it and sure wasn't alone. *"Mariner-Jupiter-Saturn 1977*, what a mouthful," aired Charley Kohlhase, almost embarrassed at even having to say it out loud. "John Casani thought we should have a contest to give the mission a better name before it departed." They got a weathered old blackboard set up in a common area where people could drift on through and leave suggestions. *Nomad. Antares.* The board slowly filled up. *Pilgrim.* Off to one side somebody had scrawled *Voyager*, which sounded appropriate enough but induced weird feelings in some.

"There was a lot of baggage in the Voyager name," explained Casani. What ultimately ended up as Viking had started life with the name Voyager—a colossally gargantuan proposal to launch Mars orbiter-lander combinations piggyback on a Saturn moon rocket. After limping along for two years, the project fell victim to budget cuts. "And it probably didn't survive for a good reason," Casani suggested. "It was big, it was ambitious, it was gonna cost a lot, and so I think the right thing happened." So in some sense, the name would be a retread.

Kohlhase recalled the concern: "Are we going to put some kind of curse on it and jinx it if we give it that name?" Time was running out. They had a logo to design, press releases to write.

"To a lot of people," continued Casani, "naming this project with the name of a project that did not survive the political process just seemed like bad boogie. You know, bad karma."

However, everyone agreed that they really did like the name. "We're not weirdos," Kohlhase affirmed. "We're scientists and we don't believe in superstitious things like that."

Casani said, "It was finally what we selected."

Voyager 2 would be launched first on the Grand Tour trajectory—a finely tuned series of aiming points capable of reaching the *publicized* selection of remote bodies; yet it was also unspeakably prepared to brush Uranus and Neptune in turn. Next would be *Voyager 1*—and if it failed to give them Titan, the other could be retargeted and to heck with the outer planets.

By mid-August *Voyager 2* stood waiting on her pad at Cape Kennedy in Florida, balancing eleven science instruments, 65,000 individual components, 5 million total parts—all on one rocket.

With the countdown clock ticking, they were about ready to go. After the Titan booster lofted their machinery to the atmosphere's edge, a burly Centaur would take over and muscle the stack into Earth parking orbit. Only after passing an electronic physical would the Centaur relight and push out from Earth, leaving one final elbowing necessary from a solid rocket motor directly underneath *Voyager*. And so at four o'clock on the morning of the first launch on August 20, 1977, Bill Shipley stood in predawn blackness on a vapid corner outside the hotel waiting for a buddy to come pick him up.

Shipley wasn't the type to particularly enjoy uprightness and ambulation during the small hours. And his friend always showed up late to everything. Waiting there like he was, Shipley's idle mind began to wander. *What if something goes haywire?* he began to agonize. *What would happen if it doesn't work?*

22. Embarking

I remember feeling like a kid again, like Tom Sawyer,
sailing to an island that he had never been on
before and wondering if there's wildlife there?

Charley Kohlhase

Bill Shipley started thinking about all the things that could go wrong, that had gone wrong; he started contemplating what he might have done differently. It gave him a shiver. Bill Shipley's quiet thoughts screamed out around him in the dark morning air, expanding, a billowing cloud of stressful memories and past machinations of days gone by, of meetings running so long the coffee went cold and they were late for lunch, late for dinner, late for everything as rediscussion after rediscussion and thought, painful methodically detailed *thought*, kept on coming. He felt a vile puff of doubt.

Yet rather quickly the cloud began shriveling back up into his head. During this brief moment of review, Shipley realized he could have done no better than what already was.

He said, "I came to the conclusion that if somebody came along and offered me another $2 or $3 million and a couple more months, I wouldn't want it." They'd worked things through about as well as anybody ever could.

Twenty minutes later when his buddy finally showed, William S. Shipley was calm again.

Charley Kohlhase figured his heart might stop any second now. Casani felt the same, standing right next to each other in the Cape's Building AE control room.

During boost, the Centaur's roll rate overstepped what was expected by *Voyager 2*'s internal gyroscopes. That is, the ship—not knowing whether she was in space or what—tumbled more than she ever thought feasible. The near headstand pushed *Voyager 2*'s gyros to their limits until her fault-protection logic ran out of options for salvaging them and tried rolling over to the backups. Undoubtedly, this would be fodder for Shipley's umpteenth notebook.

Centaur finished expending its load of fuel and blew the separation bolts. It was violent enough to jar the ship off Kohlhase's meticulously plotted course and involuntarily switch her over to the backup computer. Within milliseconds, this hiccup blipped the corrective thrusters into receiving two spurious commands in unison to fire and rectify a condition that didn't really exist—an abysmal state of affairs barely ten minutes into a twelve-year flight. As all of this occurred, the RTG and snake-in-a-can booms worked to unfurl themselves.

Down at Kennedy there was static in the headphones.

Half-unfolded, *Voyager 2* flew in Earth orbit. Nothing was actually damaged just yet; she subsisted on the backup computer, gyros maxed right out to the gills. Of course, when polled, the *spare* gyros registered an identical situation. Her intuitive brain therefore concluded that both sets had spontaneously gone bad during ascent. The primary computer went round-robin through every possible combination of good gyro–bad gyro, good thruster–bad thruster. Then, diligently moving through the list of options, *Voyager 2* requested her backup computer take the reins. That effectively rebooted the platform and set everything right. But the flapping armature of instrumentation and cameras still hadn't completely locked into place. Down on terra firma, the group chewed their nails and waited. Waited. At that moment, Charley Kohlhase's curmudgeonly father might have materialized inside Charley's head, shaking a finger as if to belt out, *See, I told you so. This is what happens when you don't do like I said.* It could be hard to keep Dad away. And the younger Kohlhase might have shaken his head like an Etch A Sketch to erase the apparition and focus on *Voyager 2*.

Ninety-seven minutes later she reported all clear.

With many years having passed, Raymond Heacock still vividly remembered the event. "You cannot imagine how long and anxious those ninety-seven minutes were to the Voyager team." And over their shoulders waited *Voyager 1*, identically built and programmed. Who's to say she wouldn't act the same way? Less than a month remained in their launch period.

All were moved by the basic primal desire to avoid failure. They'd come too far; they needed this one. Inside two frenzied, sleepless weeks *Voyager 1*'s caretakers reprogrammed her thrusters and then added springs on the boom hinge to help things snap more firmly into place. It passed every test they could dream up. No changes were made to the gyro situation, however. "We just knew that this was going to happen," explained John Casani of their final conclusion. "The spacecraft was in no danger of being damaged or anything like that. It was just behaving and reporting to us something that was totally unexpected with the first spacecraft."

But somebody did turn up a problem in the main computer; a little electronic component was intermittently causing grief. With the clock ticking down the last couple of days before launch, tension levels quickly rose. *Which lot did the part come from? How many of the others were bad? Do we have a bad lot? Are there known good ones available?* A flurry of round-the-clock phone calls ensued between Florida and California as intrepid JPL workers combed through Lab buildings testing whatever identical parts could be found scattered about. They were going through the serial numbers right through the last night before *Voyager 1* launch, and Bill Shipley went to the Cape Kennedy control center that morning having never gotten a solid answer. An hour after blastoff, he ran across one of the people who'd been working the problem.

"I wanted to tell you," the guy told Shipley, "we found another one of those parts."

Nervously, Shipley nodded.

"It was in the spacecraft we already launched."

Shipley's appetite disappeared.

"I didn't think I could explain that on the network," the other man indicated. "It would sound like a zoo if I got on the network with that story.

"So I just kept quiet."

Voyager 1's launch *had* been fraught with trouble—*lots* of it. And it was all completely unrelated to everything they'd gone scrambling for. One of her Titan main engines shut down early, leaving the Centaur responsible for more work than anticipated, in the lower and denser portions of the atmosphere. To reach orbit, *Voyager 1*'s Centaur gobbled 1,200 extra pounds of fuel, which was desperately needed at final ignition—1,200 irreplaceable pounds. During the stack's forty-five-minute coast around Earth, everybody sat doodling math, trying to figure out if the Centaur could do its job with a needle on "E."

Again Charley Kohlhase sat there with a headset on, listening to all the mathematical gibber jabber filling his earpiece between the dispersed camps of PhDs at General Dynamics, who manufactured the Centaur. Finally they decreed, "We think it's got enough." Kohlhase slid the headset off and turned to John Casani, still with headphones on, sweating bullets right there next to him. They *think*?

If the thing burned itself completely out, then it would be game over. Headsets to heads; eyes on the monitors. Without one complaint, the Centaur polished off *Voyager 1*'s injection burn just right and separated exactly on cue, leaving the final 4,500-mile-an-hour push to the kick stage. Kohlhase left to wring out his shirt. Moments later the very last drops of Centaur telemetry arrived at the Cape; the booster finished with only three gurgly seconds of fuel remaining in the tanks. Had the Voyager ships been launched with *each other's* Titan boosters, there would have been no Grand Tour.

"Flip of the coin," someone said.

Bruce Murray didn't even wait for his troops to get home after the Voyager launches; he had Bob Parks spring it on John Casani during the ride to the airport. Alone in their crappy rental car, Casani rode shotgun next to Bob Parks, who at the time functioned as Murray's assistant Lab director for flight projects.

Staring down the concrete ribbon before them, Parks spoke. He said, "Well, we got the JOP project approved in the Congress just last week." For years JPL had been pitching a large-scale "Jupiter Orbiter Probe," and finally it had gone through. Now Bruce Murray needed JOP's managerial gears up to speed for this monumentally complex flight that was still over a decade down the road. They'd put it up on the newfangled Space Shuttle, if *that* thing ever flew. In the mean time, Parks and Murray had someone in mind to crack the whips. They wanted a guy with chops.

"And so when we get back," Parks continued, "I want you to head up the project office."

Replied Casani, "Do you want me to give up Voyager?"

"No," said Parks, "what I'm gonna do is create a new organization which'll be an outer-planet organization, and it'll have two projects in it. It'll have the JOP project and the Voyager project."

Intrigued, Casani processed this new information.

Parks told him, "I want you to manage that office and, as a second assignment, also be the JOP project manager."

Casani said, "Great." To him it sounded like the most fantastic of arrangements. "I was delighted in a sense," he recalled of his feelings toward the words of Bob Parks. "He promoted me, basically, to the head of a new organization that Voyager was part of. So I had no reason to feel like anything was being taken away from me."

But really, John Casani was already off the project.

With launch past them, Kohlhase and company retired to the comparative solitude of their JPL offices to finish writing the mission rules, including all the nauseatingly detailed plans for encountering Jupiter in early '79. Prior to launch, Kohlhase had figured on the initial year and a half cruise offering him and the other mission design staff more than enough time to prop up their heels and sip a few mai tais while fleshing everything out. They were almost wrong.

A chasm of difference exists between mission design and spacecraft design. With the latter, it's all about the kinds of things pinging about back there in TOPS: radiation hardening, cameras, power. It's *all* about the hardware.

Mission design is strategizing, and it happens in two stages. *Prior* to launch, it's like assembling the pieces of a very large puzzle, spanning nearly everything from trajectories and navigation to ensuring that all of the flight and ground elements will play together just right to achieve the scientific objectives. *After* launch, mission design revolves more around planning for what you're actually trying to go out there and use the hardware *for*: imaging Jupiter's Great Red Spot and its moons; taking radiation readings and magnetometer readings; completing atmospheric radio science; and measuring charged particles and cosmic rays. Diverse groups of individuals converge to participate in hammering out these mission objectives. And often they come through the door lugging along conflicting goals.

Scientists always had the most baggage. Nominally working in isolated camps, practically the entire lot had grown complacent in having everything selfishly to themselves: the whole environment of the study, data cultivation and storage, and method and time of reporting. Not enough power on that test stand? Jack in another thousand volts. But heck, that mindset was more than understandable. They were the only fish in the pond, *their* pond. Entirely on home turf—that was the way of the scientist.

The evolving complexity of unmanned space exploration only grabbed this paradigm by the neck and gave it a good shake. In the case of Voyager, nearly a dozen onboard experiments jockeyed for attention, cocooned in an ecosystem where only so many could be operational, ideally positioned, or sending back data at any one time. The Voyager ships carried but a very limited supply of electrical power that would diminish over time. And they were not built like octopuses. Each ship only had one boom arm with one scan platform at the very end of it. And the whole of that arm could only point in one direction at a time. Swinging the cameras around hard a starboard for that must-have close-up brings everybody with. Each ship possessed one radio channel for the experiments—a solitary button on the dial, meaning no more than a single one-way conversation happened at any time. If the gamma-ray detectors were communicating their findings back home while the camera snapped away, those images had to roll onto Voyager's tape recorder and wait in limbo. When the gamma-ray uplink had finished, no further reporting was scheduled, and no commands were coming in, then the pictures could go home.

Understandably then, scientists wanted 100 percent of a spacecraft's potential. The mission was being flown for them. But that "100 percent" amounted to exponentially more than what any given flight could realistically provide. Scientists are not accustomed to working in a style of timeshare, and here is where they came into conflict with engineering. Charley Kohlhase recalled the hard lessons of Ranger and how things could quickly spiral into a mutually exclusive domino effect of wants. JPL had just launched the most sophisticated exploring machine ever built, but that very complexity required a scrupulously thought-out compendium of strict guidelines for use. Voyager needed something to separate fantasy from reality. She needed ground rules.

It's a lot like being a good parent. Children—all of us, really—operate best in environments where the limits are well defined and understood. Parents leaving too much in the gray typically raise offspring unable to succinctly comprehend just how they are supposed to act. The solution is self-evident: children perform up to what is expected of them, as long as those expectations are clear. They will learn that life is not fair; they don't always

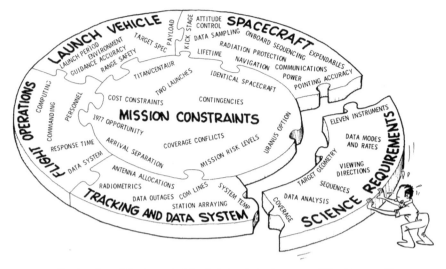

43. Shucks—how difficult can the job of mission design be, anyway? Use this as a guide. (Sketch by Richard Rackus, courtesy of Charles Kohlhase)

get the good ball or the biggest dessert or the best seat in the movie theater. This concept is easily applied to space missions, which also are in need of clear rules. Flandro and Shipley and Heacock and Schurmeier birthed Voyager. Charley Kohlhase's group would rear her.

Mission planning thus broke down each flight in turn, reducing it to an elemental timeline of planetary and satellite encounters. Into this merged the experiment needs, along with the quantifiable resources of power and computer memory and antenna time, with an eye for stretching it to the last known event. In many minds that was Neptune, whether or not it would ever come out of the closet and become official. But with a graph showing Jupiter at the front and Neptune toward the end, it could all be factored together. Objectives divided by consumables and antenna time, further divided by the number of experiments, equaled a budget. And that's how they went at it.

"Okay," began Charley Kohlhase, stepping through the process of designing a single encounter. "You can develop ten flight sequences. They're to execute between eighty days before closest approach and thirty days after." It wasn't to play hardball. Rather, this established a framework of operation such that everyone clearly understood the expectations and limits.

Kohlhase continued his example, "You shall not use any more fuel than such and such an amount, or you can't use the tape recorder any more than such and such an amount." The solitary reel-to-reel tape machine on each Voyager carried a finite amount of tape and could reasonably perform for only so long. People would call Kohlhase up and ask, "How many passes do we have, Charley? When is it going to wear out?" It could be appropriated for data or pictures; the recorder didn't care. But use of the device had to be precisely scheduled in advance. It couldn't function as a bottomless suitcase to hold every last byte of data.

Based on the compiled responses, the Sequencing Team would sculpt Voyager's command loads—the actual program for transmission to the ship. "We had a simulator to run these commands through, to make sure they did what we expected them to; and we often were surprised," said one of the engineers. "You'd think you had all the commands right, but the simulator wouldn't respond the way you expected it to."

As Kohlhase put it, "When you think about planning, you mustn't overlook any little thing like that or it'll get you, sooner or later."

The mission rules addressed every known possibility. No vital procedures were to happen with the Sun between Earth and the Voyager ships, owing to noisy radio signals. Dish time became a hot commodity as other space missions of varying nationalities would occasionally be requesting access to the Deep Space Network's array of worldwide and rather indispensable ground antennae. To what extent could each mission's critical periods be scheduled around each other? Already JPL was busy cranking out a series of diplomatic requests not to communicate on certain radio frequencies at certain times. Depending on the type of request, it went through the State Department and on to other nations.

Situation after situation, contingency after contingency, the mosaic of rules and constraints thickened into a backbreaking, massively huge binder, helpfully tabbed into chapters. One chapter was for the cruise to Jupiter, one covered the Jupiter encounter, another addressed the cruise to Saturn, and so on. Everybody got a book. The sequencing folks could orchestrate the cruise and encounter workflows, using Kohlhase's office to verify the legality of their plans. Anybody with a problem could take it right to Ed Stone, already proving himself to be more than fair in resolving conflicts.

And the spacecraft was definitely a pain to fly—really touchy. They'd been designed and built for science and no-nonsense functionality, but the sad truth was the machines were more persnickety than a Jaguar xjs. The electronics weren't working exactly the way people thought they would. The logic didn't always follow as predicted. It went on and on, transforming into hair-pulling conferences and repeated late-afternoon phone calls home to predict further missed dinners. A hundred people came on the project in just that first year alone, mostly because the ships were rougher to drive than anyone anticipated.

Fighter-jet pilots talk about a concept known as "exploring the performance envelope." See, their high-end aircraft can be as temperamental as a thoroughbred. Learning to control either one of them is so much more a feeling-out process than the issuing of rudimentary commands. It's that feel of the reins in the fingertips, the gentle tugging and what the horse is trying to communicate with it, or the sensation of the jet's control stick in the hands, the subtleties at play. With a tad too much rein on the neck, too much yaw this way and a little wing wiggle that way, you can lose it. As one of the Voyager project managers would later explain, "We were learning to operate those spacecraft for those first two years." And that is what they all began to realize: if they didn't keep it within the envelope, paying close attention to that feel of the reins, they could lose it. They'd come close to losing it a couple of times already, and the ships were just penetrating the troubling asteroid belt past Mars.

At least the belt itself was now no cause for worry. *Pioneer 10* had granted them this occasion for rest.

23. Get It

I have been trying to obtain the official credit
from JPL for over half of my life, without success.

Michael A. Minovitch

Hey wait a second. Where's Minovitch in all this? Didn't Mike Minovitch invent gravity assist and then discover the Grand Tour, all by his lonesome?

The question is tricky at best and extremely delicate when viewed from practically any angle. Depending on the person being asked, two vastly differing responses follow as to who, exactly, gets credit for these major innovations.

Begin with the broader concept of gravity assist. Anybody inquiring of Michael A. Minovitch will predictably find out in no uncertain terms that he alone is responsible—to the exclusion of basically everyone else in the recorded history of space research. He's the Man.

"I invented gravity-propelled space travel," is what he says.

Minovitch likes to talk. He also says, "It is impossible to emphasize the revolutionary impact that my invention had on the classical theory of space travel."

Hmm, notice the vernacular. It's not "I contributed to understanding the theory of . . ." or "I developed sample missions based on the principle of . . ." It's "I invented." Nothing was out there but fog and white noise until Minovitch ran his program and sprouted a halo and decreed, *Let there be gravity*. In his mind, not just the application but *the fundamental concept itself* did not exist until he came along and birthed it.

Conventional wisdom suggests introducing a morsel of perspective. The first person to sail on the ocean did not invent wind power. Isaac Newton did not invent equal and opposite reactions for every action. The Wright brothers did not invent lift. Saying someone "invented" gravity assist is like saying James Van Allen invented radiation belts. The wording doesn't compute, because there was never anything to invent—Van Allen just happened to discover this naturally occurring phenomenon.

Speaking of, what's to say of Van Allen's *own* perspective on gravity assist? *"Of course,* the general technique of so-called gravity assist has been around for a long, long time," he explained with exaggerated intonation and a hint of a smile. "It's a classical part of cometary physics, is the passage by Jupiter and the changing of the orbit. The general physics of that has been well known for a long time." And Van Allen is right.

It's been known for a long time? *Just how long?* In staking his claim, Michael A. Minovitch problematically dismisses a myriad of rather prominent figures—astronomers and theoreticians back *as far as the 1700s* who noticed the altered, sidetracking path of a comet during close passage of a planet or moon—especially when passing something huge, like Jupiter. *How is that happening?* wondered mathematicians like Parisian Jean d'Alembert as early as 1761. Seven years later another French mathematician, Pierre-Simon Laplace, also had his attention up there in the skies. He worked double duty as an astronomer. One eventful night, Laplace observed a bullied comet like d'Alembert had and wondered the same darn thing. Their pursuit of a solution led to the earliest known writings discussing the novel effect of a planet's gravity on things that flew by. When it comes to gravity assist, this is the absolute bedrock of the idea.

Later on in the mid-1800s, Urbain Le Verrier expounded on the ideas of his fellow Frenchmen from a cushy post at the Paris Observatory. He came to the same conclusions about the effect on comets as his predecessors, expanding the theory to include asteroids. When all that wasn't enough to slake Le Verrier's thirst for discovery, he went on to locate Neptune based on the odd wobble in Uranus's orbit. More than most, Le Verrier demonstrated an outstanding grasp of cause and effect.

Those French were really on to something. Astronomer Francois Tisserand stitched together a formula in the late 1880s that demonstrated an advanced, insightful, and rather comprehensive understanding of exactly was happening to all these detouring comets and asteroids. An American contemporary of his named Hubert Newton—no real relation to Isaac—even produced a math formula to gauge energy transfers between the comet and the planet. Now there weren't just observations but hard numbers to back them up. "His energy change formula is fully equivalent to that used by modern day trajectory analysts," explains Gary Flandro. And then from Tisserand and Newton, it was largely downhill.

Yet Minovitch shrugs it all off, dismissing this formative research whole scale. With a brush of his hand, it's wiped from the historical record—*his* historical record—and not supposed to factor in. According to Minovitch, none of this applies for a wide variety of increasingly convenient reasons. They didn't fully understand the concept. They were talking comets and not spacecraft. They didn't do the math right. His approach—the manner of such refutations—begins to describe an uncomfortably familiar pattern.

Most people have a blind spot, and Minovitch's is perspective. Both comets and spacecraft are obeying the same rules. No known comet carries rocket propulsion; they fly where gravity has directed them. Then moving forward, advanced astrodynamic concepts are not born overnight from whole cloth or a blank sheet of paper, or even from two pieces of orange-lined paper Scotch-taped together. They *mature* on a linear, longitudinal continuum, like Darwinism or ecology or automatic car washes. The work done by these aforementioned researchers does not pave the road and spray paint the lines and put up streetlights and do a smidgen of landscaping, but it sure does begin to lay out the road.

By the 1930s, alert science-fiction writers had picked up on the notion, embracing and extending it to their own imaginary spacecraft and situations. In the November 1939 compendium *Astounding Science Fiction*, Lester Del Ray published his fictional piece "Habit," in which the victor of a space race gets a little help from Jupiter's gravity. Ray Cummings wrote a story called "Brigands of the Moon," which appeared in the March 1930 issue of *Astounding Stories*. His spacecraft took advantage of nearby planets to reduce fuel consumption: "We were at this time no more than some sixty-five thousand miles from the Moon's surface. The Planetara presently would swing upon her direct course for Mars. There was nothing that would cause passenger comment in this close passing of the Moon; normally we used the satellite's attraction to give us additional startling speed."

German engineer Walter Hohmann counted science-fiction writers as some of his influences. Arguably operating with much less public interest or fanfare, Hohmann bridged the gravity concept into nonfictional space activities by studying orbital mechanics. His quest was to develop methods of reducing fuel consumption during planetary flight. It was all purely theoretical; Hohmann labored in the time of pre-Nazi Germany. And he was far from alone—Friedrich Tsander lived about the same time and independently worked through many identical scenarios. Both men applied gravity assist to hypothetical spacecraft understood to be realistically just over the horizon.

The list of other contributors goes on and on, but these two in particular get disregarded by Minovitch. Both men's work applied spaceship rocket power to cancel out any negative effects of planetary gravity on the intended flight path. That's propulsion from rockets, not gravity, so it's not supposed to count. In Minovitch's world people draw lines only in pencil—so if you drew in pen, then you weren't really drawing.

By 1954, with v-2s a chilling reality and Sergei Korolev's pedigree of rockets just around the bend, gravity assist began truly coming of age. An article appeared in the November 1954 issue of the prestigious *Journal of the British Interplanetary Society*. English mathematics professor Derek Lawden specifically wrote, "A number of writers have suggested that the fuel requirements of a journey between the Earth and the other planets might be reduced by taking advantage of the attractions of various bodies of the solar system, but the method of calculating such perturbing effects and the economies to be expected do not appear to be widely known."

And Lawden, by his own admission, indicated that he certainly wasn't first on the block with such an ambitious idea. His article went on to describe energy gained and lost in a planetary encounter, and it even estimated velocity increases on the spacecraft itself. He described a variety of applications of this gravitational benefit. Lawden *got it*.

An accomplished and highly regarded astrodynamicist of the time, Derek Lawden's above statement ends on a note that even plays to Mike Minovitch's benefit. In 1954 nobody *did* know how to properly lay out such a trip. There were too many unsolved factors and loose ends. Lawden said so himself.

But this is how ideas evolve.

Then at the end of Lawden's paper came this: "A series of numerical investigations of the type explained in this article would act as a very useful basis for this more theoretical research, and such work is recommended to those interested in orbital computations."

Ending it the way he did, Lawden stood there with baton outstretched, waiting for the handoff. Minovitch arguably took it and ran all the way to the IBM 7090. But how does Mike Minovitch relate his own work to Lawden's? To him, there's no connection. Lawden limited the scope of his work, Minovitch argues, by not proposing any sample missions. He also focused on the use of asteroids and planetary moons as sources of gravity. Yeah, Lawden brought up Mars at one point, but Minovitch contends that since the gravitational effects of asteroids and even Mars are pretty small, what was the point?

Derek Lawden didn't have all the answers and understood the finite limits of his work, like any good scientific researcher would. His subsequent efforts contained a whole stinky pile of uncertainties: "The best way . . . is not known," he wrote of gravity propulsion in 1955. And then again in 1958, he offered, "The way in which maximum advantage may be taken . . . has not yet been investigated." He alluded to it yet again in 1959: "Very little investigation has been carried out relating to these maneuvers." Even so, a clear evolutionary path of *Lawden's own* could be traced in just these writings—by 1959 he'd grasped the minimal impact of asteroidal gravity and wisely deleted it from his train of thought.

Derek Lawden didn't have all the answers, but he darn well understood the problem. He worked it the best he could and applied what he knew. Derek Lawden *got it.*

With this in mind, Michael A. Minovitch has, in print, argued a fairly obvious concept: "Most scientific discoveries or technological breakthroughs involve a phenomenon in nature that has been previously observed and studied."

Most anybody would agree. But then, he does something that has indelibly become his modus operandi—spinning the interpretation of a basic fact to his own advantage, introducing guilt by omission, association, or really just about anything handy: "However, if a researcher fails to recognize how the phenomenon can be used to create the scientific or technical breakthrough and regards the phenomenon as having little relevance to a basic problem, the person can obviously not be credited for discovering the breakthrough."

Minovitch thereby dismisses any prior contributions of others. *Swoosh.*
With a wave of the hand, Mike Minovitch is once again the Man.

And so it came to be, at the dawn of the space age, that this notion of gravity
assist was virtually in bloom. It was in the air, so to speak, when Minovitch
began attending UCLA in the late fifties. He'd even been exposed to the same
educational opportunities as Gary Flandro, including UCLA courses from
Krafft Ehricke in 1959.

Krafft Ehricke is to spaceflight what Colonel Harlan Sanders was to fried
chicken. A youthful Berliner in prewar Germany, Ehricke found inspiration to study rocketry and physics in part from Fritz Lang's seminal *Girl in
the Moon*, which played in theaters throughout 1928.

After Hitler picked a huge fight with Stalin, Ehricke found himself in the
midst of a panzer division on the Russian front, trying not to get his head
shot off. By 1942, as Wernher von Braun worked to assemble his Peenemünde
team, somebody there took stock in Ehricke's educational credentials. Young
Ehricke exhibited stunning prowess in nuclear physics and celestial mechanics. What in God's name was he doing out in a tank? Having been recalled, Ehricke became an important contributor to the goings-on at the V-2
base; when the Germans threw in the towel on World War II, he went into
hiding with his wife. The V-2 engineers were hotter than Joan Crawford;
everybody wanted them. Whenever someone knocked on Ehricke's door,
he'd hide and let his wife screen the visitor to see if he was American. Ehricke
only wanted to go with the Americans. One day his wife opened the door
and immediately began closing it again, delivering her stock line, "I don't
know where he is." But then she recognized the man's U.S. Army uniform
and eagerly beckoned to her husband, "He's here! He's here!" Emigration
to the United States followed.

He went from von Braun to Bell Aircraft and then to Convair, offering
his perspectives to the Atlas and Centaur booster programs. Ehricke knew
his stuff. And not long after *Sputnik 1* flew, he shifted into high gear with
theoretical research. In a 1957 paper, he demonstrated a firm embrace of all
concepts evoking gravity assist. The paper was called "Instrumented Comets," which not only foretold the advent of space probes but also niftily har-
kened back to the idea's genesis.

Ehricke also made time for university lectures. One of the stops was at

UCLA in 1959. "Mike was attending UCLA during this time period," Flandro remarked. "And I don't see how he could have escaped being influenced by Ehricke's work."

Or maybe not. "To my knowledge, Dr. Krafft Ehricke never lectured at UCLA," Minovitch says flatly. "He was a great spaceflight pioneer. But he believed that interplanetary exploration required the development of nuclear propulsion."

And what of Minovitch's eventual multiplanet flight profiles he so eagerly shared with Bill Sjogren? "These had already been proposed by Ehricke," explains Flandro, who, it should be pointed out, has never asserted credit for inventing this hot potato of gravity.

With all this in mind, ask those fellow spaceflight engineers, researchers, academics—Minovitch's true colleagues and contemporaries—who, more than any other subset of humans, are most likely to understand the context of his work, *Mike Minovitch invented gravity assist?*

"Poppycock," asserts Roger Bourke. "That's *crap.*"

"No," chirps Gary Flandro.

"I do not agree that Mike invented gravity assist," states Joe Cutting.

Gee fellas, didn't Mike have some kind of eureka moment looking at all his data?

"There were a number of people prior to Mike who recognized the gravity-assist potential," declares Cutting.

Argues Bourke, "He may have had that 'Eureka!' moment, but others did, too."

Charley Kohlhase opines, "I don't believe that Minovitch was the first to discover gravity assist—but he may have been the first to develop and use systematic procedures, albeit cumbersome, to identify several different multiplanet opportunities."

"I really feel sorry for him," Flandro offers of Minovitch. "That idea was not a new one."

And Minovitch does not mince his words when it comes to Gary Flandro: "I regard him as a scientific charlatan who wanted to achieve fame by stealing the credit for my invention."

Let the record show that these men also agree on the significance of

Minovitch's role—that he energetically twisted and tumbled and tilled the concept of gravity assist with much enthusiasm and veracity and, unlike the baguette-gnawing comet watchers of centuries past, applied it to space probes. "Nobody was disagreeing with the technical aspect of his work," begins Roger Bourke, almost in consolation. "The technical aspects of his work are fine. He *did* come up with a *method* for calculating multiplanet trajectories. And that was great! That was a contribution to spaceflight! No question about it."

But Minovitch's due credit ends just about there. The culture of JPL and engineering research in general thrives on ideas like applying gravity to a spaceship in flight. That critical thinking simply has to exist. History indicates that it did. Although no official "Multiplanet Group" existed at JPL, plenty of smart minds came in every day pursuing concepts related to advancing planetary exploration. All the pieces existed, suspended over JPL's campus—and over the rest of the world.

When Mike Minovitch presented his research in the early sixties, it sure didn't burst from the ether and it sure wasn't the be-all and end-all. His ideas were great but unfortunately not original. They fit these trajectory pieces together *a certain way* and undeniably contributed much to the knowledge base. But these are pieces that can be rearranged in *multiple* fashions, some more optimized than others. You can lay puzzle pieces next to each other and make them fit by banging on them with a hammer. But is that the best way to do things? Roger Bourke argues, "He came up with a *way* of assembling them, a very *inefficient* way. It did work. And to think that if he hadn't done this, no one else would've found it is just *crazy*! It's like, 'I'm the only person that can think in this direction!' And that's just nuts."

What of his demotion from the NASA Exceptional Scientific Achievement Award—was that part of the conspiracy?

"There are bureaucratic pigeonholes in which various things fall," explained Roger Bourke, who, it will be recalled, functioned as Minovitch's boss during the whole award episode. "And somebody decided that that fell more in that pigeonhole than another one. Now, was that a slight? Hardly. It just had to do with the definitions of these kinds of awards." But isn't what he received a real bottom-rung citation? "I don't think that one was necessarily more or less prestigious than another," said Bourke.

Minovitch tries to hang on by his fingernails, in part, because he is indeed quite close to the practical work that made gravity assist possible. There can be no doubting his involvement. Minovitch's supervisor Joe Cutting submits, "Mike made a significant contribution and deserves much credit. I don't know why he downgrades the contribution of these other people."

"I have always been saddened to watch Mike effectively ruin his career," proffers Gary Flandro, "because he cannot get past the notion that he is the 'inventor' of gravity assist."

Roger Bourke has a prediction. "He'll take it to his grave."

Michael A. Minovitch says no one else in the history of the world contributed anything whatsoever to the development of gravity assist. How could they have not?

Okay, so if Minovitch was not responsible for gravity assist, then what about discovering the Grand Tour?

Gary Flandro can't resist suggesting, "I have found it interesting that there was little contention until the Voyager mission became a reality."

The problem with Minovitch discovering the Grand Tour is multifold. First, if there was any realization of the mission's potential *at the time*, he completely failed in reporting the opportunity to his supervisor—or to *anyone* in a position to do something about it. Flights swinging by multiple destinations were apparently discussed in a chance encounter with JPL worker Bill Sjogren, but the story ends there. "I never really sat down and really talked," Sjogren said of the meeting. "I was just curious, looking at all the plots he was doing." Sjogren wasn't even in the Trajectory Group. He was in the Orbit Determination Group, working on Ranger and a completely different set of problems. "That was another bunch of guys," Sjogren explained. "I just happened to drop by the office."

Joe Cutting took over from Victor Clarke as the Trajectory Group supervisor in 1963 and specifically disclaims *any* prior knowledge of *any* Jupiter–Saturn–Uranus–Neptune flight opportunity until Gary Flandro's work during that red-letter summer of 1965. Cutting reviewed two of Minovitch's JPL Technical Reports—one in October 1963 and the other in December 1965—and says that neither one utilized what has come to be known as the Grand Tour. Minovitch's 1963 work concentrated on Venus swing bys, while the December 1965 report addressed fly bys of Jupiter. Cutting is quick to erase a line some might try to draw: "None of the Jupiter swing bys is the same as Flandro's Grand Tour trajectory," he says unequivocally.

Rather, Minovitch seemed more preoccupied with *volume*: how many planets, how many combinations, how many launch opportunities. His first calculations encompassed 1965 to 1980, but fifteen years' worth of flight options weren't enough for Mike Minovitch. He hunted for a way to get his hands on an updated *Nautical Almanac* that would extend his investigations from 1980 through 2000. Gary Flandro, alternatively, located one majestic flight option and ran with it. "I went after that one directly," Flandro explains. "That was the mission that made some sense." A trajectory *similar* to the flown Grand Tour resided *somewhere* in Minovitch's printouts. But it never saw the light of day until after *Voyager* flew.

Other events weaken Minovitch's position. When did Bill Shipley start hashing through TOPS—late in 1961, after Minovitch's holy grail paper? Nope, that didn't begin until June 1968, well after Flandro's discovery. *Wouldn't the Lab have jumped on it years earlier?* Maybe, if had they known. But nobody knew except Minovitch and Bill Sjogren.

"If Mike had already discovered the Grand Tour, then it was unknown to me, Gary, and many others," Joe Cutting says. "I first heard of the Grand Tour opportunity from Gary."

"Gary is actually the discoverer of this four-planet mission," affirms Roger Bourke.

Don't get Charley Kohlhase started: "Minovitch did *not* design the *Voyager 2* flight path!"

Kohlhase, the man *Spaceflight* magazine called "the world's leading designer of space missions," spent over a year weeding through ten thousand possible Voyager flight combinations and doesn't exactly remember doing it side by side with Mike Minovitch. "There are things that he never would have known about back in 1961!" Kohlhase insists. "We took actions to minimize any possible environmental threats, such as the near-Jupiter radiation levels and the near-Saturn ring particle hazards." And Minovitch never addressed the scientific wish list that Kohlhase and his crew had to contend with. "We even," continued Kohlhase, "made careful arrival-time adjustments to pass close to many selected moons of each outer-planet encounter, sometimes positioning key science events to occur over preferred radio telescope view periods." And wouldn't you know, he's right—Minovitch's computations never took any of that into account. "So the final Voyager trajectories were very carefully chosen," Kohlhase concludes, "with *many* more conditions applied than done by any prior researchers."

Minovitch has other claims. What about that nasty business of Victor Clarke walking off with the program? "It's his view of reality," says Roger Bourke, who knew Victor Clarke and the kind of man he was and sees no issue whatsoever. "If I had been a graduate student at the time," he begins, "and come over and said, 'Okay, well, I've come up with this neat scheme,' I would have expected that it wouldn't just be for others to admire but to actually get used there!"

Joe Cutting gets down to brass tacks, "If the program was developed using JPL computers . . . it is probably government property."

Did Mike Minovitch's dear program become a cornerstone of the gravity-assist arsenal at JPL? Did its name get changed, its true origins fuzzed over? Author scrubbed out? Was it covertly absorbed into the Pasadena regime? Gary Flandro emphatically states, "I did not use Mike's computer program in any of my work." Joe Cutting holds no specific recollection. "I don't know what software Gary used," he explains. But could Flandro have exploited the software without even knowing it? Unlikely. Flandro said his program—from Lockheed—contained a specific error that would have been watermark-style traceable back through to its beginnings. "This program had an ephemeris error, which gave the wrong position for Saturn," Flandro explained. "So I had to go back and fix my earliest calculations after the corrections to the program were made."

"There were plenty of programs which *could* have been used," Cutting says, going on to suggest that Flandro himself could have realistically even written his own.

Assuming that Flandro never touched Minovitch's software is a sensible conclusion in light of its very nature. Says Roger Bourke of Minovitch, "He had a very awkward, extremely cumbersome way of making these calculations. Not just complicated, but *cumbersome*. A very inconvenient way. And people, before very long, came up with far more efficient schemes that were plenty good to do these multiplanet trajectories."

Minovitch claims that without his program, JPL would not have "acquired the means to explore the solar system," as he puts it. But once again, it's a conditional statement. "Mike may mean," proposes Joe Cutting, "that his program was the only program at JPL specifically designed for gravity-assist transfers *at the time Mike did his studies.* At that time, JPL had many conic programs *easily adaptable* to gravity-assist transfers."

Well, even programs using vector math? For sure. That approach had been around since the late fifties, at least. The first problem Charley Kohlhase solved for Vic Clarke used vectors—in 1959. "Vector analysis was one of the most valuable courses I ever took, completing it at Georgia Tech in 1955," Kohlhase explained of his education. He utilized its power to score the first perfect 1,200 points on the Naval ROTC aptitude exam. "The use of vectors made it easy for me to solve all of the problems in the 'Bearings at Sea' and 'Navigation' portions of the test."

Michael A. Minovitch also seems to know exactly where Flandro obtained the data to create his original handmade drawing. "Cutting gave Flandro a copy of my 134 page February 1965 JPL report, where I had plotted the relative positions of all outer planets," he begins. "By visually examining these relative planetary position diagrams for the various Earth–Jupiter launch windows, it was possible to identify possible encounter sequences . . . without actually having to compute." The implication is that Flandro could easily have cribbed his drawing from Minovitch's work.

Even if Gary Flandro did refer to Minovitch's work—what, exactly, is the problem? "I believe that any competent engineer looks up past work relevant to his task," says Joe Cutting. Then he puts the nail in the coffin. "The implication that JPL was dead in the water without Mike's program is wrong."

Why would JPL do something like withhold proper credit? "They realized the importance of the invention," Minovitch explains, "and wanted NASA and the world to believe that it was 'their' invention and, as a result, obtain more NASA money to do more gravity-assist missions using my invention."

Whew.

And that's where another component of the problem comes into play. It's Minovitch's personality. He likes to talk—at length and in exhaustively meticulous detail. You don't so much have a conversation with Mike Minovitch as endure the feel of a gigantic onrushing machine. And once stuck on a topic, moving him off of it is like moving the Great Pyramid—world-class tunnel vision.

The success of complicated endeavors like planetary exploration completely hinges on professional associations and group effort. But Minovitch is not known for collaboration. To him, not only is there an "I" in "team," but there's no team in the first place. "Mike was not real good at interacting with other people," recollects Roger Bourke, "and sharing and kicking around ideas." This attitude undoubtedly contributes to his dilemma. Who doesn't benefit from a little watercooler talk or second-guessing from a colleague, from another pair of eyes? Bourke continues, "What he didn't have was what we would call today 'a network' of contacts—of people who were doing related things."

Those who worked with Mike Minovitch describe him as existing in a sort of glass bubble, where nearsightedness and self-imposed blinders hindered many important steps. Minovitch did his own thing his own way, marginally aware yet not taking full advantage of the abundant resources there virtually at his feet. Look at this business with the *Nautical Almanac*. Whereas Minovitch took it upon himself to burn weeks loading the entire thing onto punch cards, quality relations with others may have been able to save precious time—the almanac was already on computer tape. Minovitch complains about the lack of support he got from the JPL computing division in the early sixties. He didn't have a job number to bill anything to, because he wasn't pursuing an officially authorized JPL job. His important work certainly would have justified the status, but Minovitch never pursued it.

Seemingly routine events became plots to discredit him. The part-time JPL employee filled two rooms with computer printouts, then left to finish his education. They were materials carrying no official JPL sanction, deposited by a college student. Is it now JPL's fault Minovitch has no place to store them? Has it become their responsibility? When the Lab began to throw away the boxes of printouts because they needed space, it became part of a conspiracy.

Like many institutions, JPL is bursting at the seams. The place goes so far as to have an official program name for the periodic gathering and disposal of outdated documents: Records Roundup. "At least four people could have been housed in those two rooms," a Lab employee justified of the process.

Perhaps unaware of Records Roundup, Minovitch nonetheless sees this as destroying evidence that would conclusively prove his assertions.

Consider Minovitch's rejection letters from the early seventies—those from *Science, Scientific American,* and *Astronautics and Aeronautics.* Was some deep conspiracy working to keep him out of print?

The editor in chief of *Astronautics and Aeronautics* went so far as to solicit a bit of outside assistance, before replying to Minovitch with several specific comments. "I have had two exceptionally well-qualified people not mentioned in your list of references review your proposed article," the letter began. And? "They both recognize the special problems of claims of originality in this particular area of astrodynamics, and recommend against publishing your paper."

One of the impartial reviewers, who was not identified, began by recognizing Minovitch as "an important contributor to this field." However, the analysis goes on to say that "the principle of swingby trajectories was clearly understood by Lawden and Battin and undoubtedly others." Richard Battin was an MIT physicist, active in space research. As such, Minovitch's paper "appears to be somewhat biased."

"The article promises," as the editor thusly concluded, "to divert the attention of the reader from a history of ideas to a somewhat murky quest for precedents." The cover letter ended by suggesting Minovitch instead submit to a particular competition—one which, interestingly enough, "places no limits on length."

By the way, a solution to the Three Body Problem showed up only a few years after King Oscar first offered his challenge. While containing flaws and technically failing to address the specific nature of King Oscar's problem, noted French mathematician Henri Poincare nonetheless delivered an answer in 1890. Not only did it serve as a foundation for later work on the topic, but it netted him the prize. Later on down the road in 1912, Karl F. Sundman provided a solution to the problem as originally stated.

By 1967 Minovitch was back at JPL part-time and found himself on a new bent. He was going to put a spacecraft in orbit with a big tank full of hydrogen in it. An onboard laser would superheat the hydrogen and provide cheap propulsion. "He was extremely secretive about this," explained Roger

Bourke, who became head of the JPL Advanced Projects Group in the late sixties. "He hinted around that he had this invention," Bourke remembers, "but that he wasn't going to reveal it." Minovitch apparently worried about getting his affairs in order, with the creation of formal engineering drawings and the filing of patent applications. "The credit was going to be all his, and it wouldn't be stolen from him the way this multiplanet trajectory stuff had been," continued Bourke.

Yet Minovitch's life in blinders had stricken him once again. The Boston company Avco had already proposed the exact same thing.

And so with regard to Mike Minovitch, here endeth the lesson.

24. Instant Science

You don't want to go home. If you even go to the bathroom,
you miss ten or fifteen photographs that come down.

Jurrie van der Woude

April 1978 marked the one-year countdown to *Voyager 1*'s Jupiter encounter. Quietly she'd taken the lead over her sister just a few months after launch, back on December 15.

But peace was about to end. Up there in space, millions of miles from home and a Sears Craftsman deluxe 316-piece tool chest with a lifetime warranty, one paltry, little Teflon screw somehow worked itself loose from *Voyager 1* and floated right into a bunch of gears. *Screeeeeech*, and things stopped moving. Her newly damaged gearbox was designed to handle the work of articulating the scan platform to correctly point the cameras and experiments hanging off of it. Now it was stuck.

How do you fix something you can't touch?

After some harried weeks, they successfully mimicked the ship's behavior on their proof test model and agreed on a method of corrective action. The thing to do was slowly joggle the platform back and forth, left-right-left-right, until the gears chewed through the debris. It was a great solution but such an immense distraction at the same time. They'd been so busy working the problem that everybody forgot to send up the weekly command to the other ship. *Voyager 2* flipped over to her backup radio receiver, which promptly died. *Blink!* It went off the air. Suddenly they were down a man and facing dire problems exponentially greater than crud in the gears.

Everybody had always figured it would be the grandiose or obvious stuff taking them down for the count, such as booster failure, asteroids, or radiation. But this new condition occurred in an amazingly simple way. To understand what was happening, it is important to note that both ships carried dual sets of radio equipment. If one went out, the onboard computer was supposed to roll over to the other, keeping its communication link alive. It was written that way in the program code. The computer fully understood its preprogrammed instinct, and part of it went like this: *If I haven't heard from Earth in a week, then something must be wrong with the radio. So I'll change over to the other one and try again.*

Naturally, it made no sense to always be toggling between radios—that would have been like constantly putting the spare tire on a car and then taking it back off. So every week it was *absolutely critical* that some command—any trivial little instruction—go up to each Voyager ship for the express purpose of restarting that weeklong timer all over again. Sometimes the command was a "non-op"—which tells the ship, "Do nothing in regard to this command." Nothing was something. But in the heat of consternation on *Voyager 1*'s lockup foible, nobody remembered the *Voyager 2* command; lo and behold, the electronic logic kicked in. As a general rule, computers do exactly what they're told.

No ground action could be taken except to wait twelve helpless hours until the ship completed her automatic run through the hierarchy of troubleshooting steps. She got to the end of those with no luck, and the very last thing on the list was to try the original radio again. It came back up. They were in business. Thirty joyous minutes later the fuses on *that* radio blew out and darkened the ship's connection once more. Instantly, *Voyager 2* recognized her condition and began the elimination process all over again. She was their solitary—now faint—chance at the Tour. Retargeting *Voyager 1* toward Uranus and Neptune was an impossibility because of the launch trajectory.

"What can we do?" Charley Kohlhase asked, almost rhetorically. "We're on the way."

All anybody knew was what the ship had beamed back down just before cutting out entirely. And that wasn't much. A plaintive knobbing through the radio spectrum—like trolling for a good FM station—surprisingly locked on to their mute offspring. She was out there, just not on the right frequency. The signal faded . . . they turned the knob . . . and it came back again. They'd have to keep retuning, much like had been done so many years ago on board the *Glacier* to pick up *Sputnik 1*. It was another Doppler thing. Frequencies change as the ship gets farther away.

Pragmatically, it really was good timing. If the radios had conked out during a planetary flyby, they'd surely have been done for. Explained Ray Heacock, "We were lucky it happened when it did because it gave us time to solve the problem, develop a work-around, and become proficient in its use." He still had most of a year before Jupiter and the project's opening act.

In a serendipitous turn, Raymond L. Heacock had assumed Voyager's reins immediately prior to the twin ships' becoming most in need of his abilities. This Caltech-trained engineer possessed an impressive, detail-oriented head for things mechanical and had lent his narrow yet powerful talents to Bill Shipley's TOPS program beginning in 1970. Many of the widgets and gizmos flying up there had directly benefited from Heacock's influences. So when JPL director Bruce Murray decided to extract John Casani and drop him into development on the Jovian orbiter that became *Galileo*, Heacock got the nod.

What about Casani's supposed oversight of Voyager? Why wasn't that panning out like Parks had described during their car trip? As Casani explained it, "The major problem was that we were not making enough progress on developing the command sequences and everything that would be needed to do the Voyager science at Jupiter." The workload kept mushrooming. "It's also that I couldn't do both," he continued. "I wasn't able, really, to devote enough attention to the Voyager problem and also manage this Galileo thing."

So Parks said, "Casani, you focus on Galileo," and away he went.

Heacock was the world's preeminent hardware man, but this whole business with the radio fault had Murray spooked. It got him second-guessing himself about the team spread. Not sending the proper command was arguably a management situation. And if there were managerial problems as well as hardware problems, then maybe *more* problems floated above their heads that nobody knew about. The whole program threatened to buckle. At least, it did in one guy's head.

Murray decided to clean Voyager's house with brute-force manpower. He snatched up Bob Parks, temporarily installing him as project manager proper. Below him would be Heacock, administratively unchained to focus on resolving the hardware problems. Also underneath Parks and parallel to Heacock would be Peter Lyman. Lyman could specifically address the daily grind of operations, while Parks focused on righting the ship. That ought to do it. After things were back on track, the guys could break off and drift back to whatever it was they had been doing beforehand. Bob Parks hated the idea. "I recommended against the move," he diplomatically explained of his feelings. "I thought they were recovering from the problems, but Bruce wanted to take the step."

Snapping open blueprints, Ray Heacock twisted the tap on his brain to full flow. The doctor was in. Group, section, and division heads met; like Super Balls, the opinions bounced through meeting rooms, hallways, and lunch tables, ricocheting back through everything in between. And after a sweaty, long-winded process of logging the symptoms, as overtired fingers traced through coffee-stained schematics trying to isolate the fault, and with Jupiter drawing ever closer, *Voyager 2*'s caretakers finally understood their aphasic patient's condition.

Although they could still converse, she exhibited a kind of permanent hearing problem. The way it *normally* operated was like a radio with your single favorite station at a hundred different points on the dial. The signal was everywhere. But now something had happened, and the station only came in on one specific point. If *Voyager 2* was to survive and continue, that station—its frequency—had to be identified. This was easy enough. But it got tricky because, as the ship receded, its frequency would change. It would also change if the temperature in the electronics bay fluctuated by more than a single degree. So in order to maintain radio contact, the *amount* of change had to be forecast way ahead of time and continually percolated through the Lab's mainframe computers. They implemented time periods known as "command moratoriums"—quiet intervals of radio silence when no instructions would be sent from the ground until potential temperature changes had subsided, allowing them to better predict the correct frequency. It was all a simple task only in concept. "This added significant complexity," cautioned Heacock.

With the months running out before Jupiter encounter, a backup program of can't-miss observations was laboriously created from scratch and radioed on up into space. It went out to both ships, on the outside chance the same dastardly thing ever befell *Voyager 1*. The programs were almost a kind of insurance, or doomsday prophecy—something nobody wanted to use. If either *Voyager*'s radio reception went out entirely, the ships were to call up this new course and fly it entirely solo. And JPL would get whatever they got.

A project secretary came up with the idea of transforming one of the conference rooms into a picnic spot. The concept took root and blossomed into a regular de-stressing event, complete with hallway bowling using soda cans for the pins. Somebody added in Pie Day, where everyone would bring a different kind. It ran long enough to intersect 1978's Halloween, and most everyone showed up wearing a different kind of hat. A project scientist named Ellis Miner wore a huge major's hat and called himself Major Miner.

Months went by; the *Voyager 2* radio kept working. New contingency programs went up to both crafts at regular intervals, accounting for up-to-date information on their speed and position. They were getting closer. Christmastime 1978 transitioned into New Year's 1979. January became February. The Lab visibly geared up for game day. Experiments were switched on to warm. The cameras came to life; nobody had left a lens cover on. Many days were still to elapse before closest passage, but *Voyager 1* now figuratively skimmed the Jovian treetops. And no less than eighteen different programming changes would go up before she left.

Intense, pulsating waves of radiation hit next, bum-rushing *Voyager 1*, smearing the ship with over a thousand times the lethal human level. The data began streaming in. There was so much of it, raining down on the antennae that were spaced worldwide, flashing across intercontinental lines right into Pasadena and Building 264. Rain elevated to a downpour and then to flood stage. Nobody could keep up. It was processed as well as could be expected in time for the press conferences at ten in the morning. They were to be held on a daily basis during these near encounters, and the first ones were a bit skinny. Once again von Karman auditorium had been set up for the press, but not a lot of media representatives were coming in to report on this first encounter. Pioneer already went there, right? So what's the big deal? Von Karman welcomed more the likes of Star Trek groupies and science-fiction writers, plus a couple of actors.

A few legitimate reporters mingled, but they all seemed rather less than excited. Part of that had to do with the nasally monotone principal investigators up in front thumbing to life a dilapidated overhead projector—the sure precursor to another academically obtuse and bone-dry lecture.

44. Rubbing hands at far right, Bruce Murray escorts Sidney Poitier through von Karman during the Voyager Jupiter encounter. (Courtesy of NASA/JPL-Caltech)

In generations of students, the distinctively terrifying whine of an overhead projector ramping to life provokes nervous twitches and involuntary dry heaves. They hate this evil machine with its limited throw distance and uneven brightness. It puts many to sleep. Novice users can blow entire weeks twisting the focus knob in search of perfection, only to realize the machine doesn't ever truly focus. And on the day of this first press conference, its reputation was all but solidified as the onstage scientist laid out his poorly assembled, scratched-together viewgraph of abstract data meaning God knew what and began incessantly droning on. In the back of the room, several Voyager members stood there in hollow shock. "Oh, this is *awful*," they complained to one another. "This is *terrible*." As soon as the presentation ended, a couple of them dragged Ed Stone into a room and spent fifteen minutes bludgeoning him with the importance of really impressive visuals. By six o'clock that night a few hand-scribbled drafts of the next morning's graphics went out to art vendors scattered all around the LA area, who were tasked with producing high-quality color slides before a seven o'clock deadline the next morning.

The slides appeared with enough time for them to be loaded into a top-of-the-line projector for the morning's round of speakers. And instead of boring overheads, the press were now treated to a psychedelic succession of oozy black blobs that erupted from somewhere in the middle of each slide and slowly bubbled to the edges. Turnaround had been so quick that the slides were still drying, and the projector's heat was grilling them up like breakfast sausages. The on-stage science presenters liked it even less than the snickering press corps. Most every slide that didn't char over slowly wormed out of focus during the presentation, leading paranoid speakers to steal glances over their shoulders. They'd belt out "Focus! Focus!!"

These were growing pains.

Encounter lasted for many days, as *Voyager 1* threaded her way along Charley Kohlhase's meticulously charted path through Jupiter's scattering of known moons. And the press conferences did mature. Indeed, they had nowhere to go but straight up. Each typically began with Ed Stone clambering onto the podium to deliver a few opening remarks. Brad Smith tended to follow, as the head of Voyager's Imaging Team. Reporters love pictures, and over the next ten years, Brad's group would never disappoint.

Depending on the information coming down, assorted waves of scientists would then cycle through their individual at bats to explain whatever new tidbits they had on the moons, or surface details, or ring systems. A highly different and very public experience from what most scientists are used to. After the first couple of press innings, they even had a name for it: instant science.

Think about how the scientific process normally operates. Most learn the basics in elementary school: What is observable? What is repeatable? Forming hypotheses and testing theories consumes months if not years, and sometimes decades. Scientists like to be fairly positive about the nuts and bolts of something before standing up in front of a bunch of reporters to talk about it.

But *Voyager 1*'s cascading discoveries left them all with no time. Now performing better than most probably anticipated, the hungry space machine expelled such an onslaught of information that staying on top of it all required constant vigilance and off-the-cuff analysis . . . mingled with intermittent sleep. It was a real schedule-clearing kind of environment. Many lived out of the vending machines and in days-old clothes, interpreting whatever they could so as to have it in some semblance of order for the next press conference—attended by increasingly aroused packs of reporters now as hungry for information as *Voyager 1* herself. But what should be mentioned? The wow factor always mattered. Any basic knock-your-socks-off stuff—like a ring around Jupiter—was prime justification for mention, while the grittier details were perhaps kept as close to their chests as they dared for later analysis. Only the best, or most prevailing, theories came out at the press conferences. But they'd been formed at a record pace.

On March 8 the final Jovian press briefing was held. Afterward, everyone would take a long shower and somehow try to unwind during the two-year jaunt to Saturn. Now with a fairly saturated roomful of reporters and other curiosity seekers in von Karman, Larry Soderblom first took the podium to speak. A thin man with businesslike attire, hair that would not have been out of place on Paul McCartney, and rather British features in general, Soderblom had been a student under Bruce Murray only ten years before. He liked grilling hamburgers and training his Irish setters. Now he counted himself a member of Voyager's Imaging Team.

Soderblom was a research man, about as much public speaker as he was kung-fu master, and his first remarks came out a bit disheveled and confusing. To the swollen crowd he explained how he'd been discussing this impending press conference the previous evening with Torrence Johnson and how in the world was he going to describe the four newly imaged Galilean moons in the little time available. Although Jupiter shepherded many other moons, four of the largest ones had initially been eyeballed by Galileo Galilei prior to his telescope's confiscation by the Catholic Church. Since his time, they've always been known as the Galilean satellites. And *Voyager 1* deftly showed how each moon was different almost beyond comprehension. First had been the pockmarked Callisto, seemingly the oldest of the four, displaying craters on top of craters. Then came Ganymede, practically overrun with fault lines that resembled "tire tracks in the desert," as one person described it.

Next? Fabled Io demonstrated apparent chemical seepages hither and yonder, which served to generate bizarrely uneven surface temperatures. Just a few years back, Ames's *Pioneer 10* had actually been the first to recategorize Io from cold, lifeless moon to geologic curiosity, effectively pushing it onto *Voyager 1* planning's radar. But *Pioneer 10* hadn't noticed *anything* like what a JPL employee was unwittingly about to discover. Finally then, to round out the quartet, Europa sported what looked to be a fractured, paper-thin surface in a constant state of upheaval. And underneath that surface? The million-dollar question. Instant science couldn't touch it.

Approaching for his turn at the microphone was a bleary-eyed Edward Stone. "I think we have had almost a decade's worth of discovery in this two-week period," he summarized. "And I think that all of the people who have been talking to you feel the same saturation of new information which has occurred." The press, in receipt of their sound bites and bullet-pointed factoids, thusly packed up and checked out of the Pasadena motels and returned home to file stories.

They also wanted pictures. *Lots* of pictures—they wanted so many that the Lab's photographic division could hardly keep up. A given print allocation had long been planned, and money set aside with which to use. But Voyager was now the hottest thing around. And when the requests escalated to altogether unforeseen levels, there wasn't a person in the department who could say no. "Everything just went out the window, and we started producing things," explained photo lab supervisor Robert Post. "It was so nerve-racking."

Red safety lights on and enlargers humming, they began printing nonstop, 24-7 for two weeks solid. There were twelve press releases every day during encounter, generating six hundred to seven hundred prints per release. They had no idea *Voyager 1* would be so popular. "It almost blew us out of the water," Post remembered of the media's thirst for imagery. "It was too much, way too much; it was almost chaotic. It almost snapped the system." Duplicate negatives, reprints. A one- to two-sentence description accompanied each shot. They were typed up and handed to volunteers, who commandeered every spare desk and open bit of work surface within a large radius of the photo lab. Scissors in hand, the captions had to be cut out and taped onto the back of *every single print* before it could go out.

"Everybody was so exhausted," Robert Post continued. "You're not even sure where you are, but you keep going because you don't want to miss anything. I still to this day don't understand how we did it."

Emanating from the twin ships, pictures arrived on Earth essentially as a stream of numbers shoehorned into a radio wave. When reconstituted into a proper image, most required a bit of "help." A strip of detail might be missing, or the colors in need of calibration. The cleaned-up shot then went onto magnetic tape to be transferred to 35 mm film for traditional developing. From there it all worked like regular photography, only to preposterously exacting standards. Every single print had to utilize the same color reference. Nobody was much interested in the press coming back and asking, "Which orange is the correct Jupiter?" Over thirty years later, they still print to the same dead-on colors, using reference images stored more lovingly than the crown jewels.

"Black Friday" is the term now used to describe that first week of the Jupiter encounter. The press got their images and their slides and their captions, but it almost cost JPL their entire photo squad. "I have seen people laying in the hallways of the photo lab sound asleep," recalled Jurrie van der Woude, who worked in the division up to 120 hours a week at the time. "They literally worked until they fell at their posts, but the world never knew about it."

Handing out the shots could turn into a feeding frenzy. To help create order, van der Woude brusquely imposed a very necessary hierarchy onto the photographic distribution. First off, the main press briefing had to be over and the question-and-answer period underway before he handed out squat. Waiting was only fair to the speakers. The grouchy wire service people got their photo packets first. They were always grouchy because for them the deadlines never went away. Images could get on the evening press in Europe or early morning in the Far East if the shots got on the wire soon enough. So those folks grabbed their packets and sprinted out the door at full speed. Everybody else had to hang on until the press conference was totally over. No exceptions. Then the newspaper and television people could approach for their materials, while any magazine reps were forced to hang back. They always came last.

"That's the way my world ran," explained van der Woude, who correctly assumed that his touches of procedural administration would mercifully help things along. "I stuck to that."

On the day of that final press conference, Linda Morabito got up and came to work in Building 264 to begin another fourteen-hour shift. Yeah, the

Jupiter encounter had ended. For any reporter the magic was therefore all over for a good two years until Saturn, but true Lab insiders understood how the trickle of magical moments would inevitably persist for many weeks at the very least.

Linda's heartstrings were plucked by the romantics of optical navigation, involving analysis of the growing stacks of *Voyager 1* photographs to figure out just exactly where the ship was. Everyone had a fairly good idea of course, but tracking the precise flight path remained a crucial and ongoing task. It could be divined by measuring distances between key stars and the centers of specific bodies—like Jupiter's moon Io. Morabito later elaborated, "No star catalogue existed anywhere in the world that had the stars' celestial coordinates to the accuracy that the Voyager mission needed." It was like having a city map that showed the highways but no side streets. "So Voyager commissioned production of its own star catalogue." And in this phase of the flight, Linda Morabito's particular hurdle mandated a determination not of where *Voyager 1* was going but of where she *had been*. This, combined with radio tracking data, made for a more accurate overall plot of the heavens, which in turn would incrementally simplify this whole process the next time around.

Morabito entered the work area only to be hit in the face with a blast of chilly, refrigerated air. At least she knew it was coming. Not every computer was as huge as the IBM 7090, which incidentally was no longer in use at JPL. Morabito's electronic machinery didn't require the space of a backyard—only that of a sizeable bedroom. It was up on a grid of removable tiles with network lines and air-conditioning fixtures located down below the floor. "It was cold, noisy, and an interesting environment in which to work," is how she put it. *Voyager 1*'s data stream came in to her from one of the Deep Space Network's three gigantic dishes spread in equidistance around Earth's circumference. If the one in California pointed away from *Voyager 1* at the time of communication, then the one in Spain or Australia collected the signal and pushed it on to JPL. This automated process was several orders of magnitude removed from the days of James Van Allen personally retrieving data tapes and plotting them out in a motel room with drugstore school supplies.

Up in Morabito's chilly office, giant reels of thick magnetic tape spooled through her MODCOMP IV minicomputer, boasting a whopping thirty-two

45. Linda Morabito at JPL, 1977. (Courtesy of NASA/JPL-Caltech)

kilobytes of core storage. All day, and typically all evening, she processed one grayscale image after another. They weren't selected for beauty or composition or any of those things. Navigation didn't even need color shots—just ones of the Jovian moons with star patterns in the background.

On March 9, the day after Ed Stone's punch-drunk summation, Morabito sat before another parade of gray images. They'd been taken as *Voyager 1* looked over her shoulder on the way out, glancing back at Jupiter's receding entourage of moons. On each shot, Morabito jacked up the brightness level—one of her little tricks to ferret out stars from an otherwise dim background.

Picture after picture went by, but then she stopped on one of them. What was that? The brightness was still up; she dialed it back down. Now it wasn't there any more. Crank it back up again—*yep*, there it was, barely visible. Some kind of goober artifact maybe? A little halo—there was an arc, like a little cloud formation, just off the edge of Io.

A cloud on Io?

It didn't seem right. Io held no atmosphere; everybody knew that. It had to be something else. Voyager cameras weren't perfect. The design was almost ten years old already, lifted from the *Mariner 10* camera. And Linda

Morabito knew most all of the handful of distortions and anomalies they presented. But this wasn't one of them. Could it be some newly developed flaw? Fingers hovering, the tendency lingered to punch a few keys and move on to the next shot, but something in Morabito's keyed-up craw prevented her from doing so. She felt like she was on to something.

Over the next six hours, in frequent consultation with a revolving slew of guests and their opinions, Morabito eliminated most every possibility out there and circularly kept coming back to where she'd began—that cloud. She ran it past the camera engineers to confirm it wasn't some heretofore unseen defect in the lenses or something with the image processing or with the transmission. Nope. She verified it wasn't some new moon—entirely within the realm of possibility. Continual discoveries of new moons in every direction were now a Voyager given.

The regular workday activities began tapering off. People drifted out; it was a Friday after all. JPL's parking lots emptied, and traffic levels on I-405 built to their usual rush-hour proportions. The Sun dropped along with the temperature. Even Ray Heacock had piled it in. People like him were already working marriage-killing shifts of thirteen hours on, eleven hours off during the two or three months of prime encounter time—and that was when things ran smoothly.

Soon only Morabito and one other man, Peter Kupferman, remained in her department, staring at the screen and that cloudlike arc. Even through slackened, glazy eyes both knew they weren't seeing things. At one point Peter tiredly offered, "The more you work with it, the more you think it's real." And between them grew an eerily likelihood that it might end up being a cloud after all.

At this late hour with no dinner, no snacks, and no breaks, after dwindling down a short list of remaining possibilities, it might have been beyond painless to just give up. All the easy answers were discarded and gone. They tried approaching it another way, correlating the feature with Io's puckering black acne. And . . . something clicked. It was so beyond the scope of comprehension, though, as to be exceedingly ridiculous. But as Sherlock Holmes once aptly observed in *The Sign of Four*, "when you have eliminated the impossible, whatever remains, however improbable, must be the truth." Linda Morabito knew what she had, and she knew it was the truth.

Morabito headed out, departing the Lab as she drove straight toward her parents' house to join an overly late evening meal already in progress. There, over the uniquely soothing atmosphere of family and food, she explained the long day's goings-on, eliminating the impossible, and her improbable remnant.

Morabito's dad looked at her kind of blankly for a minute and then suggested, "Do you realize you may have discovered the first volcanic activity outside the earth?"

He was right; that's exactly what Morabito had. The arcing white cloud was nothing less than a plume of debris rising from one of those blackened pinpoints on the surface of Io. Ed Stone later put this into perspective by saying, "I guess I would pick the volcanoes on Io as the first major surprise. That really told us that the outer solar system was not a bland reproduction of the same phenomena over and over again." Other planet's moons could never again be assumed to resemble our own gray ball o' dirt. Io had thrown that paradigm in the garbage.

Once Linda Morabito nailed the volcano connection, it all made sense. Io is small, but it's also treacherously close to Jupiter and other moons. It's the pickle in the middle. The endless gravitational tug of war causes Io tidal bulging on the order of *three hundred feet*. Compare this with the typical Earthly variation of thirty-six inches.

There's more. Io barely clings onto its very self. Every second one to two *tons* of sulfur and oxygen are gravitationally stripped from the moon by Jupiter, contributing to a doughnut-shaped feature encircling the host planet. It actually glows in ultraviolet light. On occasions when Io crosses Jupiter's magnetic field, bad things happen. An enormous amount of electrical energy—something on the order of 1 million amps—whips up in response to the duo's intimate proximity. The unfortunate moon becomes a defenseless recipient of heinous storms and, yes, volcanoes.

By the time *Voyager 1* stopped photographing Io, no less than nine active volcanoes had been discovered. One of them is so large it has a name— Pele—whose size approaches the limits of reason.

Remember the Krakatau explosion? In 1883 the volcanic South Pacific island of Krakatau exploded with a sound loud enough to be heard not only two thousand miles to the east in Australia, but also three thousand miles to the west on the island of Mauritius. The blast equaled two hundred megatons of dynamite. Ensuing shock waves rippled around the globe a total of seven full times. Tsunami waves crashed all the way through the English Channel. Ash ballooned fifty miles up. Six-inch slugs of pumice fell up to ten miles away. And more than fifteen cubic miles of soot and ash hung in the sky for weeks. The Krakatau explosion generated the loudest sound ever made on Earth. It has nothing on Pele. When this feature of Io erupts, debris is thrown *170 miles up*—thirty times the height of Mount Everest. On Earth, Pele's fallout zone would bury France.

Four brisk months were all that remained before *Voyager 2*'s blind date with Jupiter. Almost two years into flight, many of the personality issues had sorted themselves out. The right people were—on the face of it—settled into their roles of day-to-day ship operations. A core group of flight controllers was very obviously emerging—people with the right kinds of constitutions to sit there day after day and work the situation, work the problem, whatever came at them. They were people who didn't get uptight or panicky when something didn't go right or when some little nagging detail still hadn't been resolved. JPL engineer Dick Laeser commented, "There were a few cases where we had to ask people to leave, because they kind of wanted to stay, but they didn't fit personality-wise in the kind of team environment that we were trying to build." Expert dog trainers talk about how the dog's owner always needs just as much training as the dog. In the context of Voyager, this paradigm held. Successfully running these ships depended as much on the right blend of personalities holding the leash as it did on healthy mechanical innards.

Millions of miles away, two brilliant spacecraft had almost trained their handlers.

And after Jupiter, one of the Voyager managers looked at another to say, "I think we're beginning to learn to operate these things."

Esker Davis hadn't been on the job very long when Ray Heacock showed him a letter that had just pitter-patted in through the NASA channels. It was the response to Heacock's projection of a Voyager cost overrun in the neighborhood of $17 million by the time both ships rolled through Saturn. That was an extra $17 million on a program already lightening U.S. taxpayers by some $24 million a year. Compared to running a war, it was peanuts, but somebody upstairs in Washington DC was barking heavily about it all the same.

Trouble like this, in part, had precipitated the regime change. Heacock was a straight-up hardware man, never holding much interest in working the finances. He liked to build stuff. When *Voyager 2* had all its difficulty with the radio receiver, Bruce Murray pulled his executive coup d'état installing the triumvirate of project managers. As the problems abated and Jupiter approached, Heacock returned to his role of sole manager. But then the money troubles began pecking holes in everything. Heacock thus galloped off to spend his days on how the U.S. might best examine Halley's comet during its 1986 flyby. That left the reins (and the accounting hell) for Esker Davis to choke on.

Until shortly before, "Ek" had served as deputy project manager under Heacock, patiently living in the shadows while learning Voyager's many idiosyncrasies. Davis hadn't formally applied for project manger; it didn't work that way. Nobody sat down and filled out a tear-off sheet like getting a job at Burger King. "When I was notified that I was the top candidate," Davis recalled with kindergarten-teacher enthusiasm, "I couldn't believe it was real; I couldn't believe how fortunate I was." And soon thereafter the budget mess fell in his lap.

NASA's response to the cost projection was decidedly south of enthusiastic. The frazzled space agency, still two painful years away from actually flying their ungainly new manned spacecraft, spent most days wiping Shuttle doo-doo off their feet. They were constantly stepping in it. Shuttle's planned off-the-shelf pedigree had quickly devolved into an alarmingly huge list of big-ticket show stoppers, like blown engines and heat shield tiles failing to stay glued on. Nobody wanted to hear about an anticlimactic Saturn flight running a little short on cash. NASA cautioned JPL, "Not another dime"; to say the least, they remained in a huff over the Lab's fiscal estimates.

"We were pretty jolted by that," explained Davis of the reaction. "To what extent NASA was playing hardball with us, I guess we'll never know." But Esker Davis possessed a richly developed skill set from years of budgeting air force projects. That kind of experience figured strongly into his appointment. So Davis reset his synapses to view the problem as a business owner would—if you're 17 million in the hole and the client says "you're SOL," where can you sneak in a little liposuction?

"You must have a plan," Davis preached. He overhauled the whole project structure top to bottom, stripping it down to the skivvies and rebuilding. Three wholly different teams worked on processing data; Davis consolidated them into one and sliced off $3 million right there. He learned that many of the outside subcontractors had been overpaid. They were holding onto JPL funds as obligations against any outstanding, last-minute finishing costs. The Lab wouldn't get their money back until a final audit, which might take years. So with great fury, Davis unleashed a team of hell-for-leather accountants who anaerobically descended on the porked-up subs to extract refunds.

Somebody made a discovery about the data tapes. In those days, when the science data came in, it was all laid down on endless winds of fat magnetic tape. A full reel of the stuff measured a foot across and weighed thirty-five pounds, easily capable of fracturing the metatarsals on any butterfingered assistant. But the scientists always insisted on using new tape, right from the box. They went through it by the pallet; Voyager was burning $750,000 a year in boxes of new tape alone. "But we got the scientists to accept recertified tape," described Davis, explaining how the used stuff could be bulk erased and scrubbed and then given a clean bill of health.

Penny-pinching ideas snowballed, until everyone's heads got into a rhythm of not only saving money but also outwardly looking for ways to slim down. Voyager's $17 million gap was closing like the Red Sea. Things were going so well that the U.S. government's General Accounting Office flew some people out to have a little look-see. They monopolized Davis's time by asking how he had done this. Ek Davis merely explained how, like any competent business manager, he'd scrutinized and restructured to make Voyager's home office run more like a for-profit than not. "It was recognized as one of the best project plans NASA had ever seen," Davis proudly alleged. His latest balance sheet looked like JPL would right well finish both Saturn encounters under budget. And so with a touch of uncertainty, the polite inquiry finally went out as to a possible continuation of the mission—to Uranus and Neptune.

They were in.

Esker Davis had saved Voyager. And most of the fiscal details got wrapped up quickly enough to rejoin the program already in progress. It was time for the ringed beauty.

Years down the road when both ships ended their strings of encounters, Ed Stone would finally shelve the last of thirty-eight notebooks, heavy with ink, covering his day-to-day dilemmas and decisions and deeds. As *Voyager 1*'s Titan encounter was being planned, though, Ed was still only loading up the first ten. Some of his jottings addressed the conflicting scientific agendas and personalities he dealt with nearly every day. And all of them competed for priceless time on the spacecraft. Charley Kohlhase had the big picture formally nailed in place, but it could be modified slightly depending on which exact way they desired to cross a specific body. And the headache of the moment these days centered on that small, unassuming Saturnian moon.

Titan was considered important enough so as to almost eclipse Saturn in the magnitude of its planning. The orange ball had some kind of a dense atmosphere, and it might have some kind of liquid on or near the surface. Planetary scientists had been clawing their eyeballs for years trying to learn more, and now *Voyager 1* offered a clean shot. To that end, her navigators came up with a few different choices in the basket for how to swing on by. One option positioned the ship so that it flew past Titan *before* reaching Saturn. Or Saturn could come first, and *then* it could fly over to Titan. Which was better? Did it even matter? Kohlhase needed a ruling.

Ed Stone lumbered back and forth between the two. It seemed almost like a toss-up, and that's what buggered him. Somewhere in all those reports and analyses sat a good reason to favor one over the other. It just had to be found. He got to thinking about safety. With Saturn coming first, the ship could ignorantly head into ring particles larger than anyone figured. Or it might discover an entirely new ring the hard way, as it were. Hmm. Stone looked up from his work and felt visiting Titan first certainly would appeal more to the crew.

But if they did it that way, then the flyby of Saturn's rings wouldn't be as dramatic—or as scientifically productive. The radio science folks pined for visiting Titan *after* the planet, in part because the ship's radio transmissions could be directed through the rings and shed valuable light on their makeup. Stone ultimately went home scratching his head, trying to make a decision based purely on scientific merit that would somehow keep everyone happy. Who knew when the next real chance to study Titan would come? The final decision rested with Stone.

"It was clear to me that 'Titan before' made a lot of sense for safety reasons," he explained of his final conclusion. It threw the radio science team a jump ball. They went off in all directions to hopefully extract a bright side from the lamentable decision. Soon one of them jauntily reported back to Stone that he'd found a possible flight path using "Titan before," allowing a nearly ideal analysis of *half* the ring system. And that, the man explained, would be just dandy enough for them. The maneuver came off as planned, and Stone breathed a sigh of relief. "You don't know until it's all over whether it's really going to work," he said. "There have been a number of things like that along the way."

The subsequent imaging of Titan disappointed everyone on a large scale. The place was socked in by a smoggy, orange haze, albeit in several discernable layers. Then some good news came again—maybe the best news of all. Titan might actually have *lakes* of liquid methane on its surface. And years later when a tiny parcel of cameras and instrumentation known as *Huygens* was actually put down on Titan, that's exactly what it found.

Afterward came the ringed mama herself. Saturn's rings follow a million-mile-long circular path. They're made up largely of ice particles—some the size of dice and others the size of minivans. Ring thicknesses vary, but most are on the order of two hundred yards or less. And some weren't perfectly smooth all the way around. In places, they showed bends or kinks. The only real explanation called for small, yet-unnoticed shepherding moons folded in there somewhere, truckin' along in the ring path while exerting a measure of gravitational bending. The bizarre yet plausible theory persisted for years until later unmanned flights indeed found these diminutive shepherd moons.

Voyager 1's angling through the Saturnian bodies yielded much useful information on the system, but it also doomed the ship's future course. The flight path was constructed such that Titan's gravity unceremoniously yanked the ship upward—like rapidly jerking up on some fishing rod while trying to land a muskie fighting the hook. *Voyager 1* headed north, up out of the flat horizontal plane on which our planets all orbit.

And after that, she went off into nothing.

Over eight months later Esker Davis still found himself awestruck by the imagery. *Voyager 2* was now commencing its own survey of Saturn—although from different angles than her sister and with a separate list of targets. Volcanoes? Methane lakes? Both were hand-rubbing moments of amazement. And what would come *after* those? They'd just have to wait and see. Davis had trained as a geologist. It was all so great. Sometimes he'd just sit there in the evenings immobile, staring at a television monitor as the images rolled in one after another. Once, his wife Kay phoned to ask what he was doing. "Just watching the pictures come in," he told her. It gave him goose bumps being the initial human to view a picture of one of Saturn's moons, rich in unanticipated details. It was cause for wonderment.

"I was there and saw that very first picture."

Shortly after the *Voyager 2* Saturn encounter, Charley Kohlhase was scheduled to give a small talk in Caltech's Beckman Auditorium. Seven hundred science teachers from everywhere in the nation were coming through on what amounted to a glorified field trip. Kohlhase lankily entered Beckman Auditorium from the rear, with the knot of his necktie askew, futzing with the papers in his hands as he walked. People heard the man coming and then saw him. They rose as one. He got closer to the podium, and the clapping began. It swelled. Feeling a rush of emotion, Kohlhase slowed, unconsciously stretching out the moment. This full-blown standing ovation celebrated one humble man who gave himself to Voyager. He had simply done his job; even so, the tears welled up in Charley Kohlhase and ran down his cheeks.

"You never forget that," he said, reminiscing.

Perhaps the memory of his exacting father entered Kohlhase's head at that moment. His dad had wanted him to study mechanical engineering in college, but Charley switched over to physics. "He told me it was the greatest mistake of my life, but somehow I stuck with the decision.

"I did very well at it, I think, because he could not compete with me in that area."

And that afternoon, as Charley walked teary-eyed up on stage to applause, the ever-present haunting shadow of his father disappeared at long last. Charley'd done good.

With the ringed sphere behind them and a mystifying Uranus better than four years distant, some Voyager crew began reporting a similar experience—broken-record comments. "Well," their friends would suggest, "now you don't have anything to do for four years." But these were people who didn't understand.

The mission's laundry list was endless. Nobody had done much of the detail work for what was to come. Meticulous encounter sequences had to be designed. Eleven scientific entities still had eleven different experiments running. Big decisions had to be made over these four years: who got what and when, where compromise would occur, which known moons would be targeted, and where the ship might browse for new moons. Ground equipment had to be upgraded. The big dishes of the Deep Space Network required new construction; some were going to be rebuilt up to fifty feet larger in diameter to accommodate the increasingly weak signals. Each ship would receive new programming to apply a method of stringing its computers together *in series* to generate enough horsepower for operation out at Uranus.

Minor little things like rain also became meeting-stopper topics to address. Voyager took advantage of a radio spectrum called X-band frequency for its mega-long-distance data transmission. Despite X band's abilities, it was also susceptible to rain, which would innocently mix loads of noise into the signal. And if an important planetary encounter was happening with *Voyager 2* beaming live results down to, say, the Madrid antenna, a simple little rain shower could wreck the party. There would be no second pass— no second chances. "What in the world can you do if there's rain in Spain?" asked Charley Kohlhase, rhetorically throwing up his hands. "Aren't you just out of it?"

Mission design could—and did, ultimately—adapt to this seemingly unpreventable hitch. Critical data beaming home could simultaneously find temporary safe harbor on *Voyager 2*'s tape recorder, to be replayed later on command. The strategy also worked in case the ground received an incomplete transmission that was seemingly nonsensical and wanted a complete do over. The tape recorder could play back the same stuff ten times if need be.

But the tape recorder giveth, and the tape recorder taketh away. Repeatedly boosted by gravity, *Voyager 2* now flew tens of thousands of miles per hour faster than she had been going at Jupiter or even Saturn. And the amount of sunlight out by Uranus was practically nil; the camera shutters would have to stay open much longer to nab most anything. So in sum, if the photographs were taken in identical fashion as before, every one of them would come back all blurry—it would have been like trying to get a shot of a cow while going 140 miles an hour on the interstate. It just wouldn't work.

Good photographers know they can essentially "stop" a fast-moving subject by panning along in sync as it whips by. Applying the trick to *Voyager 2* called for trading roles since *she'd* be the one moving instead of her subject. It just had to be worked out by pivoting or canting the scan platform somehow, just ever so slightly during each photo. And this is where the tape recorder problem entered the equation. At the start of every picture, the recorder flipped on, jiggling the entire ship. Then, confound it, there was another jiggle when the recorder stopped. What happens if a photographer gets his elbow bumped right when clicking the button? He gets blurry pictures.

The jiggles were a given—unavoidable consequences of design. They couldn't be stopped, but could they be compromised somehow? Cancelled out? It was time for another back-bending jaunt through the possibilities.

And don't forget this: nobody knew *precisely* where Uranus and Neptune were. Of course they had a fairly good idea, but nailing down the painful details couldn't happen until the ship drew closer. An accurate flyby of Uranus, executed so as to continue on to Neptune, is afforded only a single chance for success. "You know when you're coming into a planet that this is it," explained Robert Cesarone, who calculated these final trajectories. "I guess it's like landing a glider; there's no second chance. There's a three-quarters-of-a-billion-dollar mission riding on the fact that you designed this procedure properly and you're executing it properly at three o'clock in the morning."

Some months before the 1986 Uranus encounter, Charley Kohlhase happened to look up the Shuttle manifest and noted that a launch was scheduled for January 20. It rubbed his fur the wrong way—closest Uranian approach was only four days later and definitely set. His worst-case spider-sense hummed to life as Kolhase contemplated how any major emergency on board Shuttle could saturate the Deep Space Network. And *bing*, just like that, there'd be nothing from Uranus.

After conferring with a few others, he took the potentiality across campus to senior JPL management. They agreed that, under the circumstances, slipping the Shuttle launch a few insignificant days made the most sense. In turn they forwarded their request to NASA headquarters. It should have been a rubber stamp; what did the Shuttle mission care? *Voyager 2's* motion dictated Uranus encounter dates—the Shuttle was only going up to lob out a dumb communications satellite that could have been orbited much more cheaply with an unmanned booster. They could fly it any old time. But JPL's request went all the way up the chain to NASA administrator James M. Beggs, who promptly gave it the thumbs-down even though virtually everyone underneath him considered the delay to be a wise one.

Apparently he yapped, "The White House doesn't want the launch slipped."

Knowledge of Uranus's tenure as a planet dates precisely back to March 13, 1781. On that night, British astronomer William Herschel had been surveying the cosmos with his eighty-four-inch-long telescope. It extended the power of his eyes by 227 times. And at one point that evening, he thought he'd spotted a comet. Like everything else, it went into his notebook for review.

The subject came up about a month later at the next meeting of the Royal Society of London. During the proceedings and discussions, Herschel had to confront the sticky issue of two other astronomers having recorded observations of this feature *before* him. Nonetheless, he received accolades as the discoverer. Herschel promptly named his new baby after King George III, which held for a while. Herschel had actually used the term *Georgium Sidus*, which literally translates to "George's Star." Oops. The name quickly transmogrified to "Georgian Planet."

Nobody outside of England offered much in the way of name acceptance. French astronomer Joseph Lalande suggested using "Herschel" and went so far as to incorporate a prominent "H" in the new planet's astrological symbol. The design caught on but not a specific moniker, which begat a rash of suggestions: Astraea, Cybele, or Neptune—yes, Neptune. Andrei Leksel of Russia evoked compromise by offering George III's Neptune and, as a deferential alternate, Great Britain's Neptune.

But there was no consensus whatsoever. It dragged on: Transaturnis, Minerva, Austraa, or Hypercronius. Finally, Berliner Johann Bode coughed up "Uranus" and threw it over the fence just to see how everyone would react to it. He'd taken the name Ouranos, Greece's god of the sky, put it through some kind of mental grinder, added a dash of Latin, and crazily got "Uranus." Not long afterward, the name began popping up in charts of the heavens, and it was as good as solidified—except with those stubborn Brits, for whom *Her Majesty's Nautical Almanac* did not actually budge from Georgium Sidus until 1850.

Sunny days are rare on Uranus. It gets *one four-hundredth* of the light blanketing Earth. It'd be hard to get a tan out there. It's mighty cold, too—just a touch above absolute zero. And things are crooked. Uranus was already known to operate on a peculiar ninety-degree tilt—as if some titanic, Earth-sized chunk of ice happened along a billion or so years ago and whacked it off-kilter for all eternity. It rotates over to under, as it were, versus Earth's side-to-side spin. Of course, the cocked angle means Uranus's moons orbit perpendicular to the rest of the solar system's. Once *Voyager 2* arrived, who knew what else they'd find.

Press coverage wasn't what it used to be. So many reporters wanted into von Karman Auditorium that JPL needed to adapt. Staffers cleared out its museum annex, stripping it bare, while another group followed along right behind slapping down desks one after another to generate writing space for as many reporters as the fire code would permit. All unconsumed real estate went to banks of television monitors, which fed in the substance of any JPL news conference in progress next door.

Reporters saturated the phone lines with foreign and domestic tongues, drained the vending machines, and clogged the toilets. They were welcomed with open arms. Most questions dealt with *Voyager 2*'s health, the health of her operators, or what might be expected in the Uranian system. One reporter then stood up and asked about the Deep Space Network dish in South Africa and if its existence there was an indication of JPL's support of apartheid?

Uranus was going to be Dick Laeser's first encounter as project manager, and he'd spent part of the past four years emotionally bracing himself for disappointment. "It really looked like it was going to be dullsville," he matter-of-factly offered. "It was not a very interesting-looking planet."

Balding and experienced, Laeser had clambered his way up through the ranks of Voyager. An electrical engineering degree from MIT led to four job offers right out of college. One of them came from JPL. It promised the absolute least pay, but it also offered the most specific detail on what, exactly, he'd be doing. "The others were vague," he remembered, "they were

. . . jobs. The vagueness of them turned me off. I was intrigued by doing something very specific, and it sounded like fun." Back when Jack James was first launching Mariner, Laeser spent a lot of time plugging away at the Deep Space Network. Later on, Bud Schurmeier had needed a man running point on that for MJS'77; Laeser went for it. The role expanded into work on the other necessary ground systems around the time the Jovian radiation problems were just coming into play. From there it leapfrogged to managing operations, and then to acting as deputy project manager under Esker Davis.

And now he found himself looking at picture after picture of bleak Uranus rolling across the television screens. On approach, *Voyager 2*'s cameras only made out so much. The planet itself was mostly a blobby, blue washout. There were too many clouds moving in a . . . check that again . . . *hmm*, yeah, in a surprisingly *fast* way. High wind speeds out this far? It didn't make sense.

Appearances aren't everything, however. In short order, the ship's bandolier of additional instruments picked up the slack. They described radiation belts around Uranus—thick and strong, making Earth's look like chump change. The wind got special attention. Wind speeds were running up to *three times as fast* as those on Earth. And near Uranus's equator, the winds impulsively and quixotically reverse direction. How was instant science going to address that one?

In due course, the scientists forged a link between Uranian wind and what's *not* happening at the planet's core. No internal heat source could be located—therefore, no atmospheric turbulence exists to slow or otherwise affect the winds. They just keep on a-going. Ed Stone described the enthralling and unforeseen discovery: "As we went further out in the solar system, we found higher speed winds rather than lower speed winds. There is less solar energy input, yet the winds are faster." Nobody knew yet that Neptune's winds would run faster still.

The first pictures came out razor sharp. The secret to fixing the jiggle problem centered on *Voyager 2*'s little thrusters, which hung all over the equipment bus. In dead on synchronization, they'd give a little toot as the recorder jumped to life and perfectly cancel out the movement. Tooting 'em again at the end completely solved the problem.

Then the images arrived from Uranus's moon Miranda, and nobody knew how to respond. They stood there attempting to grasp what displayed on-screen. "I about jumped out of my skin," recalled Dick Laeser. "Some people told me my behavior was a bit bizarre." Miranda resembled nothing ever seen before. Was it feeble moon, or monster? Had the thing been accidentally dropped at some point? Jumbles of shattered parts barely clung together. An enormous, jutting cliff decorated one hemisphere. Other areas looked as if they'd been hurriedly spackled over in some freshman attempt at repair. Part of the surface lay cratered. Part of it sported huge, parallel grooves waywardly cutting across the surface—like an overgrown farm implement had gone berserk. Some of the furrows measured up to twelve miles deep—pretty remarkable for a moon only three hundred miles across. Even Iowa is bigger.

So much for dullsville. While practically leaping up the steps to von Karman's stage, the more clinically reserved scientific descriptions evaded most of the speakers. Everybody stood there for a bit watching Miranda's imagery all spooling out on live television, and their imaginations picked up the slack:

"A chevron."

"A race track."

"A layer cake."

That's all they had for this particular moment of instant science.

On January 28, 1986, von Karman Auditorium prepared for another onslaught of reporters. That morning, it was relatively quiet. In ninety minutes, the stage lights would brighten for JPL's final round of instant science and parading imagery from the Uranian system. Off to the side, wearing a plaid button-down with sleeves discreetly rolled to the elbows, Jurrie van der Woude stacked up his press kits and prepared to watch over them like a presidential bodyguard. Nobody knew that a singular happening down in Florida would soon redirect everyone's attention. At 11:38 a.m. Florida time, seventy-three seconds after launch, one of Space Shuttle *Challenger's* booster rockets encountered some trouble way down low in back, right next to the gigantic rust-colored tank of liquid hydrogen and oxygen. This was the flight carrying Christa McAuliffe, the first member of President Reagan's Teacher in Space Program. One of her ceremonial tasks, incidentally, was to be the triggering of a non-op command for *Voyager 2*.

These identical white boosters, which flank the huge orange tank, are the most powerful solid-fuel rockets ever flown. Neither booster is comprised of a single tube. They are made of four major segments bolted together; that day, immediately after ignition, a hellish puff of black smoke shot out from between two of the segments. A seal had failed.

This allowed a focused and insanely hot fountain of gases to torch directly through the booster wall and eat a hole in the external tank. Searing flames ignited the liquid oxygen held inside, and up it went, the concussion detonating the vessel of liquid hydrogen above it and blasting the external tank to smithereens. *Challenger* had not been designed to handle such forces. It broke apart, and pieces began raining down on the South Atlantic. The shuttle did not actually explode.

When the news flashed through von Karman, three-quarters of the reporters grabbed their stuff and ran for airplane flights to Florida, leaving JPL and their Voyager at Uranus. Having stolen the Grand Tour's money ten years before, *Challenger* now stole its thunder. The activities in von Karman were postponed.

Shortly after Uranus receded into *Voyager 2*'s rearview mirror, Dick Laeser solemnly moved out to Virginia and opened up a little JPL office to support the impending space station. He would decommission it only five years later; it was a dead-end, unfulfilled assignment soaking up twice as much money *per year* as Voyager did over the life of its entire mission.

Laeser handed over the driver's seat to Norm Haynes, who would see Voyager through to the last port of call. JPL was facing another long haul to Neptune—over three full years. Hopefully that would be enough time to finalize the ship's trajectory, design her encounter sequences, reprogram her computers, and expand the network of humongous receiver dishes.

Those three years might well pass quickly owing to the endless questions and desires that kept materializing. Everybody lusted for a good wing past Triton. It orbits *backward* relative to Neptune's own orbit; due to this wacky retrograde motion, the moon is in a constant mode of stress. Neptune actually strips energy away from Triton, ripping it off like a hot body wax job, slowly yet progressively decaying the moon's orbit. Many, many years in the future, Triton will either impact Neptune outright or cleave into shards and form a ring of debris.

Triton and Neptune both? What a combination. Pulling it off demanded careful aim over the top of Neptune's North Pole, letting gravity fling *Voyager 2* down past Triton as if the ship were riding the ultimate playground slide. It was a breathtaking yet dicey option forwarded by Kohlhase's team to the scientists. But *Voyager 2*'s experiments bore differing recommendations for how close to buzz these targets. Pictures are always better when taken closer up, and everyone loved the pictures. An atmosphere, however, is best sampled from a vantage point some distance farther away. It became another judgment call for Ed Stone. "We had a real tradeoff," he explained, nearly stumped. He said, "If we flew too close, we might lose radio science data on Neptune's atmosphere, but we'd get better images of Triton." The ship wasn't going into orbit; there could be only one pass. The closer *Voyager 2* flew to Neptune, the sharper its flight path would bend, and, therefore, the closer it would pass to Triton. Both were inextricably connected. Stone continued, "However, if we backed off far enough to be safe for the radio science occultation of Neptune, we'd lose high-resolution images of Triton." So what was a scientist to do?

Some folks on the navigation team approached him with a surprising option. Although still hammering out the details, they'd stitched together something of a Hail Mary, go-for-broke kind of option. A nearly perfect course suiting *both* objectives was actually feasible—provided a large number of critical and potentially fatal events all came off without a hitch. They would have to fly *through* the Neptunian ring plane and literally skim the top of Neptune's pole—not more than 1,800 miles above the clouds. Nobody was exactly sure how big the planet was or how high up the atmosphere went above it—or what all might happen if ring debris peppered the ship.

If those two minor details resolved themselves, getting both planet and moon depended on one other teensy mitigating factor. *Voyager 2*'s precise arrival time at Neptune would have to be determined . . . to within *one second*. As Ed Stone explained, "The spacecraft was moving so fast and so close to Neptune that if you're off by five seconds where you're pointing the spacecraft, you're pointing in the wrong direction."

Their big reward would come only by way of the tightest of margins.

25. Circles of Gold

What would we want them to know about us?
And how would we communicate it to them?

John Casani, concerned about extraterrestrial ice-breaking

To understand why Jon Lomberg spent part of his life trying to think like an alien, flash back to long before the discovery of chevrons on Miranda and moments of instant science . . . before twin machines were even christened *Voyager* . . . all the way back to December 1976. MJS'77 machinery was still in testing; JPL was neck deep in the Mars landers from their Viking program, which had already been on red soil for months. During Carl Sagan's Pasadena visit in support of Viking, he bumped into John Casani. Sagan was always welcome. He did good things for planetary exploration—like preaching its benefits to the masses on Johnny Carson's *Tonight Show*. Sagan had a way of translating scientific babble into a common language that garnered widespread appeal.

Quietly, the mustached Casani drew Sagan aside and asked for a bit of help. "We really ought to have something on there," he explained to the Cornell astronomer of his wish for the MJS'77 ships.

Most JPL offspring ended up orbiting the Sun, by and large, or dishearteningly impacting the very planet they'd been sent to investigate. Conversely, these new Grand Tour girls awaited uncommon and exceptional fates. If the missions came off properly, both ships would eventually leave the solar system and drift on through the galaxy. Although distances would become too great to maintain radio contact and electrical power would ultimately give out, the pair would effectively, probably, last forever. Really? Yeah. Since *forever* tended to be quite a long time and since some*thing* some*where* out there might one day intercept a ship, JPL very much wanted to have a sort of extraterrestrial greeting aboard.

Would Sagan be willing to assemble something? Or as Casani put it to him at the time, "Would you wanna give it some thought?"

"Let me put together a proposal," Casani remembers Sagan volunteering. "I'd like to line up a couple of other people to help me on this, and I'll come back to you."

Off the top of his head, Carl Sagan figured on what he termed a "modest extension" of that which was currently riding on a couple of other spacecraft. Before Ames had launched *Pioneers 10* and *11*, Sagan was instrumental in the creation of small metallic plaques—one fastened to each ship. In pictograms, they described our location relative to thirteen pulsars and included a line drawing of people waving hello. From this otherwise satisfying recipe, he could perhaps expand and improve. Take the plaque; sprinkle on some detail of our chemical makeup, some protein structures maybe, or a pinch and a dash of biology; and give a brief stir. Maybe that would do the trick? Quickly, Sagan banged on a few doors, soliciting advice from the likes of physicists and philosophers. To round things out, he also hit up a few prominent science-fiction writers: Isaac Asimov, Robert Heinlein, and Arthur C. Clarke, who are the types to contemplate meeting extraterrestrials.

What kind of message? The air filled with great swirly clouds of ideas and options and then even more ideas. A celestial sphere? A galactic map? The pulsar thing again? What about a string of binary numbers that could be gridded out to make a picture? A spool of magnetic tape with images on it?

One month later, Sagan was in Hawaii for a gathering of astronomers. He shared a cottage with Frank Drake, a buddy of his from Cornell. It was not your average cottage. Located at the Kahala Hilton, this one opened onto a large saltwater pool, serving as home to a happy pair of dolphins. At night the men slept with the windows open, listening to the sleek animals outside happily trolling around their watery quarters. But Sagan wasn't getting the sleep he wanted or needed. He felt continually preoccupied by John Casani's impending flight. More than once, Sagan had confessed to Drake how they should just settle for a repeat of what went on *Pioneers 10* and *11* and call it finished. He was busy enough as it was. No matter what their choice, it would probably get back to some complainer who felt slighted enough by a minor omission to blast editorial cannonballs at the *Omaha World-Herald*. Drake was still holding out for pictures. Sagan rhetorically asked, *Which ones?* Any plaque could only hold a few. What were they supposed to say about Earth and its people in just a few pictures?

One day at the conference, Frank Drake offered a pivotal suggestion: how about a phonograph record? It really did seem like a great idea. A record provided for the obvious inclusion of music and sound. But, when properly translated, images could be encoded into the audio spectrum as well. A phonograph record could actually be made to hold a fair number of pictures—many more than any simple plaque.

Frank Drake scribbled out a tentative photographic rundown, including "DNA," "Times Square with automobile," "Sydney Opera House," "Indian Music." He also tossed in "Milky Way showing Sun" and "Magellanic clouds." Based on what they knew right then, both men figured each picture might consume as much as three-minutes worth of runtime—equaling relatively few images, although certainly more than by any other method. They could put in a dozen or so and still have room for plenty of music. When later tests demonstrated that each picture required only three seconds, the capacity soared to over a hundred images. Earth and humanity could be much better represented.

Records released to stores for consumer purchase were made by pressing blobs of warm vinyl into a mold. However, sending vinyl made little sense. *It* sure wouldn't last forever; the harshness of space would eventually do it in. But Sagan learned how the originating mold was first created from a "mother" of nickel or copper—definitely a sturdier material. Well then, how about sending a mother outright? Phone calls with Casani determined that, unfortunately, a nickel disc might interfere with the shipboard instruments; so the men settled on a copper-mother record that would also be gold plated. At the standard speed of thirty-three and one-third revolutions per minute, a regular twelve-inch disc provides some twenty-seven minutes of playback per side. One of the sides could be all pictures, encoded into the audio spectrum. That left just twenty-seven minutes for music. Jeepers, what could they do with only that much? It wouldn't even hold two movements of a symphony, and that was discounting all the nonmusical snippets they wanted to include: talking, laughing, and so on.

What about *two* records. Could they put in two? Casani shook his head. JPL planned to mount it on MJS'77's ten-sided bus, and the increased thickness of two records would affect the temperature of the electronics. Either they had to come up with something else or confront the reality of including less material.

On the *outside* of the bus? Wouldn't it then just be open to the ravages of space? Offered Casani, "We wanted to locate the record on a place that if someone were to recover the spacecraft, the record would be externally and readily accessible. So that meant the record had to go on one of the outside surfaces." They decided to cover it with a protective shell of titanium.

By this point, Carl Sagan counted less than six months to assemble a cross-sectional essay of human culture, encode the thing, and deliver. He didn't have any contacts in the recording industry. He was a scientist, not Elton John.

He had six months to encapsulate humanity.

Fortunately, Carl Sagan wouldn't have to go it alone. Half a dozen assistants—all volunteers—dropped whatever they were in the middle of to each head up one particular section of the record. And John Casani backhanded a few dollars into their court that he had managed to carve away. "It was project money," Casani explained of his manna from heaven. "We have a little budget to do public outreach—things that will engage the public in the project." This meager funding pushed the idea incrementally closer to being real, and that's when the JPL legal team figuratively cleared their throats. "The lawyers got very interested in it at the time, and their concern was that we would make copies of the record and sell 'em," Casani remembered. "They wanted to avoid any of that." He gave his assurance that it would never happen.

Jon Lomberg, a writer and artist who previously had illustrated Sagan's *The Cosmic Connection*, bravely volunteered to lead up the imaging effort. His sole direction came from Frank Drake's original list of possible subjects and the accompanying instruction, "Your job is to find the photographs and construct diagrams if necessary." No specific images had been chosen—just generalities. And with that flimsy prep, Lomberg was off and running with six weeks till deadline. To clarify, there were six weeks until NASA needed to receive the final disc. In reality, Lomberg enjoyed far less time to actually gather his collection.

Almost immediately and with recurring frequency, Lomberg started playing the alien game with himself. He'd look at candidate pictures and try to wipe the schema from his head, pretending that *this* was the first time he'd laid eyes on what sat before him. What was that creature holding? What was it wearing? What was that off to the side?

The exercise always wore him down. Another bugaboo was how to call attention to specific things in the pictures. One tactic he hit on was the use of silhouettes, presenting two images side by side. One would be the original, and the other would be a high-contrast, black-and-white version outlining the key subject. As Lomberg wrote in his explanation, "It's a way of saying, 'This is what we want you to see in this picture.'"

The final list redefined "eclectic," and it started with arithmetic. Binary notation is equated to numbers, numbers are pieced together into simple math problems, and math is used to delineate scale. "We cannot imagine," Carl Sagan wrote of the approach, "a civilization for which one and one does not equal two." The mathematics lesson gave way to other pictures: Earth from space, cross sections of the human body, fetal gestation, mothers nursing children, school, games, laughter and sadness. One picture tries to portray human error by showing a crawler-tracked snow vehicle stuck in a crevasse. Other shots contained trees, animals, and oceans. There were images of seashells, the Great Wall, a track meet, and children playing. All pictures were consciously slanted toward the uniqueness of Earthly life. Each required copyright releases. One particular shot of a diver among fish had to be left out when Lomberg and the photographer couldn't agree on the royalty payment.

The musical selections on the record are Earth's greatest hits, presented in hopes that they might be appreciated by alien intelligence. But the pictures that Drake and Lomberg chose are not intended as Earth's greatest images. Rather, they are a deliberate attempt to depict our world in a way that ET might understand.

After the pictures had been selected, actually encoding them into an audio sine wave brought to light one small glitch—nobody anywhere made a machine to do that. Every audio engineer had told them it *could* be done, but, heck, JPL? NASA? The Defense Department? *Nobody* had a box that could do it. Although simple in concept, not even the mighty American television networks had created such a beast. But someone managed to find Glenn Southworth at Colorado Video in Boulder. The holder of some ten patents on unique video processing and transmission devices, Southworth had as of late been perfecting his latest creation—born from his wholehearted conclusion that any day now the television and news industries would be in need of a device to send video over phone lines. "Our business is primarily the design and manufacture of video instruments for research laboratories," he once said. "But now and then we come up with a device that is sheer fun!" Southworth loved his work, loved finding new applications for the medium. If a television engineer encountered some weird request, some totally oddball conversion that needed to happen to a piece of video, the best thing to do was get down on all fours and pray for Glenn Southworth.

On his interest in the record project, Southworth commented, "I was delighted. It seemed something like being one of the stone masons chiseling hieroglyphics into the Rosetta stone but with far wider implications!"

Shortly prior to deadline, a large group convened on Central Avenue in Boulder for the transfer session, which lasted almost until midnight. "I can't remember if we had any supper or not," recalled Southworth. Each picture, formatted on a slide, was projected into a high-quality black-and-white video camera. This fed into Southworth's magic box, transmogrifying the image into the audio spectrum. In turn, the resultant audio spooled onto a special data tape recorder specifically loaned to the project by Honeywell. They made four copies. Half walked out with the record team, and Colorado Video sent the others later after the first ones safely touched down.

Most any VCR these days comes out of the box accessorized with instructions in roughly eight languages. The record needed something different. Something universal—like pictures. While the playback stylus would be stashed adjacently, the record itself would wait out eternity underneath its titanium cover. On the face of that cover, the group decided to lay out an engraved progression of simple line drawings—a user manual. Painstakingly, the drawings indicated how the stylus fit over the record, which direction the record turned, how individual scan lines built up to a complete picture, and what that very first picture would be: a circle inside a rectangle, the "calibration frame." Doing his very best to think like an alien, Jon Lomberg counted on the recipient comparing this engraving on the cover with the first reproduced image off the record itself to verify sameness. Then they'd know it was being operated correctly.

Around the edge of the stylus-on-record picture is listed the proper rota-

46. In the foreground, Jon Lomberg attends the launch of *Voyager 1*. Wendy Gradison is to his immediate left, who assisted in gathering the images. (Courtesy of Jon Lomberg)

tion time for correct playback. This and a couple of the other cover drawings include annotations written in binary code. Everyone felt the binary approach translated better; it's based on a system of two instead of our own nonsensical ten. Aliens might not possess ten fingers. But how do you explain sixteen and two-thirds rotations per minute to a civilization unschooled in our measurement of time?

Near the cover's edge is one final engraving. It's a simple representation of hydrogen in its two lowest states, with "|" placed between them. This cue of the momentary shift from one to the other indicates a scale of time. Being the most abundant thing in the universe—there's even more of it than junk mail—hydrogen serves as the common denominator. The amount of time it needs to transition states is constant everywhere in the universe, and here it equals "1." The speed of sixteen and two-thirds rotations per minute is therefore depicted in binary code, which in turn has been derived from the speed of a hydrogen atom. The group figured that was about as fundamental as anyone could get.

One last feature graces the cover. Electroplated on it is a dab of nearly pure uranium with a known decay time. Whoever or whatever finds the ship need only analyze this in order to figure out when this curious machine began life.

Everyone on the record team witnessed both launches in person; all experienced great emotional release at the twin crafts' departure and the completion of such a magnificent gesture to Earth and the universe at large. Hugging, tears, and joy.

The Voyager record is not a time capsule. Time capsules are sealed into a new building's cornerstone or buried to mark the occasion of a town's bicentennial. The intent is for them to be dug up and pried open in a hundred years by the same sorts of Earth-dwelling, festival-organizing human communities who buried them in the first place. Most of the detail inside those boxes could have been found in any nearby library, and without risking dirty fingernails. Time capsules are a localized curiosity, because you can *relate* to what went in.

What *Voyagers 1* and *2* have are entirely different. Embedded within its grooves are attempts to describe ourselves, to demonstrate our capabilities for thought and communication. Along with Lomberg's imagery, people of the world voice their hellos in fifty-four languages; some say "Peace," while others go further. Volcanoes erupt and babies cry and humpback whales sing. These are the sounds of Earth. Bach is aboard, alongside Senegalese drums, an Indian raga, and "Johnny B. Goode." Taken as an entire composition, these records bear warm, respectful greetings from one fervent species living on an unremarkable blip in the western arm of the Milky Way galaxy. Their attitude is enthusiastic, celebratory, gracious, and dear. They are perhaps the finest example of mankind's good intentions.

They are a celebration of the human spirit.

They are our calling card.

They are *we*.

Millions of years will slip away before the record begins to stand even its most remote chances for discovery. Until then, humanity would have to sit down and be patient for a very large amount of time—only twelve modest years had, to date, elapsed since *Voyager 2*'s launch. A dozen down, she was just about to Neptune.

And 930 miles off course.

26. Last Light

They've nailed it right down the middle.

Norm Haynes, to the reporters, August 23, 1989

They came from everywhere to see her off. Late August 1989 showed mighty fine weather to commune with an old, old friend and an even older planet in very new ways.

Here, rolling into town, came Bud Schurmeier with Bettye Jo at his side, although they were not in the orange dune buggy. "I don't think I could have had a more interesting or fun career," Schurmeier offered. "It was always fun to get up and go to work." Now retired, kids grown, the Schurmeiers had moved away from Pasadena long before and started growing avocadoes on their scraggly two acres in Fallbrook, roughly between Anaheim and San Diego. Avocadoes were about the only suitable plant for the local turf. Running the irrigation system costs so much money they sell off most of the pickings to cover the water bills.

"So you're in the avocado business whether you want to be or not," Schurmeier explained, moderately amused by the situation.

Roger Bourke couldn't make it. "I was in another part of JPL working diligently on stuff that was completely unrelated," he said, more than a bit remorseful. "We were at an off-site location. I wish now I'd gone."

Where was John Casani? "Jesus, I don't remember!" he exclaimed. His *Galileo* launch was less than two months away and kept the man fairly preoccupied.

The Blue Fox, unfortunately, was no more. "Whoever owned the property wanted to do something else with it, and they tore the thing down," Bud Schurmeier elaborated. "But it was a great hangout."

Mike Minovitch came up from Los Angeles dressed in his Sunday best,

47. Mike Minovitch, left, shakes hands with Charley Kohlhase during JPL's *Voyager 2* Neptune encounter, August 1989. (Photo by Richard Dowling, courtesy of Michael Minovitch)

consisting of a discreet brown business suit and tie. Near JPL's front entrance, he ran into Charley Kohlhase. The men hadn't seen each other in a while. The last time Kohlhase could recall was when he'd sat in on Minovitch's droning 1963 seminar for the engineers. Both shook hands and momentarily chatted in front of JPL's prominent billboard depicting the current locations and odometer readings of *Voyager 2* and a couple other in-progress missions.

As Minovitch remembers it, this chance encounter between the two carried more meaning than just some inane handshake. "When Charles Kohlhase greeted me at the main entrance to JPL," he said, "he was congratulating me for the *invention* of gravity-propelled space travel."

The big three-year coast was just about over. For the last nine days, the old girl had steadily crept up on her final target, while the JPL people had steadily been sleeping less and less. Brigades of reporters crammed themselves into every corner of von Karman Auditorium. So many people wanted in that JPL security had to crack down on a syndicate of bogus Neptune encounter press passes in circulation.

Up on von Karman's stage that Monday, August 21, Ed Stone welcomed the reporters and the crowds. "I suspect many of you are here for your sixth and final Voyager encounter," he earnestly began, then changing gears to wax a bit sentimental: "In a certain sense, this is the final movement of the Voyager symphony of the outer planets, which began with Jupiter ten years ago. And I think in true symphonic form, the tempo appears to be accelerated as we approach encounter." He could only smile in satisfaction at the breadth of the crowd before him, as well as at his splendidly moving analogy.

Four hours by radio and 2.5 billion miles away, *Voyager 2*'s twin cameras whirred to life, handily demonstrating that our solar system wasn't totally done serving up the surprises. Long-observed Neptunian ring arcs progressively resolved into—*click*—extraordinary detail that was expected to depict extremely fine yet complete rings. How fine? *Click*. Like particles of smoke. Four arcs ultimately closed their gaps, but the stubborn refusal of some to fill out only complicated this planet's already baffling qualities. One pip-squeak ring arc was—*click*—summarily unmasked as a completely new moon. Another ring arc never did make it all the way around. Years later, tiny, unforeseen moons would be discovered, their presence gravitationally limiting the ring material.

And Neptune itself? Instead of a frozen, carbonized world, the planet eerily resembled Earth, down to the rich blues and hazy white clouds. "It looks homey," opined one JPL worker. Those blue tints owe themselves to all the abundantly flowing methane revealed to be present.

Hard numbers kept rolling in even with *Voyager 2* still days away from closest encounter. They'd picked up some winds blowing down there at four hundred miles an hour. Everyone was keen for a closer look at the odd anomaly christened the Great Dark Spot, noticed months before as a kind of sister feature to Jupiter's own Red Spot. Was this another storm like on Jupiter?

Invited back for the last big show, Glenn Southworth once again selflessly configured his specialized video equipment to disseminate imagery the moment it came in. Back at the time of the Uranus encounter, he'd managed to get in on that action, too. Completely at his own expense, Southworth had installed his custom equipment at five locations, including MIT and the Smithsonian's Air and Space Museum. As pictures came through JPL, Colorado Video encoded and retransmitted them over special phone lines to the five, who received the images for display. MIT went so far as to set up a round-the-clock screening room for students to drop in and check out the feed. As a television geek, Southworth could only chuckle to himself. "Ten thousand people far from Pasadena were able to see the Uranus images at the same moment as the scientists at JPL," he gaily explained. So this prior success only led to a mushrooming of demand. For Neptune, AT&T collaborated with Colorado Video in setting up a phone bank of special 1-900 lines. Anybody in the world using Southworth's equipment could dial in during the encounter to access the images in real time.

"Colorado Video picked up the tab for both encounters," he mentioned of the pictures-on-demand setup. "We didn't make any money, but we gained a lot of friends."

Thirteen-year-old Troy eagerly walked into the JPL Space Flight Operations Facility and took a look around. Then his father ambled up beside him to begin sharing the cavalcade of encounter. "My dad asked me about six months in advance if I wanted to go," Troy said years later. "Needless to say, my response was a very enthusiastic 'Yes!'"

They dipped in and out of von Karman to catch the press conferences and got a bite in the cafeteria. Troy had an opportunity to finger the spare Voyager ship. They made it to the JPL gift shop, which enjoyed a bustling business of selling patches, pins, mugs, and pretty much anything else the Voyager logo could be screened on. Troy got a keychain.

"Excitement about Neptune was so thick, it was almost tangible. Journalists and reporters were everywhere." It was all the lad could do to soak it in.

Bud Schurmeier greatly enjoyed the festivities. "They had it nice and organized, so that you weren't just sitting there doing nothing," he indicated. "You could have something to eat, or you could talk with people you hadn't seen for a long time."

Norm Haynes told a story of himself laughing at a dusty, twenty-year-old astronomy book he had on the shelf in his office. Mostly it was filled with all kinds of detail about Venus and Mars, and it had a bunch of photographs of them taken through big telescopes on the ground. As the book got into Jupiter and Saturn, the particulars dribbled off at an exponentially rapid pace. Then by the time it got to Uranus and Neptune, the facts went off a cliff. Both planets were covered in about three paragraphs discussing how far they orbited from the Sun and how long those orbits took. That was it. "We didn't know anything about them," he said.

Norman R. Haynes stood in charge on that August day. He'd been at JPL since 1959, back in the cro-magnon dark ages of *Pioneer 4* and rockets that blew up most of the time. How things had changed. Cutting his teeth on programs like Mariner, Haynes gradually ascended through the Lab's hierarchy. He labored over plans for Galileo until Dick Laeser got pulled from Voyager late in 1986 and opened up the spot. Haynes got the offer.

"Wonderful!" he proclaimed.

Soon Haynes assumed responsibility for the aging machine nearly 3 billion miles away. Unlike Ray Heacock or Esker Davis, he'd never worked as a Voyager underling in any capacity; his attention and energy had always been someplace else, on other projects. And so in a very real way, Norm Haynes was altogether unacquainted with the spacecraft, its evolution, and its team. But he inherited that team—the well-oiled machinery of people, now down to a few hundred, who had long mastered the pulleys and levers of *Voyager 2* puppetry.

His first task?

"I realized right away that the best thing for me to do was to not screw up a good thing."

One item with the potential to turn into a giant rat tail had been the main flight computers. A gaggle of engineers had rushed Haynes with the idea of switching over to *Voyager 2*'s backups and lobotomizing the primary computers' memory to insure a pleasing freshness for Neptune. Just as many people came back and told Haynes how stupid that would really be. For God's sake, they pleaded, the main computers worked fine. The backups were only used in testing. What if there was some problem with them that nobody knew of after twelve years of flying? What if they got halfway through reprogramming and the things died?

"The few areas where I really had to make some fundamental decisions were almost always related to how much risk we wanted to take," explained

Haynes. "Somebody had to make those choices." And in this scenario of ping-ponging between computers, he considered the overall risk and ended up deciding not to change a thing.

Making his way through the oddly numbered Lab buildings ambled a silvering Gary Flandro, thoroughly enjoying all the festivities. Although never running into Minovitch, he did get to chatting some with former coworkers about those long-distant epochs and the mission's formative accretion. The many hordes of press reporters crawling the joint were generally unaware of the flight's true genesis. Flandro was only too happy to mention that seminal 1965 day with him and two sheets of graph paper.

To one reporter he said, "I must tell you, I expected the opportunity to be wasted."

Radio link? *Check.* Spacecraft health? *Check.* Strutting, humming with life, almost smiling, *Voyager 2* glided perfectly toward Neptune. Right on the money.

Of course, it didn't exactly line up that way by chance. News articles over the years have characterized the overall *Voyager 2* flight as, in one example, comparable to the sinking of a two thousand–mile golf putt. And yeah, that's not far off the mark. Either is a pretty long way to go with such accuracy. But that golf ball doesn't carry little thrusting rockets aboard to alter course; it is affected by only one field of gravity, and it sure doesn't have buildings full of really smart people meticulously plotting and goosing its position along the way. Oh, and on top of all that, the cup would of course have to be *moving*.

Voyager 2's Neptunian advantage took the form of the pack of disheveled engineers who shuffled into Room 461-B a week before this final encounter—right up there on the second floor of Building 264. Ed Stone was there with Norm Haynes and some others—and here came Ellis Miner, voraciously chewing through a defenseless apple. They assembled in that room at nine at night to evaluate the ship's position and assess whether or not she required one final nudge. On the far wall at one end of the room, a trio of monitors quietly displayed Neptune's beauty in a series of staccato images, updating almost as quickly as they splashed down into the receiver dishes.

Engineers prefer to not deal in uncertainties; with regard to the ship itself, *Voyager 2*'s whereabouts were quite certainly known. Speed, for example, is determined to within a millimeter per second by measuring the radio signal's Doppler shift. Timing those same radio signals marks her distance to within forty feet or so—not bad at all for a 3 billion–mile excursion.

But at nine o'clock in Room 461-B, the issue was this: however certain they all were of *Voyager 2*'s speed and position, her approaching *target's* exact location—and for that matter, its true size—was the subject of a near-continuous series of evaluations, the guesses improving slightly over time. Nothing had ever visited Neptune before. Imaging analysts were leaping on *Voyager 2*'s pictures as soon as they arrived, grab and tackle, poring over them Morabito-style, yanking the brightness and contrast this way and that to better establish Neptune relative to the unchanging background of stars. Up in Navigation, they refer to this as the location "settling down." Six months before, the Neptune error was 930 miles, which one of the group described as "alarming." Neptune wasn't *precisely* where they thought it was.

Then the following Monday, around the same time Ed Stone got up in von Karman to welcome the reporters back, *Voyager 2* received and executed a small, agreed-upon maneuver to gently adjust her position. "We really diddled around for awhile," confessed one of the men at that meeting in 461-B of the caustic uncertainties as they debated how best to fine-tune the ride. It represented the last time anyone would, in a manner of speaking, turn her steering wheel.

Heading right down the middle of the road, Neptunian gravity took over, pulling her down, math equations spinning through the arcs of her descent, prancing and dancing and leading the way. It was a perfect approach. Carefully and brilliantly, Navigation had delivered Ed Stone his magical one-second leeway, the ultimate Hail Mary that scooped Neptune and Triton both—just like they said they would. To finalize that one decision alone took over a year.

Again, data began flooding in—big-picture stuff. Neptune dizzily rotates almost twice as fast as Earth. And those zippy winds? Forget about it. The highest natural wind speed ever recorded on Earth measured 231 miles per hour. Even this was nothing more than a brief gust on Neptune. Nobody was prepared to hear of calculated wind speeds reaching a full *1,200 miles an hour and more*—hell for bicyclists yet heaven for windsurfers. Only slightly off pace roiled Neptune's cyclonic and peculiarly enigmatic Great Dark Spot, blasting along at 760 miles an hour, able to wend around the planet in eighteen brisk hours. What begets a Great Dark Spot? Nothing less than a raging, pulsing storm—varying its shape and whirling off minor squalls as it goes.

Gary Flandro ran into Charley Kohlhase; the two hadn't spoken in a long time. Heartily shaking hands, Flandro turned and introduced his son, Troy. *This is one of the men who got us to Neptune,* explained Gary.

"He is a fascinating scientist," the younger Flandro has said of Kohlhase. "It was an honor to meet a man who made such a wonderful contribution to the Voyager missions." He was indeed a man who helped it persevere through the years.

And the generations.

Gathered in von Karman, Ed Stone and Brad Smith and some of the others groggily held themselves together on stage during the latest press conference. Bedraggled and stubbly and telling of Neptunian revelations, they described their findings to the undulating sea before them of journalists silently jotting notes on pads of paper, or recording the proceedings on video, or dictating the new story via telephone. Out there at the solar system's frozen edge, *Voyager 2* had the world on a leash, and she wasn't even in high gear yet. More distance remained to close on her quarry.

They opened up the floor, media shouting their questions. *Tell us again, how many words of memory are there on the spacecraft?*

"Well, there's sixteen thousand words of memory," someone up on stage replied.

All the reporters laughed. "Gosh, that's the funniest thing," they said. And it *was* a funny thing. By 1989 even a floppy disc held more.

Another question—*How are the Imaging Team members responding to the pictures?*

"Literally jumping up and down," Brad Smith responded.

Are you ready for the closest encounter?

Norman Haynes took that one. "We sent up the near-encounter command load six times," he explained. "And the spacecraft received it six times. We will send up a few tweaks later today."

Questions kept coming. There's no way any of them up there on the podium could even begin to have all the answers. They were beyond saturation with raw data pouring down on them like it was, hard rain changing to sleet. And tomorrow, Friday, August 25, was to be the enthralling day of closest approach.

"This is it," Ed Stone concluded, wrapping the day's press conference. "This is the twenty-four hours we've been waiting for." He stood there, hands on the now-weathered podium, outlined by television lights heating up his eyeglass frames, gazing out over the commingled wash of faces and expressions and clothes rumpled much like his own. Those reporters out there—they wanted the facts. But they also liked Ed Stone and anxiously hovered in absolute silence, writing utensils poised, waiting for him to offer something profound.

Ed finally told them, "I like to compare *Voyager* to a highly trained athlete, performing right up to its limits and not beyond.

"We couldn't be happier."

Two new moon discoveries came in just before midnight on the twenty-fourth. *Voyager 2* was now within slapping distance of Neptune, dropping down on final approach, almost at the treetop level, catching the edges of Triton on the sensors. Then she snapped a few pictures of it. Just a shade over four hours later the signals reached Earth, plucked from the cosmic chaff by an eighty-meter dish and zapped on to Pasadena.

All around the country, JPL satellite feeds pumped the complete unedited saga—pictures and press conferences and the whole ball of wax—into planetariums and schools and pretty much everyplace else that wanted them. A guy in Texas invited the public to his own personal downlink and packed the joint to standing room only. People brought their children. In a decidedly pre-Internet era, cable television's USA Network offered a special *Neptune All Night* program that provided live imagery backed up with interviews and analysis. It was hosted by JPL scientist Rich Terrile, who'd labored untold hours over Neptune's ring arcs in his attempts to close them by locating previously unseen ring material. Someone drew up a cartoon of him holding what looked to be a hula hoop with one part of it missing. Standing inside it, Terrile was saying, "Hi. I'm Rich, but I'm still trying to make ends meet."

That all was just inside the United States. Glenn Southworth's equipment worked perfectly in near-instantaneous *worldwide* distribution of the images to similarly organized parties round the globe. At one point, someone on the JPL campus accidentally hung up the special phone line and cut everyone off in the blink of an eye. Collective sighs arose from Grenoble, France, and the one thousand space buffs assembled before television monitors jacked into the feed. Missing nary a beat, Southworth tap-danced through the blackout by repeating some of the best pictures and at one point tossed in one of his cat. "The next day, I received a magnificent bouquet of flowers from Grenoble," he said.

Already, Triton shaped up to be another great brainteaser for deduction, a cosmic Rubik's Cube for the ages. The pictures weren't yet complete, showing only snippets of color in little swatches. The details yielded themselves in agonizingly slow fashion as *Voyager 2* continued her plunge. Nobody knew yet what to make of the rugged terrain and varied colors. And there were all these wispy, little brushlike marks everywhere. "Triton will upstage Neptune," came one prediction.

Someone else offered up, "It's the last time we're seeing a planet for the first time. We've completed the reconnaissance of the planets, except Pluto."

It was one o'clock in the morning.

It was time for the end.

She went barreling on through the ring plane first, going sixty thousand miles an hour and then sixty-one thousand, peppered by debris at three hundred hits a second on the sensors, nattily buzzing Neptune proper just twenty-one breathy miles up. Heck, people in balloons have ascended more than twenty-one miles above Earth.

Bull's-eye.

Then she went down behind the planet, moving, shooting the rapids, looking, learning. Crisp photos, nice and razor-sharp, resulting from *fifteen-minute-long exposures* that combined in perfect synchronicity with thruster toots and recorder jiggles and the delicate twistings of the spacecraft. Euphoria, delight, accomplishment, the sound of champagne corks popping— all on a four-hour delay. It was one o'clock in the morning on Friday, August 25, 1989. Anything past Saturn was supposed to be gravy. But deep down, every single person on Voyager had well understood all along how true success would only be experienced by picking up Uranus and Neptune, the impossible cosmic seven-ten split. That would be the only way.

Congress be damned. The budget be damned. Only Voyager mattered. And they'd done it, bucking the funding issues and the politics and Jim Fletcher and that damnable Space Shuttle and the Russians and the radiation and just about everything else that threatened them on practically a daily basis. Damn them all. They would complete the 2.8 billion miles after twelve years of flight and twenty-four years of evolution. They'd succeeded through the generations; Troy Flandro was proof. Four planets, one determined spacecraft, eight project managers, and tens of thousands of women and men who contributed to *Voyager 2*'s success—all were symbolically riding aboard, with the remainder of Earth's population nicely encapsulated on one twelve-inch circle. In a sense, we were all there, leaving a piece of ourselves to the cosmos. It was all on her, the logical extension of Mariner, of our minds, the heir to the throne, the queen for the ages.

Lounging in front of a monitor that night, James Van Allen could only chuckle when arriving data indicated the existence of a radiation belt around Neptune.

Plenty of details were still en route from the ship, arriving hours later as sunlight pushed the horizon, trash cans overflowed, and pizza and gritty government coffee streamed through the Lab's arterial hallways and offices nearly as fast as the incoming data. Whipping past Triton, *Voyager 2* athletically wound her scan platform on command to backhandedly collect imagery of this poorly understood moon.

Triton gob smacked everyone. "It's in the wrong orbit," stuttered Brad Smith up at the von Karman podium to the zillion reporters. "It's going backwards around Neptune." Like Uranus, Triton had taken a hit sometime long ago. And most of the moon's features seemed understandably icy—definitely not rock, but rock-hard *ice* measuring just a few dozen degrees above absolute zero. Then came the real kicker: those wispy, little brushlike marks were every bit alive as Io. Close, painstaking analysis tentatively identified them as *liquid geysers* erupting from the polar caps. It didn't make sense and it didn't end there. Triton also contained lengthy fault lines *oozing ice*, along with gigantic basins filled with *ice lava*. It was upending, mind-bending stuff right out of the comic books. "This is a crazy idea!" bellowed Larry Soderblom as he manhandled the podium. "It's probably wrong, but it's the best we have at a current time." He was exhausted; his eyeballs hurt.

One of Voyager's crew later noted something remarkable about that day. He said that at midnight nobody knew anything about Triton, but by five o'clock the next morning the world had better imagery of it than the guy's own home town. "My God," he remarked, "to go out with that planet and such a moon!"

And *Voyager 2* withdrew, receding, waving over her shoulder, beginning the slow exit. The train was leaving the station, *hiss-hiss*, as she blew kisses in a final goodbye. They only cheered her on, applauding, whistling, shouting her name, urging her like they would an underdog runner winning the Olympic marathon. Emotion overcoming exhaustion, fists pumping, tears flowing, and applause, applause, applause cascaded into waves— *waves*! Ringing . . . fading deep into the echoes of the solar system, as she proudly marched off into nothing.

27. Continuum

Near perfection was marginally adequate.

Esker Davis on the Voyager Philosophy

In mid-December of 2004 *Voyager 1* hit the termination shock after reaching 8.7 billion miles from home. This is where Konstantin Gringauz's million-mile-an-hour solar wind abruptly dawdles. The shock's presence was hypothesized but never encountered until right then.

The solar wind blows a bubble, as it were, around our planetary system—the pressure of which is basically holding out material from the rest of the universe. You live inside it. Every star creates its own bubble, and nobody knows for sure how big they get. It really depends on the star. Where the wind begins to slow denotes this termination shock. Going *through* the shock isn't like popping the bubble or like going through a door. The transition is deep, progress gradual—sort of like a movie with a long dissolve to change scenes. This scene-changing region is what's known as the heliopause, and once you're through it, there's flat-out empty deadheading all the way to the next star system. By June 2005 *Voyager 1* was calmly describing what it's like to be neck deep in this environment as she cruises ever onward and outward, heading for the vast nothingness of true interstellar space to chart the uncharted oceans. And JPL is in touch with both ships every day. A nearly indescribable number of years from now, in approximately 40,176 AD, one of the *Voyagers* will have traveled closer to a star other than our own. And by 296,000 AD, *Voyager 2* will make its closest approach to Sirius.

So many are unfamiliar with Voyager's "Solar System Family Portrait," JPL's "Picture of the Century," captured on that magical Valentine's Day in 1990. In comparison to something like Miranda or Triton, the "Family Portrait" is abstract and uninspired—especially if you can't accept how all

48. As the Family Portrait comes together behind them, Candy Hansen makes an important point to Ed Stone. On the right is JPL scientist Torrence Johnson. (Photo by and courtesy of Anita Sohus)

of humanity is hunkered on that flyspeck of a dot. But that doesn't hamper the image's impact. "For many years we had the mosaic of the solar system put up on the wall in the von Karman Auditorium," explained Candy Hansen. "Every year they would have to replace the picture with Planet Earth, because people just couldn't resist touching it."

Each Voyager ship has worked better than anyone reckoned. Surely without intending to, JPL proved that restricting a spacecraft design, in the end, actually makes good sense. Mission parameters require careful, precise definition, not open-ended catchalls leading to a watering down of the device's original intent. That only makes planning balloon into obesity as designers hedge against every imaginable contingency. Van Allen called it "Engineers Gone Wild"—presumably without all the nudity. A perfect example of this is Ranger. Standardizing on a modular frame and bus was a smashingly good idea. But then what happened? As soon as NASA and the Defense Department realized that ship could hold pretty much anything, the bitter bickering began with regard to exactly what Ranger was supposed to *do*. Before long the entire program was awash in complexity and perplexity as people were summarily removed from their jobs for trying to stick with the original plan.

"It was a huge bite we were trying to bite off," said Bud Schurmeier, speaking in yeah-wasn't-that-crazy tones of the initial TOPS-based Grand Tour proposal. "Huge amount of new technology." In the end, JPL—and science—actually benefited from NASA's rejection of the titanic project. "I've said that many times," agreed Schurmeier. "Man, we were bitin' off a huge chunk, and we were probably, as I said, saved from having to swallow that by the cancellation of the original Grand Tour."

Gary Flandro specifically notes the Lab's ability to continually exceed expectations when denied money or otherwise tethered in some fashion. "I think JPL always does better when it's kinda struggling," he opined. Back in the sixties and seventies, money evaporated like steam into manned flights. The resulting drought threatened to shutter the oddly numbered collection of buildings in Pasadena and send everyone packing. "They had to fight tooth and nail every step of the way just to keep their funding," Flandro explained of JPL's low-man stature. "Cause the manned program was just eating it alive."

Ironically, then, robotic space exploration demonstrates a superior return of information as compared to manned flight. "I am biased of course," began Flandro, "but standing back and looking at what's been achieved, and trying to put some kind of a value on it, I'd have to say that we gained much more from the unmanned program than we have from the manned program. In terms of just about everything you can imagine."

Soviet physicist Roald Sagdeev agreed wholeheartedly. Back in the seventies, leaders of the Soviet Union were all about squishing research into their prestigious cosmonaut flights. As the former head of the Soviet Space Institute, an overwhelmed Sagdeev burned endless hours trying to reconcile a respectable scientific agenda with the flight crew's needs. And *people*, quite understandably, always came first. "The itinerary of the crew in space is completely controlled by the need to survive," he pointed out, which is not always obvious to politicos. A few of the science packages could not be left on while the crew was sleeping, eradicating 30 percent of the mission right there. Experiments requiring high voltages or dangerous substances were out. The crew needed more breaks to eat, to update their journals, and to take pictures.

Finally Sagdeev concluded that experiments could be operating only about 2 percent of the total flight time. As he later wrote, "To keep the 'right stuff' in orbit was an extremely inefficient way to do science."

Voyager, Luna, Mariner, Sputnik—none of the ships in these programs cared whether it was Miller time. Obediently they attempted whatever was asked of them. James Van Allen stated, "There's been an enormous advance in sophistication of scientific instruments, telemetry capabilities, ruggedness, and the capability of all our scientific sensors. It's absurd to think of trying to do these things with human beings." And what of America's plan to return people to the Moon and then hopscotch on to Mars? Van Allen offered, "I think it's just pure insanity myself."

His own graduate student concurred. As George Ludwig put it, NASA offers "as one of the justifications for the manned program, all of the great science that they're going to do. And yet if you look through the published papers, very little of it comes from the manned program."

Suggested Van Allen, "The repair and the upgrading of the Hubble telescope is probably a bright spot in the history of human tending of spacecraft." Beyond that, he steadfastly maintained a belief that crewed spaceflight is essentially obsolete.

"I'm not too sure where the space business is going," admitted Dick Laeser. Without question, the process has definitely changed, starting at the highest levels and dripping all the way down into the teeny cracks of graduate students and university machine shops. "It was a very freewheeling society," Van Allen smiled of the good ol' days. Way back when, a lot of things were done more on a handshake basis. Entire programs were decided over a single phone call. It was true for manned flight as well; Abe Silverstein birthed the entire Mercury program with a scribbled memo that read, "OK— Abe." But today a penny-wise NASA wants their programs and projects all on spreadsheets with no cell left unfilled. They're supposed to be budgeted right down to the Post-it Notes. There's too much on the line anymore. Funds are tight, and the potentially embarrassing details of their use hang before the public like soiled underpants. Politicians want *quantitative* scientific return for their expenditure, as if the researchers are clairvoyantly able to divine their final results, as if more money equals more science. They want *qualitative* deliverables for an irreversibly jaded public—like the glamorous imagery of Martian ice to splash on the covers of magazines and JPL bookmarks for the kids.

Such increased demands along with high-level accountability have created a very different environment for those frontline troops still hunkered in the trenches. Bud Schurmeier seemingly harbored no regrets over his decision to leave. "When I talk to guys up there," he mentioned of today's JPL, "it isn't as much fun as it used to be, they say." So what *exactly* is different? "The bureaucracy sets in. There are much more requirements and procedures and things. We didn't have any of those when we started."

The sales pitches are entirely different. Van Allen used to follow his own thread of research, based on nothing more than personal interest. He'd lift details from his over-thumbed binder of problems and then crank a few pages through his typewriter to send off to Washington DC. "All of us were doing our research with a great deal of independence," said George Ludwig. "We each had our own projects."

Today even this banal course of action is comparatively bloated and ponderous. "Most proposals these days have multiple authorship and multiple institutional participation," Van Allen began, describing the approach NASA currently finds to be attractive. "A particular proposal might have six different institutions at universities and research centers participating in maybe six or ten investigations. It gets to be so the list of investigators is as long as our proposal!" He laughed, knocking the table in punctuation, but the shape of his eyes told a weary tale.

"There's so many more people in the field these days."

They begin the long string of phone discussions, conference calls, and face-to-face meetings to hammer out the details of who's doing what. It's design by committee, often reaching over several time zones and languages. Considerable effort goes into merely slicing up the labor requirements. As Van Allen explained it, "Mr. X is going to do this chapter, and Y is going to do the part that has to do with neutrons or something. And the principal investigator tries to blend all this together into a coherent proposal." His chin dropped—a funny taste suddenly in his mouth. Clearly, Van Allen didn't like such an environment, but he could only shake his head and concede, *what else are we supposed to do?*

The scientists now assemble in Washington DC for selection committee meetings, where they make their emotionally invested presentations. They play carnival barker. Then, they all go home and wait months. Sometimes word comes back that NASA has decided to cancel the mission or prescribe some nonsensical delay. "That happens quite frequently, mind you, these days," Van Allen cautioned. Far too often, the originally proposed mission is judged overambitious for the budget. "So everybody's sort of back on his heels, so to speak." Tension drags along for months, perhaps years. Some iteration of the project *may* be resuscitated—with enough advocacy. The experimenters start all over, repeating the whole process. Again and again, scientists are forced to play Dr. Frankenstein with their experiment's purpose, budget, and instruments. These are not people who enjoy working the phones and selling themselves. They want to *learn*.

Approval finally comes through, but these individuals step into a landscape far removed from what it was. A spacecraft used to go from drawings to launch within a few years, but they were simple. The entire flight transpired over a comprehendible amount of time—in the case of Ranger, sixty-six quick hours elapsed between liftoff and impact. Some people working the mission didn't even bother with sleep.

Complexity has swollen in a form commensurate with the mission itself. Now look at a protracted flight like that of the *Cassini* probe, as big as a minivan, launched toward Saturn in 1997. Simply commuting to the job site required seven whole years. In mid-2009 the craft began its fifth year of in situ operations, projecting a total lifespan of roughly two more years beyond that. And when was it born? *Cassini*'s origin traces back to at least 1988. If a couple were to have had a daughter on the day blueprinting started, they would have attended her kindergarten orientation while Cassini was still being built. Launch would have happened amid third-grade book reports and field trips. She would have gotten asked to junior prom around the time of Saturn approach and will graduate college before the program even ends.

John Casani's *Galileo*—the comparably sized Jupiter-bound ship—required $300,000 worth of custom scaffolding to surround it during assembly and testing. Closer in size to Volkswagen Beetles and proportionately less capable, the Voyagers' only cost $80,000. And *Explorer 1* came together on a secondhand lab bench. But that's the price of progress.

Galileo was downright huge. Seventeen feet tall and weighing over five thousand pounds, the ship carried eighteen experiments dispersed throughout its humongous girth and was topped by a collapsible wire-mesh antenna. Its computer program gobbled over 2 million lines of code. Part of the super-structure actively rotated to gather data from all angles, while the remainder of the ship held steady for cameras and the other experiments. In sum, this was the kind of machine that could *not* be built by graduate students.

"Around here we used to have *lots* of graduate students, even *undergraduate* students, building the equipment," Van Allen said, waving his arm to indicate the greater Iowa campus. "You know, actually *building* the equipment. And debugging it. And you know, sitting in front of an oscilloscope and seeing how the pulses are coming through right." This modus operandi is no longer possible. Most universities now engaged in space research maintain squadrons of professional engineers on staff. The students themselves have practically nothing to do with the actual construction. "Our graduate students these days here are working with the data from missions they never saw the instrument for." A light smile broke across Van Allen's face. "Not even *saw* the instrument, much less had anything to do with development. Because the instrument was built before they even were in graduate school! And launched!" He gargled a laugh and slapped his meaty hands together.

Arguably, *Galileo* and *Cassini* could be about the most convoluted spacecraft that will ever be fashioned. So why all this complexity and rigmarole? Why make spaceships with spinning halves and nonspinning halves? Why can't they just make 'em like they used to?

Quite simply, we've learned all the easy stuff. Only a handful of decades ago, next to nothing was understood about the environment above Earth's atmosphere. Nobody knew anything about radiation or magnetic fields. "We were in the exploratory phase," Van Allen remarked. People like him could blaze through the asteroid belt on four-page proposals because they didn't need much else. "To investigate the environment," the broadly worded document could read.

But today the big bullet points have already been identified and measured and properly tallied up. We know which planets have rings, which ones have magnetic fields, which moons have atmospheres. We know what the atmospheres are made of and which the most promising candidates are for harboring primitive life. These prior discoveries mean investigative energy must now delicately tune in to finer details. And as such, the payloads are now more complicated, larger, and more expensive.

"I really feel that I'm pretty well outstripped by the new missions and the new instruments, so I have trouble thinking of anything I can really contribute." And Van Allen let loose a huge sigh, elbow on the table, resting his chin in a palm. "Life is a lot more sophisticated these days."

From an apartment south of Griffith Park in Los Angeles, surrounded with books, Michael A. Minovitch continues a tireless pursuit of the credit he feels to be long overdue. That NASA Exceptional Scientific Achievement Medal is out there somewhere, suspended in the ether, containing an elusive, magical citation for "inventing gravity propulsion" with a blank space where the name should go. More than anything, Minovitch wants *his* name on it yet figures JPL already has cooked up a scheme for this very important sheet of paper. "They plan to fraudulently give the award to some other person after I die with some blatantly fraudulent explanation. That is what the evidence shows. They can hardly wait to do it."

In 1994 Minovitch sighted in on Richard Battin of MIT, responding to what he perceived as Battin's attempt at harvesting credit for gravity assist. Easily a major contributor to the field of astrodynamics, Battin had taken issue with Minovitch's long-winded presentations to the International Astronautical Federation (IAF) conferences some four years before. There, in dozens of pages of extremely small print, Minovitch laid out his own history of "inventing" gravity assist, along with a person-by-person refutation of basically everyone else's contributions. Battin's 1959 efforts to plan a Mars trip received slight disrespect.

But throughout all those pages the would-be gravity king never mentioned MIT's work on flights to more than one planet. See, the prestigious MIT Instrumentation Lab had apparently conducted some gravity-assist work of its own *back in 1958*, enabling Battin to publish only one year later. In a 1959 paper, the MIT professor explored "propulsion-free round trips to the planet Mars." And over the course of many detailed pages, Battin offered mathematical justification for his curious voyage while specifically noting that "a digital computer program has been prepared that mechanizes the preceding computational procedure." Of course, Battin could never have co-opted a program Minovitch wouldn't spawn for another two years. And funny thing, Minovitch cited this source in many of his original IAF abnegations, but he didn't mention the program. Later in January 1961 MIT proposed a combination Venus–Mars journey, trumping Minovitch's holy grail paper by a firm seven months. Battin never laid claim to "inventing" gravity assist—he just applied some math to the concept and reported his findings. "The vehicle will," Battin mentioned in that January work, "receive from the Venusian gravity field alone a velocity impulse, sending it in the direction of Mars." Sound familiar?

None of this two-planet stuff came to Minovitch's attention until Battin presented a minor retrospective in 1994 and formally published in 1996, calling attention to MIT's omission from the Minovitchian historical perspective. That wasn't the basis for the whole paper, but it did appear near the end. Minovitch didn't care much for the heat. And when he failed to win the Nobel Prize around the same time, Minovitch put two and two together and sued the hell out of Battin. Fully seven counts—libel, slander of title, deceit, and others—were asserted by Minovitch's lawyer and summarily dismissed in 2005 by the Massachusetts Superior Court. Closing remarks in the court's decision contained this: "It is strongly arguable that the type of dispute relating to scientific discovery between competing academics rests ultimately within the judgment of the community of their peers."

And the peers have spoken.

Despite pushing well into retirement age, Gary Flandro remains employed at the University of Tennessee Space Institute, occupying the Boling Chair of Excellence in Space Propulsion. "I really look forward to going to work and working with students and doing scientific and technical work. I just love it!" he enthusiastically declared, like every day was Christmas. "Teaching two classes every term at least, maybe sometimes even three. And yet a full load of research and the whole bit!" We all have our hot buttons.

NASA saw fit to bestow its Exceptional Achievement Medal on Flandro in 1998 for, according to their citation, "Seminal contributions to the design and engineering of missions, including the Grand Tour opportunity for the epic *Voyager* explorations." Who nominated him? Charley Kohlhase.

Unofficially retired, Kohlhase enjoys taking his camera to photograph wildlife in Alaska, Yellowstone, and anyplace else that appeals to him. As of 2009 he still consulted for JPL about two days a week, primarily focusing on Mars efforts but keeping his fingers sticky with half a dozen projects. At Kohlhase's 1998 JPL retirement party—if it can really be called that—the host whipped out a little tape machine and hit Play. Carl Sagan died in 1996,

49. Three Voyager teammates reunite at the mission's thirtieth anniversary, September 2007. *From left*: Robert Cesarone, Candy Hansen, and Charley Kohlhase. (Photo by Ralph Roncoli, courtesy of Charles Kohlhase)

but his voice suddenly, pleasingly rang through the room. From the afterlife, Sagan implored, "CHARLEY KOHLHASE'S NAME SHOULD BE AS WELL KNOWN AS MICHAEL JORDAN'S!"

"Something my children will always remember," said Charley Kohlhase.

Like Flandro and Schurmeier, former NASA tyrant Oran Nicks was also a sailplane pilot. In 1998 he crashed a sailplane of his own design and construction and did not recover. Despite Nicks's confrontations with Jim Burke during Ranger, the two maintained professional respect through the later years and managed to find common ground—Burke also piloted sailplanes.

"I have often marveled at my good fortune in surviving the problems with Ranger and remaining as program director," Nicks remarked years before he died. "My vulnerability as 'coach' of a losing team that was so much in the spotlight at the time had made me continually aware of the debt I owed to colleagues like Bill [Cunningham] and Bud [Schurmeier] and to my supportive superiors, particularly Ed Cortright."

Jim Burke offered, "Nicks, if at times not lovable, was never unprofessional." The two had flown together and broken bread together. Their relationship matured. "The Ranger experience led him and me, later in our careers, to achieve a friendship of the kind only reached through adversity." Burke even retains good words for things like Vela Hotel, which did so much to frustrate him back then. "Today we are witnessing a huge payoff from that work. As a result of the Vela Hotel instrument developments, a whole new field of astronomy has opened with the discovery of gamma-ray bursts in the cosmos. At the Moon, the discovery of excess hydrogen near the poles has set off much excited discussion and mission planning to see if there may be deposits of ices in polar cold traps." And now he refers to the Los Alamos people as "heroes." Things change indeed.

In retirement Jim Burke has spent an inordinate amount of time attempting to clean his garage. "Once," he recalled, "a well-meaning friend invited us to one of those find-yourself retreats." The guru asked each of them to confess their deepest desires. Burke answered, "To clean out the garage," and the guru was insulted.

"No, dammit," Burke shot back at him, "you asked and I answered seriously. *Seriously*, that is my deepest desire." But the guru wasn't buying it.

"So much for self-discovery."

The Ludwig family is alive and well. Still fiercely in love with Rosalie, George and she built their stable to five children. Through the eighties at least, the family continued to make a tradition of summer camping trips. The Ludwigs are tent campers. Rosalie went back to school and got her nursing degree, adding to the current family roster of two nurses, two PhDs, and one MD. They're never sick.

"I have an extremely good stereo system for listening to classical music," Ludwig explained of his never-ending fascination with electronics. "And I had a very, very good turntable that quit operating here a while back. And it dates from the seventies, but it's built like a battleship and is still a classic today." He maintains an electronics workshop in the basement of a home the Ludwigs built in Virginia. "It has been impossible to find the matching parts," he continued about the latest project with the turntable. "So I've had to sort of reverse engineer it and figure out what the transformer characteristics had to be and get parts and try them out. I've got it working again now, and I just have to finish it up."

50. For Van Allen's ninetieth birthday, the University of Iowa threw him a party so grand that even George Ludwig came to see it. Reunited in 2004, Ludwig is on the left, and Van Allen on the right. (Courtesy of the University of Iowa)

Over the years, many have asked George Harry Ludwig about the apparent slighting by Van Allen when it came to the discovery of radiation belts and their subsequent naming. Should they be called the Van Allen–Ludwig Belts?

"I've done a lot of thinking about that question," Ludwig quickly responds. "In terms of the radiation belts, I think it's entirely appropriate that they carry Van Allen's name and his name only. He was the experienced cosmic ray researcher in our department. He was the one who had the best general background. He was the one who, from that background, proposed the satellite experiment." Van Allen was the physicist and put the physics into the results. It's that simple. There was not a hint of resentment in Ludwig's voice whatsoever.

"On the other hand, I do feel that I stake a claim to being chief instrumentor for those satellites," Ludwig continued. "That, I think, was my primary contribution. And in spite of the fact that I did work on the physics to some extent, my primary contribution—the area in which I felt most comfortable and most knowledgeable—was in developing and building the instruments."

With that, Ludwig gets back to his stereo.

The night after Sergei Korolev's death, his senior staff numbly assembled to come to terms with their leader's untimely demise and figure out a successor. Strictly speaking, they worked outside official lines, because the Central Committee was supposed to handle these things. But Korolev's people wanted no governmental lame ducks or other intrusive riffraff. Working through the night, the group finally drafted a letter to the Central Committee explaining their wishes for the capable Vasiliy Mishin to succeed Korolev. The committee already had their own guy in mind, but relented after five months and installed Mishin. A whip-smart engineer, Mishin had for many years already labored shoulder to shoulder with Korolev on his infinite projects. He was Sergei Korolev's First Deputy. "He knew the business," recalled a coworker of Mishin, "and, most important, could screen options as fast as a computer."

Unfortunately for Soviet space exploration, Mishin's diplomatic skill lagged far behind his predecessor. He was never the wheeler and dealer that Korolev could snap into being at a moment's notice. He failed in swaying the Communist Party like Korolev had. Mishin tended to become blunt and obtuse in some situations, clinging to his guns, whereas Korolev would have somehow negotiated a way out. "He deferred to no authority," Mishin's coworker continued, "as long as the authority in question came up with solutions that defied logic and common sense to serve a hidden agenda. That is why he was not popular." But Mishin was their man all the same, his name a secret until the collapse of Communism in the Soviet Union. Having escorted the Soviet space program through a failed manned lunar program and the advent of space stations, he was ousted from such work in 1974 and died in 2001.

After being transferred away from Korolev's design group, Gleb Maksimov joined the Soviet Space Research Institute and earned his PhD. His association with Russia's great chief designer lasted only a decade. "Those ten years with Korolev were the most difficult and the happiest years of my life," he said. Later Maksimov taught at the Moscow High Technical School and died in 2000.

"I am a sort of scoutmaster around here," remarked James Alfred Van

Allen of his relationship with the students that came and went and trundled the halls of his beloved physics department. Over the years, Van Allen continued his dogged pursuit of unlocking the universe. "I'm a sort of a problem-oriented sort of a guy, like to gnaw away at it." Right through most of 2006 he still rose every work day, made his own lunch, then steered onto Dubuque Street in his '85 Jeep Cherokee. Dubuque Street has a hill that cyclists can hit forty miles an hour on going down. Every day Van Allen roared up it, morning Sun to his left, Iowa City growing from the horizon as he scanned for a good parking spot near that building with his name out front. On a wall of his office an ancient, drab metal shelf collected the three-ring binders as they accumulated: Problems I, Problems II, and so on. "Those are solutions. Those are interesting problems to which I've found the solution." He kept another small leather-bound notebook on his desk proper to hold the current ones.

"I love to teach; I love to explain things," was one of the last comments he offered of his sixty years in the field. "I love to work. I just love to solve problems."

One of the bugaboos that was never quite solved is, ironically, the one that got him shooting off rockets in the first place. As George Ludwig put it, "The 'Cosmic Ray Problem' hasn't been solved to this day! We know more about them, but not some of the very basic questions, like where do they come from, and how do they attain such high energy?" The mystery continues, but that's the kind of thing that always got Van Allen up anyway.

He retained the evening work habits at his worn corner desk, putting in additional thought about magnetospheres, or cosmic rays, or cluster missions, or gravitational focusing, before switching off his desk lamp and heading to bed.

On August 9, 2006, he didn't get up.

"Planet" is the Greek word for "wanderer." The ancestors of James Van Allen and Konstantin Gringauz and Charley Kohlhase are the wanderers of times long gone. They were the forefathers who left the safety of caves to forge pathways where none had existed, to break trail and investigate and discover. Their methods have changed but their mindsets have not. We as humans are preprogrammed to explore. And unmanned spacecraft are wondrously able to share vicarious exploration with *all of us*.

Other great adventures—easily found in history books—are celebrated by the masses. Any high schooler who paid attention could describe how Hannibal took fifty thousand men and thirty-seven elephants across the Alps to meet Roman invaders on Italian soil. They will know Magellan went around the world or that Ernest Shackleton lost his ship in Antarctica, only to give Death himself the finger and return without losing a man.

But it's also fundamentally important for any one person to understand how, in the last decades of the twentieth century, a steadily evolving pedigree of exploring machines veritably stuck their toes outside Earth's atmosphere to proclaim, *come on in, the universe is fine.*

And to understand who sent them there in the first place.

Sources

Books

Allen, Richard Hinckley. *Star Names: Their Lore and Meaning.* New York: Dover, 1963.

Batson, Raymond M. *Voyager 1 and 2 Atlas of Six Saturnian Satellites.* NASA SP-474. Washington DC: NASA, 1984.

Bille, Matt, and Erica Lishock. *The First Space Race: Launching the World's First Satellites.* College Station: Texas A&M University Press, 2004.

Burgess, Colin, and Chris Dubbs. *Animals in Space: From Research Rockets to the Space Shuttle.* Chichester, UK: Springer-Praxis, 2007.

Burrows, William E. *Exploring Space: Voyages in the Solar System and Beyond.* New York: Random House, 1990.

Byers, Bruce K. *Destination Moon: A History of the Lunar Orbiter Program.* NASA TM X-3487. Washington DC: NASA, 1977.

Chaikin, Andrew. *A Man on the Moon: The Voyages of the Apollo Astronauts.* New York: Penguin Books, 1994.

Chertok, Boris. *Rockets and People.* 2 vols. NASA History Series. Washington DC: NASA, 2005–6.

Dunlop, Storm. *Collins Atlas of the Night Sky.* London: HarperCollins, 2005.

Dyer, Davis. *TRW: Pioneering Technology and Innovation since 1900.* Boston: Harvard Business School Press, 1998.

Ehricke, Krafft. *Space Flight.* Vol. 2, *Dynamics.* Princeton: D. Van Nostrand, 1962.

Fimmel, Richard O., and William Swindell. *Pioneer Odyssey.* NASA SP-349/396. Washington DC: NASA, 1977.

Fimmel, Richard O., James A. Van Allen, and Eric Burgess. *Pioneer: First to Jupiter, Saturn, and Beyond.* NASA SP-446. Washington DC: NASA, 1980.

Gatland, Kenneth. *Robot Explorers.* London: Blandford Press, 1972.

Gray, Mike. *Angle of Attack: Harrison Storms and the Race to the Moon.* New York: W. W. Norton, 1992.

Griffin, Michael D., and James R. French. *Space Vehicle Design.* 2nd ed. Reston VA: American Institute of Aeronautics and Astronautics, 2004.

Hall, R. Cargill. *Lunar Impact: A History of Project Ranger.* NASA History Series. Washington DC: NASA, 1977.

Hall, Stephen S. *Mapping the Next Millennium: The Discovery of New Geographies.* New York: Random House, 1992.

Harford, James. *Korolev: How One Man Masterminded the Soviet Drive to Beat America to the Moon.* New York: John Wiley and Sons, 1997.

Harland, David M., and Ralph D. Lorenz. *Space Systems Failures: Disasters and Rescues of Satellites, Rockets and Space Probes.* Chichester, UK: Springer-Praxis, 2005.

Harvey, Brian. *The New Russian Space Programme: From Competition to Collaboration.* Chichester, UK: Springer-Praxis, 1996.

———. *Russia in Space: The Failed Frontier?* Chichester, UK: Springer-Praxis, 2001.

Kelly, Thomas J. *Moon Lander: How We Developed the Apollo Lunar Module.* Washington DC: Smithsonian Institution Press, 2001.

Kloman, Eurasmus H. *Unmanned Space Project Management.* NASA SP-4901. Washington DC: NASA, 1972.

Koppes, Clayton R. *JPL and the American Space Program.* New Haven CT: Yale University Press, 1982.

Lemaire, J. F., and K. I. Gringauz. *The Earth's Plasmasphere.* Cambridge: Cambridge University Press, 1998.

Levy, David H. *Skywatching.* New South Wales, Australia: Weldon Owen Pty, 1995.

Maxfield, Clive, and Alvin Brown. *Bebop Bytes Back: An Unconventional Guide to Computers.* Madison AL: Doone, 1997.

McNab, David, and James Younger. *The Planets.* New Haven CT: Yale University Press, 1999.

Meltzer, Michael. *Mission to Jupiter: A History of the Galileo Project.* NASA History Series SP-2007-4231. Washington DC: NASA, 2007.

Moore, Sir Patrick. *The Great Astronomical Revolution: 1543–1687 and the Space Age Epilogue.* West Sussex, UK: Albion, 1994.

Morrison, David, and Jane Samz. *Voyage to Jupiter*. NASA SP-439. Washington DC: NASA, 1980.

Murray, Charles, and Catherine Bly Cox. *Apollo: The Race to the Moon*. New York: Simon and Schuster, 1989.

Naugle, John E. *First Among Equals: The Selection of NASA Space Science Experiments*. NASA History Series SP-4215. Washington DC: NASA, 1991.

Neufeld, Michael. *Von Braun: Dreamer of Space, Engineer of War*. New York: Alfred A. Knopf, in association with the National Air and Space Museum, Smithsonian Institution, 2007.

Newell, Homer. *Beyond the Atmosphere: Early Years of Space Science*. NASA History Series SP-4211. Washington DC: NASA, 1980.

Nicks, Oran. *Far Travelers: The Exploring Machines*. NASA History Series SP-480. Washington DC: NASA, 1985.

———. *A Review of the Mariner IV Results*. NASA SP-130. Washington DC: NASA, 1967.

Perminov, Vladimir. *The Difficult Road to Mars: A Brief History of Mars Exploration in the Soviet Union*. NASA Monographs in American History 15. Washington DC: NASA, 1999.

Rhea, John, ed. *Roads to Space: An Oral History of the Soviet Space Program*. Translated by Peter Berlin. New York: Aviation Week Group, 1995.

Ryan, Craig. *The Pre-Astronauts: Manned Ballooning on the Threshold of Space*. Annapolis MD: Naval Institute Press, 1995.

Sagan, Carl. *Murmurs of Earth: The Voyager Interstellar Record*. With F. D. Drake, Ann Druyan, Timothy Ferris, Jon Lomberg, and Linda Salzman Sagan. New York: Random House, 1978.

———. *Pale Blue Dot: A Vision of the Human Future in Space*. New York: Ballantine Books, 1994.

Sagdeev, Roald Z. *The Making of a Soviet Scientist: My Adventures in Nuclear Fusion and Space from Stalin to Star Wars*. New York: John Wiley and Sons, 1994.

Siddiqi, Asif. *The Soviet Space Race with Apollo*. Gainesville: University Press of Florida, 2003.

———. *Sputnik and the Soviet Space Challenge*. Gainesville: University Press of Florida, 2000.

Sloop, John L. *Liquid Hydrogen as a Propulsion Fuel*. NASA History Series SP-4404. Washington DC: NASA, 1978.

Swift, David. *Voyager Tales: Personal Views of the Grand Tour.* Reston VA: American Institute of Aeronautics and Astronautics, 1997.

Tomayko, James E. *Computers in Spaceflight: The NASA Experience.* Contract NASW-3714. Washington DC: NASA, 1988.

Ulivi, Paolo. *Lunar Exploration: Human Pioneers and Robotic Surveyors.* With David M. Harland. Chichester, UK: Springer-Praxis, 2004.

U.S. Air Force. *Soviet Aerospace Handbook.* Washington DC: U.S. Government Printing Office, 1978.

U.S. Army Space Institute. *Army Space Reference Text.* Fort Leavenworth KS: U.S. Army Space Institute, 1993.

Van Allen, James A. *Origins of Magnetospheric Physics.* Iowa City: University of Iowa Press, 2004.

Wise, David, and Thomas B. Ross. *The U-2 Affair.* New York: Random House, 1962.

Wolverton, Mark. *The Depths of Space: The Story of the Pioneer Planetary Probes.* Washington DC: Joseph Henry Press, 2004.

Periodicals and Online Articles

Alibrando, Alfred P. "Manned Venus–Mars Fly-by in 1970 Studied." *Aviation Week and Space Technology,* March 4, 1963.

"ASK Talks with Tom Gavin." *ASK,* April 2004. http://appel.nasa.gov/ask/issues/17/17_interview.php.

"Atlas-Able IV Instrumentation Detailed." *Aviation Week,* December 14, 1959.

Baldwin, Hanson. "U.S. Atom Blasts 300 Miles Up Mar Radar, Snag Missile Plan, Called 'Greatest Experiment.'" *New York Times,* March 19, 1959.

Battin, Richard H. "The Determination of Round-Trip Planetary Reconnaissance Trajectories." *Journal of the Aero/Space Sciences* 26, no. 9 (September 1959): 545–67.

———. "On Algebraic Compilers and Planetary Fly-by Orbits." *Acta Astronautica* 38, no. 12 (1996): 895–902.

BBC News. "1966: Soviets Land Probe on Moon." On This Day: February 3, 1966. http://news.bbc.co.uk/onthisday/hi/dates/stories/february/3/newsid_4063000/4063471.stm.

Breus, Tamara K. "An Unforgettable Personality." *Journal of Geophysical Research* 102, no. A2 (February 1, 1997): 2027–34.

Broucke, Roger A., and Antonio F. B. A. Prado. "Jupiter Swing-by Trajectories Passing Near the Earth." *Advances in the Astronautical Sciences* 82, no. 2 (1993): 1159–76.

Burke, James D. "Personal Profile." *Spaceflight* 26 (April 1984): 178–83.

Cesarone, Robert J., and Andrey B. Sergeyevsky. "Solar System Stellar Flybys." *Journal of the British Interplanetary Society* 38, no. 11 (November 1985): 527.

Cesarone, Robert J., Andrey B. Sergeyevsky, and Stuart J. Kerridge. "Prospects for the Voyager Extra-planetary and Interstellar Mission." *Journal of the British Interplanetary Society* 37, no. 3 (March 1984): 99–116.

Chaikin, Andrew. "Hard Landings." *Air and Space Magazine*, June–July 1997.

Charny, Vitaly. "Semyon Ariyevich Kosberg." *Belarus SIG Online Newsletter*, no. 14 (March 2004). http://www.jewishgen.org/Belarus/news letter/Kosberg.htm.

Clarke, Arthur C. "Planetary Fly-By Orbits." Letter to the editor. *Acta Astronautica*, February 12, 1997.

Grinter, Kay. "45 Years Ago: Pioneer V Gives a Regal Performance." *Spaceport News*, March 4, 2005.

Heacock, Raymond L. "The Voyager Spacecraft." *Proceedings of the Institution of Mechanical Engineers* 194, no. 28 (1980): 214–224.

Lange, Volker. "Journey to the Gas Giants." Morgenwelt. http://www.mor genwelt.de/futureframe/001023-planets5.htm (accessed December 2, 2004; site is now discontinued).

———. "Metallic Snow and Methane Rain." Morgenwelt. http://www .morgenwelt.de/futureframe/001127-methanerain.htm (accessed December 2, 2004; site is now discontinued).

LePage, Andrew. "The Great Moon Race: The Commitment." Pt. 2. *Saguaro Astronomy Club SACNEWS*, no. 190 (November 1992): 1–4.

———. "Recent Soviet Lunar and Planetary Program Revelations." *Saguaro Astronomy Club SACNEWS*, no. 199 (August 1993): 1–5.

Marcus, Gideon. "Pioneering Space." Pts. 1 and 2. *Quest* 14, no. 2 (2007): 52–59; 14, no. 3 (2007): 18–25.

"Mariners to the Red Planet." Interview with William Momsen. *Astrobiology Magazine*, April 19, 2004. http://www.astrobio.net/news/print .php?sid=930.

Momsen, Bill. "Mariner IV—First Flyby of Mars: Some Personal Experiences." http://home.earthlink.net/~nbrass1/mariner/miv.htm.

NASA Quest. "Transcript of the Pioneer 10 Virtual Conference." http://quest .nasa.gov/sso/cool/pioneer10/general/amonetxt.html.

Office of Management and Budget. "Department of Defense." The White House. http://www.whitehouse.gov/omb/budget/fy2005/defense .html.

"100 Years Ago in Scientific American: The Riddle of Mars." *Scientific American*, June 11, 2007.

Parker, P. J. "Grand Tour Spacecraft Computer." *Spaceflight* 13, no. 3 (March 1971): 88, 120.

"Pioneer VI Designed for Moon Orbit." *Aviation Week*, September 12, 1960.

Poe, Margaret. "Iowa's Epic Voyager." *Daily Iowan*, September 7, 2006.

"Pre-election Moon Shot Still Possible." *Missiles and Rockets*, October 3, 1960.

"Project Vanguard: Why It Failed to Live Up to Its Name." *Time*, October 21, 1957.

Rayl, A. J. S. "The Stories Behind the Mission: Personal Reminisces from the Voyager Team." Planetary Society. http://www.planetary.org/ explore/topics/space_missions/voyager/stories.html.

"Reach into Space." *Time*, May 4, 1959.

Rickard, Doug. "Memoirs of a Space Engineer." ABC Science Online. http:// www.abc.net.au/science/slab/memoirs/default.htm.

Rigby, Mark T. "Island Lagoon Tracking Station." Woomera on the Web. http://homepage.powerup.com.au/~woomera/tracking.htm.

RoadsideAmerica.com. "Sputnik Crashed Here: Manitowoc, Wisconsin." www.roadsideamerica.com/story/12959.

Sagan, Carl, Linda Salzman Sagan, and Frank Drake. "A Message from Earth." *Science* 175 (1972): 881.

Schiaparelli, G. V. "La Vita sul Pianeta Marte" [Life on planet Mars]. Translated by Paolo Ulivi. *Natura ed Arte* 4, no. 11 (1895): 81–89.

Selivanov, A. S., V. N. Govorov, A. S. Titov, and V. P. Chemodanov. "Lunar Station Television Camera." Translated by Reilly Translations. *Tekhnika kino i televedeniya* [Techniques of film and television], no. 1 (January 1968): 9–17.

Snopes. "Nyet!" http://www.snopes.com/disney/parks/nikita.htm.

Teplov, I. B., and A. E. Chudakov. "Sergei Nikolaevich Vernov." Lomonosov Moscow State University, Skobeltsyn Institute of Nuclear Physics. http://www.sinp.msu.ru/eng/maininc/vernov.html.

Van Allen, James A. "What Is a Space Scientist?" *Annual Review of Earth and Planetary Sciences* 18 (1990): 1–26.

———. "What Is a Space Scientist? An Autobiographical Example." University of Iowa. http://www-pi.physics.uiowa.edu/java/.

Verigin, Mikhail I., and Norman F. Ness. "Obituary: Konstantin Iosifovich Gringauz." *Space Science Reviews* 65, nos. 1–2 (March 1993): 1–3.

"Voyager at Neptune: For 10 August Days the Science Flowed Instantly." *Countdown*, November 1989.

"Washington Countdown." *Missiles and Rockets*, May 11, 1959.

"Wesley Merle Alexander, Ph.D." *Baylor University* CASPER *(Center for Astrophysics, Space Physics, and Engineering Research) Newsletter* 6, no. 1 (Winter 2004): 15.

Yaffee, Michael. "Scientists Analyze Lunar Orbit Failure." *Aviation Week*, October 3, 1960.

Yefimov, V. "Anniversary of the Space Television." *Tele-Sputnik* 5, no. 3 (March 1996): 50.

Interviews and Personal Communications

Battin, Richard H. Telephone conversation with the author, January 25, 2007, regarding Minovitch lawsuit.

Bille, Matt. E-mail messages to the author, September 8 and 9, 2007, regarding origin of the term "the Deal" in relation to *Explorer 1*.

Bourke, Roger. Interview by the author, August 13, 2006, and related correspondence.

Burke, James D. Interview by Jeffrey Kluger, May 6, 1997.

———. Interview by the author, February 20, 2006, and related correspondence.

Cahill, Lawrence. Interview by the author, August 31, 2005, and related correspondence.

Carpenter, Don. Interview by the author, April 20, 2006, and related correspondence.

Casani, John. Interview by the author, April 24, 2007, and related correspondence.

Cesarone, Robert. Personal communications with the author, November 30, 2007–January 18, 2008, regarding *Voyagers 1* and *2* trajectories.

Cortright, Edgar M. Interview by Rich Dinkel, August 20, 1998. NASA Oral History. http://www.jsc.nasa.gov/history/oral_histories/CortrightEM/EMC_8-20-98.pdf.

Cutting, Elliot "Joe." Interview by the author, December 10, 2006, and related correspondence.

Dumas, Larry. E-mail message to the author, September 4, 2006, regarding Minovitch lawsuit.

Ferrari, Kay. Personal communications with the author, 2004–8, regarding JPL culture, policies, personalities, and anecdotes.

Ferris, Timothy. Personal communications with the author, June 18–July 8, 2008, regarding creation of the Voyager record.

Flandro, Gary. Interview by the author, February 9, 2005, and related correspondence.

Flandro, Troy. Interview by the author, March 26, 2007, and related correspondence.

Fomenko, Alex. Personal communications with the author, 2006–7, regarding Soviet launch facilities, Soviet military culture, and Russian culture in general.

Freund, Edmund, and Marcella Freund. Interview by the author, October 3, 2005.

Gilruth, Robert. Interview by David DeVorkin and John Mauer, March 2, 1987. Glennan-Webb-Seamans Project for Research in Space History. National Air and Space Museum. http://www.nasm.si.edu/research/dsh/TRANSCPT/GILRUTH6.htm.

Hall, R. Cargill. E-mail message to the author, March 24, 2006, and related correspondence, regarding circumstances surrounding replacement of Jim Burke as Ranger project manager.

Hansen, Candice. E-mail correspondence with the author, November 18, 2007–January 1, 2008, regarding Voyager imaging and program in general.

Kluger, Jeffrey. Telephone conversation with the author, November 6, 2006, and related correspondence, regarding Kluger's interviews with Jim Burke and William Pickering about NASA Ranger program.

Kohlhase, Charles E. Personal communications with the author, November 10, 2006–July 24, 2008.

Koppes, Clayton. E-mail message to the author, September 23, 2007, and related correspondence, regarding the origins of the term "the Deal."

Lewicki, Chris. Interview by the author, January 23, 2005.

Lomberg, Jon. E-mail correspondence with the author, October 2, 2007–August 3, 2008, regarding the Voyager record.

Ludwig, George. Interview by the author, August 23 and 30, 2005, and related correspondence.

Minovitch, Michael A. Interview by the author, July 21, 2006, and related correspondence.

Mueller, George. Interview by Martin Collins, February 15, 1988. Glennan-Webb-Seamans Project for Research in Space History. National Air and Space Museum. http://www.nasm.si.edu/research/dsh/TRANSCPT/MUELLER4.htm.

Pickering, William H. Interview by Jeffrey Kluger, May 8, 1997.

Ramo, Simon. Interviews by Martin Collins, June 27, 1988, and January 25, 1989. Glennan-Webb-Seamans Project for Research in Space History. National Air and Space Museum. http://www.nasm.si.edu/research/dsh/TRANSCPT/RAMO1.htm and http://www.nasm.si.edu/research/dsh/TRANSCPT/RAMO2.htm.

Richter, Henry. E-mail message to the author, December 20, 2007, regarding playing gin rummy at JPL and the origin of the term "the Deal."

Schurmeier, Harris "Bud." Interview by the author, July 6 and 19, 2006, and related correspondence.

Siddiqi, Asif. E-mail message to the author, April 10, 2007, regarding Ryabikov's January 30, 1956, decree extending permission for Korolev to launch a satellite and how this level of permission compared with Khrushchev's, later given to Korolev during an in-person visit to Korolev's design bureau. Also, e-mail message to the author, December 18, 2007, regarding Korolev's attitude toward military space projects.

Sjogren, William. Interview by the author, September 10, 2006, and related correspondence.

Sonett, Charles. Interview by the author, March 13, 2006, and related correspondence.

Thomsen, Michelle. Telephone conversation with the author, July 3, 2007, and related correspondence, regarding preparation of Van Allen's Jovian magnetosphere model.

Van Allen, James A. Interview by the author, May 3, 4, and 5, 2005, and related correspondence.

VonDelden, Hugh. E-mail messages to the author, January 16 and 17, 2008, regarding Voyager record mounting, names of technicians, and location of record stylus on spacecraft.

Webb, James. Interview by Martin Collins and Allan Needell, November 4, 1985. Glennan-Webb-Seamans Project for Research in Space History. National Air and Space Museum. http://www.nasm.si.edu/research/dsh/TRANSCPT/WEBB9.htm.

York, Herbert. Interview by Martin Collins, January 24, 1989. Glennan-Webb-Seamans Project for Research in Space History. National Air and Space Museum. http://www.nasm.si.edu/research/dsh/TRANSCPT/YORK2.htm.

Other Sources
Ames Research Center, Palo Alto CA

"Pioneer 11 to End Operations after Epic Career." Ames press release, September 29, 1995, 95–163.

California Institute of Technology, Pasadena CA

"Ronald Scott Dies: Designed Soil Scoop for Early Unmanned Moon Mission." Media relations release, August 19, 2005.

James A. Van Allen, private collection

"James Van Allen Computation Book." October 1957.

Jet Propulsion Laboratory, Pasadena CA

Bourke, Roger. "IOM 393.1-213." JPL interoffice memorandum to Michael A. Minovitch, August 13, 1971.

Cesarone, Robert. "Near Star Encounters for Voyager-2 Neptune North Polar Trajectory." Results of computing run, September 12, 1989.

———. "Near Star Encounters for Voyager-2 Neptune Polar Crown Trajectory." Results of computing run, July 20, 1987.

Chahine, M. T. "Summary of JPL Credit." Letter to Michael A. Minovitch, November 19, 1997.

Clarke, Victor C., Jr. "Clarification of Comments." Letter to Dr. Norriss S. Hetherington of the University of Kansas, July 22, 1974.

———. "Interplanetary Round-Trip Program." Memorandum to J. Scott; CC C. R. Gates, T. W. Hamilton, W. E. Bollman, R. Roth, M. Minovitch, W. Hoover, S. Dallas, F. Jordan, J. Detlef, W. Kirhofer, and W. G. Melbourne, June 21, 1962.

Cutting, E. "Joe." "Cover Letter for Preliminary Copy of Report." Letter to Michael A. Minovitch, May 21, 1965.

———. "Proposed AIAA Paper." JPL interoffice memorandum to Michael A. Minovitch; CC V. C. Clarke Jr, January 21, 1964.

"Final Imaging Plans for Mariner IV." JPL press release, June 28, 1965.

Hobby, George L. "Summary of Spacecraft Sterilization Program." Letter to Oran Nicks of the NASA Lunar and Planetary Programs, January 25, 1962.

James, J. N. "Significance of 19 June 1963." JPL interoffice memorandum to All Concerned, June 19, 1963.

"Mike Minovitch Will Continue His Discussion of 'Multiple Planet Trajectories' Today." JPL interoffice memorandum, February 5, 1963.

Minovitch, Michael A. "A Method for Determining Interplanetary Free-Fall Reconnaissance Trajectories." JPL technical memorandum, no. 312-130, to Section 312 Engineers, J. F. Scott and W. Scholey, August 23, 1961.

———. "NASA Exceptional Performance Award." JPL interoffice memorandum, no. 393.1-202, to Paul D. Lehman, August 2, 1971. Includes attachment.

———. "NASA Exceptional Performance Award." JPL interoffice memorandum, no. 393.1-202, to Paul D. Lehman, November 2, 1971.

———. "NASA Exceptional Performance Award." JPL interoffice memorandum, no. 393.1-213, to R. D. Bourke; CC C. R. Gates, P. D. Lehman, and W. H. Pickering, August 11, 1971.

———. "Unexplained 11 Month Delay in Getting JPL Clearance for the Paper Entitled 'The Development of Gravity Thrust Space Trajectories.'" JPL interoffice memorandum to R. D. Bourke; CC C. R. Gates, T. W. Hamilton, and R. R. Stephenson, May 17, 1972.

"Proposed Citation—NASA Exceptional Scientific Achievement Award—1971 for Dr. Michael A. Minovitch, Jet Propulsion Laboratory." Mission Analysis Division.

"Solar System Mosaic." JPL press release, June 6, 1990.

Space Explorers. Pamphlet. Pasadena CA: JPL / California Institute of Technology, July 12, 1958.

Stavro, W. "Origin of 'Gravity Assist' Trajectories." Interoffice memorandum to R. V. Morris and E. Cutting, April 20, 1971.

Temperature Control in the Explorer Satellites and Pioneer Space Probes. JPL External Publication No. 647.

Michael A. Minovitch, private collection

Abelson, Philip H. Letter to Michael A. Minovitch, May 5, 1966.

Flanagan, Dennis. Letter to Michael A. Minovitch, July 28, 1966.

———. Letter to Michael A. Minovitch, June 16, 1971.

Kohlhase, Charles E. "Misinterpretation of Telephone Conversation." Letter to Michael A. Minovitch, September 23, 1996.

Los Angeles Section of the American Institute of Aeronautics and Astronautics Presents the Thirteenth Annual Western Region Student Conference, May 2 and 3, 1963. Pamphlet. Los Angeles: AIAA Western Region Headquarters, n.d.

Minovitch, Michael A. "The Development of Gravity Thrust Space Trajectories." Undated Composition.

———. "Effect of IAA 90-630 Paper." Letter to Charles E. Kohlhase, October 25, 1990.

———. "Gravity-Propelled Interplanetary Space Travel." Home video lecture.

———. "Hetherington's Paper on the Historical Development of Gravity Assisted Trajectories." Letter to Victor C. Clarke Jr., June 10, 1974.

———. "The Invention of Gravity Propelled Interplanetary Space Travel: A Technical and Historical Presentation to the Jet Propulsion Laboratory." October 1997.

———. "JPL's Readiness to Acknowledge My Invention." Letter to Charles E. Kohlhase, September 4, 1996.

———. Letter to Dennis Flanagan, June 14, 1966.

———. Letter to Dennis Flanagan, May 29, 1971.

———. Letter to John Newbauer, May 28, 1971.

———. Letter to Philip H. Abelson, March 22, 1966.

———. "Making the Recognition Official." Letter to Charles E. Kohlhase, October 30, 1989.

———. "Michael Minovitch Notes." Computation book excerpts, February 20–April 6, 1962.

———. "Pending Legal Action if Hetherington Article Is Published." Letter to William Pickering, June 10, 1974.

———. "Regarding Our Recent Telephone Conversation." Letter to Charles E. Kohlhase, February 20, 1990.

Newbauer, John. Letter to Michael A. Minovitch, January 3, 1974. Enclosure: Uncredited review of proposed Minovitch article.

Pickering, William. Letter to Michael A. Minovitch, July 22, 1974.

Ruppe, Harry O. "Response to IAA Paper 90-630." Letter to R. L. Dowling, November 26, 1990.

Subotowicz, Mieczyslaw. "Nomination of Michael Minovitch for 1992 Nobel Prize in Physics." Letter to the Nobel Committee for Physics and Chemistry, August 21, 1991.

National Aeronautics and Space Administration, Washington DC

Dohnany, J. S. *Dust Cloud Produced by Luna-5 Impacting the Lunar Surface—Case 211.* NASA CR-156595. Washington DC: Bellcomm, August 3, 1965.

"Mariner–Venus '73 Flight Genesis." NASA news release no. 70-112.

NASA. *Highlights of the Press Conference at the Moscow House of Scientists on 10 February 1966.* NASA ST-PR-LPS-10448. Washington DC: NASA, February 17, 1966.

———. "National Aeronautics and Space Administration Organizational Chart." October 24, 1958.

———. "National Aeronautics and Space Administration Organizational Chart." March 23, 1959.

———. "National Aeronautics and Space Administration Organizational Chart." November 1, 1961.

———. *Soviet Press Communiques and the Pravda Editorial on "Luna-9" Soft Landing on the Moon.* NASA ST-PR-10445. Washington DC: NASA, February 10, 1966.

———. *Surveyor Spacecraft System, Slides Used in Quarterly Technical Status Review on 24 July 1963.* Report. Washington DC: NASA, July 29, 1963.

"Remarks by von Braun and Pickering upon Return of Explorer I." NASA news release no. 70-49.

University of California, Los Angeles

Hollander, Frederick H. "The Determination of Interplanetary Reconnaissance Trajectories for Free-Fall Space Vehicles." Request for access to UCLA Computing Facility, April 2, 1962.

Murphy, Franklin D. "Invitation to Represent UCLA at the All-University Spring Festival." Letter to Michael A. Minovitch, March 20, 1963.

UCLA Computing Facility. "Problem MAII." Memorandum to Peter Henrici and Michael A. Minovitch, January 18, 1962.

University of Iowa, Iowa City

All materials listed in this section are from the James Van Allen Papers, Special Collections Department, University of Iowa Archives, Iowa City.

Candee, T. W. "Termination of Contract #950345." Telex to Ray Mossman, July 11, 1963.

Chinburg, Dale. "Disassembly of SUI Experiments." Telex to Don Schofield, November 19, 1963.

———. "Formal Authorization of Orders." Letter to A. V. Doble, November 6, 1963.

———. "Forwarding of Parts." Letter to Don Schofield, October 21, 1963.

———. "Lack of Screened Components." Letter to R. L. Heacock, June 19, 1963.

———. "Mariner Memo #026: JPL Parts Screening." Letter to James A. Van Allen, December 10, 1963.

———. "Removal of Oil Film." Letter to Vincent F. Molonea, October 15, 1964.

————. "Revised Monthly Cost Report." Letter to V. F. Molonea, June 9, 1964.

Crawford, J. A. "Non-payment of JPL Vouchers." Memorandum to V. M. Enson, March 22, 1965.

Hones, Edward W., Jr. "Demise of Your Good Friend Explorer I." Letter to James A. Van Allen, April 1, 1970.

Kalisman, Herbert. "Authorization to Publicize Use of Anton Equipment for Explorer I." Telegram to George Ludwig, February 4, 1958.

Ludwig, George. "A Chronology of SUI Space Instrumentation." March 1960.

————. "Clarification of Final Schedule." Memorandum to James A. Van Allen, January 22, 1958.

————. *Computation Book* 56, no. 1. April 25, 1956–March 5, 1957.

————. *Computation Book* 57, no. 1. March 5–September 10, 1957.

————. *Computation Book* 57, no. 2. September 10, 1957–June 30, 1958.

————. "Finalizing Schedule." Letter to James A. Van Allen, January 22, 1958.

————. "George Ludwig Journal." January 20, 1958–August 23, 1960.

————. "Review of Test Schedule and Equipment Delivery." Letter to Homer Newell, April 16, 1957.

————. "Two Very Mild Regrets." Handwritten personal note, March 17, 1981.

Ludwig, George, and James A. Van Allen. *Status Report on the SUI Cosmic-Ray Instrumentation by George Ludwig and James Van Allen.* Iowa City: State University of Iowa, April 19, 1957.

Molonea, Vincent F. "Concern over Reported Costs." Letter to Dale Chinburg, September 16, 1964.

NASA. "Document #PT-204 Pioneer F Asteroid Analysis." With handwritten annotations by James A. Van Allen. "Probability of Pioneer F and G Asteroid Encounter."

Nowicki, Richard. "GMI Proposal 11353: Six Tape Recorder Mechanisms." Letter to George Ludwig, September 21, 1956.

Pickering, William. "Division of Responsibility for Explorer I Experiments." Letter to Major General Otis O. Benson Jr., March 26, 1958.

Ray, Ernest C. "Granting of Authorization to Publicize Use of Anton Equipment for Explorer I." Telegram to Herbert Kalisman, February 6, 1958.

Simpson, John. "Alarming Situation Regarding Data Tapes." Letter to R. K. Sloan, July 1, 1964.

———. "JPL Timing for Release of Scientific Data." Letter to Jack James, June 16, 1964.

"Six IGY Proposals." SUI physics department internal memorandum, May 27, 1956.

Stuhlinger, Ernst. "A Toast to Explorer I." Letter to James A. Van Allen, April 6, 1970.

University of Iowa. *National Science Foundation Quarterly Financial Report, IGY Expenditure Basis.* Form NSF-636A, Grant #Y/32/1/147, March 31, 1957.

Van Allen, James A. "Cosmic-Ray Observations in Earth Satellites." Application for Research Project for the U.S. IGY Program, IGY Project 32.1, August 6, 1956.

———. "Details of Tape Recorder." Letter to Warren Berning, November 21, 1956.

———. "Electrically Wound or Driven Wristwatch." Letter to Hamilton Watch Company, January 7, 1957.

———. "Final Tabular Form of Explorer I Data." Letter to Hugh Odishaw, February 1, 1961.

———. "Poor Job of Getting the Data Out." Handwritten personal note, February 4, 1958.

———. "Possibility of Flying Scientific Apparatus on Earth Orbiter." Letter to Ernst Stuhlinger, February 13, 1956.

———. "Proposal to Collaborate with the ABMA and JPL in the Conduct of Further Satellite Studies." Letter to Ernst Stuhlinger, May 5, 1958.

———. "Request for Substitute Budget." Letter to Richard Porter, January 23, 1956.

———. "Request of NRL-Designed Items." Letter to Homer Newell, November 20, 1956.

———. "Status of Explorer I Instrumentation." Handwritten personal note, February 2, 1958.

———. "Termination of Pioneer 10's Mission." Memorandum, February 20, 2003.

———. "Validity of Large Radiation Intensity above Altitude of 1000 Kilometers." Letter to W. K. H. Panofsky, April 9, 1958.

Waterman, Alan T. "Conveyance of Congratulations." Telegram to Virgil M. Hancher, February 1, 1958.

Whipple, Fred. "Private Technical Discussion of Possible Satellite Vehicles." Letter to James A. Van Allen, June 10, 1955.

The White House, Washington DC

Brooks, Overton. "Recommendations Re the National Space Program." Letter to Lyndon B. Johnson (identified as "Chairman, National Aeronautics and Space Council"), May 4, 1961.

Johnson, Lyndon B. "Evaluation of Space Program." Memorandum to John F. Kennedy, April 28, 1961.

Kennedy, John F. "Memorandum for Vice President: What Do We Have a Chance of Beating the Soviets With?" Memorandum to Lyndon B. Johnson, April 20, 1961.

Von Braun, Wernher. "Attempt to Answer Some of the Questions about Our National Space Program." Letter to Lyndon B. Johnson, April 29, 1961.

Ahlquist, Diron. "[OK-Lawmen-Outlaw] The First 6 Months of 1897." E-mail to RootsWeb mailing list, October 26, 2003, under "The Younger Pardon Fight." http://archiver.rootsweb.com/th/read/OK-LAWMEN-OUTLAW/2003-10/1067198033.

Bellan, Josette. "The Jet Propulsion Laboratory Space Exploration: Past, Present, and Future." Invited plenary lecture, 33rd Israel Annual Conference on Aviation and Astronautics, Haifa, Israel, February 24–25, 1993.

CollectSpace. http://www.collectspace.com.

Corneille, Philip. "Mars Literature." http://mars-literature.skynetblogs.be.

Dowling, Richard, William J. Kosmann, Michael A. Minovitch, and Rex W. Ridenoure. "The Effect of Gravity-Propelled Interplanetary Space Travel on the Exploration of the Solar System—Historical Survey, 1961 to 2000." IAA-99-IAA.2.1.08, 50th Congress of the International Astronautical Federation, Amsterdam, Netherlands, October 4–8, 1999.

———. "Gravity Propulsion Research at UCLA and JPL, 1962–1964." IAA-91-677, 42d Congress of the International Astronautical Federation, Montreal, Canada, October 5–11, 1991.

————. "The Origin of Gravity-Propelled Interplanetary Space Travel." IAA-90-630, 41st Congress of the International Astronautical Federation, Dresden, GDR, October 6–12, 1990.

Encyclopedia Astronautica. http://www.astronautix.com.

Ferris, Timothy. Interview by Johnathan Heuer, May 8 and 15, 2002. *The Human Chorus.* KALW Radio.

Flandro, Gary A. "From Instrumented Comets to Grand Tours: On the History of Gravity Assist." Paper 2001-0176, presented at the 39th AIAA Aerospace Sciences Meeting and Exhibit, American Institute of Aeronautics and Astronautics, Reno NV, January 8–11, 2001.

Grahn, Sven. "Sven's Space Place." http://www.svengrahn.pp.se/.

Her Majesty's Nautical Almanac Office. http://www.nao.rl.ac.uk/.

International Business Machines Corporation. *IBM Data Processing Division Press Technical Fact Sheet.* Distributed October 4, 1960.

————. *Operator's Guide for IBM 7090 Data Processing System.* White Plains NY: International Business Machines Corporation, 1959.

McIlwain, Carl. "Discovery of the Van Allen Radiation Belts." Presentation made during the Invited Lecture Series on James Van Allen Day, Iowa City IA, October 9, 2004.

McNab, David, dir. and ser. prod. *The Planets.* BBC Television / A&E Network coproduction. Executive producer, John Lynch. 1999.

Minovitch v. Battin. Civil Action No. 00-5159(J), (Mass. Super. Ct. December 11, 2002). Kirk Y. Griffin. "Amended Complaint and Jury Trial Demand."

————. Civil Action No. 00-5159(J), (Mass. Super. Ct. February 6, 2003). Jeffrey Swope. "Answer of the Defendant Richard H. Battin to Amended Complaint."

————. Civil Action No. 00-5159(J), (Mass. Super. Ct. April 22, 2005). Jeffrey Swope. "Defendant's Motion for Summary Judgment on All Counts."

————. Civil Action No. 00-5159(J), (Mass. Super. Ct. April 22, 2005). Peter G. Hermes. "Plaintiff Michael A. Minovitch's Opposition to Defendant, Richard H. Battin's Motion for Summary Judgment."

————. Civil Action No. 00-5159(J), (Mass. Super. Ct. July 15, 2005). Thomas A. Connors. "Memorandum of Decision and Order on Defendant Richard H. Battin's Motion for Summary Judgment."

Mitchell, Don P. "Mental Landscape." http://www.mentallandscape.com.

NASA, JPL, and Caltech. "Voyager: The Interstellar Mission." http://voyager .jpl.nasa.gov/index.html.

National Space Science Data Center. http://nssdc.gsfc.nasa.gov.

Preusch, Aaron, prod. Untitled James Van Allen Memorial Video Feature. 2006.

Rudd, R., J. Hall, and G. Spradlin. "The Voyager Interstellar Mission." IAA-96-IAA.4.1.03, 47th International Astronautical Congress, Beijing, China, October 7–11, 1996.

Soviet Press. "Destination—Venus." February 12–March 3, 1961. Compiled and translated by Joseph Zygielbaum. Astronautics Information, Translation No. 20. Communiqués and Papers. JPL. Caltech. Pasadena.

S. P. Korolev Rocket and Space Corporation Energia. "Energia." http:// www.energia.ru/english/.

Thurman, Sam W. "Surveyor Spacecraft Automatic Landing System." AAS 04-062, 27th Annual American Astronomical Society Guidance and Control Conference, Breckenridge CO, February 4–8, 2004.

"The Unexpected Universe." Space Age season 1, episode 3, WQED Pittsburgh and NHK Japan, in conjunction with the National Academy of Sciences. 1991.

Zak, Anatoly. "Russian Space Web." http://www.russianspaceweb.com.

Index

In the Outward Odyssey: A People's History of Spaceflight Series

Into That Silent Sea
Trailblazers of the Space Era, 1961–1965
Francis French and Colin Burgess
Foreword by Paul Haney

In the Shadow of the Moon
A Challenging Journey to Tranquility, 1965–1969
Francis French and Colin Burgess
Foreword by Walter Cunningham

To a Distant Day
The Rocket Pioneers
Chris Gainor
Foreword by Alfred Worden

Homesteading Space
The Skylab Story
David Hitt, Owen Garriott, and Joe Kerwin
Foreword by Homer Hickam

Ambassadors from Earth
Pioneering Explorations with Unmanned Spacecraft
Jay Gallentine

To order or obtain more information on these or other
University of Nebraska Press titles, visit www.nebraskapress.unl.edu.